Springer Undergraduate Texts in Mathematics and Technology

For further volumes:
http://www.springer.com/series/7438

Glenn Ledder

Mathematics for the Life Sciences

Calculus, Modeling, Probability, and Dynamical Systems

 Springer

Glenn Ledder
Department of Mathematics
University of Nebraska-Lincoln
Lincoln, NE, USA

ISSN 1867-5506   ISSN 1867-5514 (electronic)
ISBN 978-1-4939-4452-1   ISBN 978-1-4614-7276-6 (eBook)
DOI 10.1007/978-1-4614-7276-6
Springer New York Heidelberg Dordrecht London

Mathematics Subject Classification (2010): 92-01, 92B05, 00-01, 00A06, 00A71, 37-01, 60-01

To my wife, Susan, for her inexhaustible patience

Preface

Science is built up with facts, as a house is built with stones. But a collection of facts is no more a science than a heap of stones is a house.

Jules Henri Poincaré

There are several outstanding mathematical biology books at the advanced undergraduate and beginning graduate level, each with its own set of topics and point of view. Personal favorites include the books by Britton, Brauer and Castillo-Chavez, and Otto and Day. These books are largely inaccessible to biologists, simply because they require more mathematical background than most biologists have. This book began as lecture notes for a course intended to help biologists bridge the gap between the mathematics they already know and what they need to know to read advanced books. The only prerequisite for the course was the first semester of the calculus sequence. Topics included mathematical modeling, probability, and dynamical systems. My original notes included a brief review of calculus, which I subsequently expanded into the first chapter of this book so that it could be used for courses that do not require a calculus prerequisite or by biologists whose calculus experience is but a distant memory. Most students will probably find this book to be more challenging than the typical calculus book, albeit in a different way. I do not make as many demands on students' computational skills, but I require a greater conceptual understanding and an ability to harness that conceptual understanding for service in mathematical modeling.

A Focus on Modeling

In its early days, science consisted of careful observation and experimentation, with a focus on collecting facts. However, as eloquently stated by the French mathematician, philosopher, and scientist Henri Poincaré, this is not enough to make science work.

In contrast with science, mathematics is a purely mental discipline focused entirely on structures that we create in our minds. It can be very useful in science, but it has to be connected to science carefully if scientifically valid results are to be achieved. The connection is perhaps best made by a metaphor:

The muscles of mathematics are connected to the bones of experimental science by the tendons of mathematical modeling.

As you read through this book, you will see that mathematical modeling goes far beyond the "application" problems that mathematics text authors include to make mathematics appear relevant. The problem is that what little modeling work appears in these problems is generally done by the author rather than the students. At best, the experience of doing these problems only benefits science students if their science instructors are also good enough to do the modeling work for them.

This book is written from a modeling perspective rather than a mathematics or biology perspective. The lack of modeling content in the standard mathematics and science curricula means that the typical reader will have little or no modeling experience. Readers may find the modeling skills of Section 1.1 and Chapter 2 to be difficult to learn, but the effort to do so will be well rewarded in the remainder of the book and in any subsequent attempts to read biological literature with quantitative content. While it is unreasonable to expect readers of this book to become expert modelers, my primary goal is to make them sufficiently comfortable with mathematical modeling that they can successfully read scientific papers that have some mathematical content.

Pedagogy

There are a lot of connections between mathematics and biology, yet most students—and even many mathematicians and biologists—are unaware of these connections. One reason for this situation is that neither the historical development nor the pedagogical introduction of either subject involves the other.

Biology grew out of natural philosophy, which was entirely descriptive. Modern biology curricula generally begin with descriptive biology, either organismal or cellular. The mathematically rich areas of genetics, ecology, and physiology make their appearance in advanced courses, after students have come to see biology as a non-mathematical subject.

Calculus and calculus-based mathematics were developed to meet the mathematical needs of physics, and it remains standard practice to use physics to motivate calculus-based mathematics. Other areas of mathematics, such as game theory and difference equations, were motivated to some extent by biology,[1] but these topics appear in specialized courses generally taken only by mathematics majors. Probability is another mathematical topic with strong connections to biology, but it is generally encountered in statistics courses that emphasize social science applications.

> *The basic premise of this book is that there is a lot of mathematics that is useful in some life science context and can be understood by people with a limited background in calculus, provided it is presented at an appropriate level and connected to life science ideas.*

This is a mathematics book, but it is intended for non-mathematicians. Mathematicians like to have a mathematical definition for a concept and consider the meaning of the concept to be a consequence of that mathematical definition. Contrary to that plan, I prefer to begin with a functional definition and then present the mathematical definition as the solution of a problem. For example, in probability I define each distribution according to its purpose rather than its mathematical representation and then present the mathematical representation as a result. This is pedagogically appropriate; there are infinitely many functions that satisfy the mathematical definition of a probability distribution, and we should only be interested in those that have some practical value. The context should precede the mathematical definition.

---

[1] Leonardo of Pisa, more commonly known as Fibonacci, developed his famous sequence as the solution of a difference equation model of population growth.

A mathematics book for non-mathematicians needs to be clear about the extent to which rigor matters. A colleague of mine once started a talk to undergraduates with a joke: "An engineer, a physicist, and a mathematician are traveling in a train in Scotland when the train passes a black sheep standing along the track. The engineer concludes that sheep in Scotland are black. The physicist concludes that there is at least one black sheep in Scotland. The mathematician concludes that ..." Mathematicians have no trouble finishing the joke: the mathematician concludes that there is a sheep in Scotland that is black on one side. This insistence on rigor is both a strength and a weakness. It was long the common practice in calculus books (and such books are still popular) not to introduce the logarithmic function until after the definite integral, even though the students have seen logarithmic functions in precalculus. This example and others support my contention that "Mathematicians are people who believe you should not drive a car until you have built one yourself."

It is my aim to provide a balanced approach to mathematical precision. Conclusions should be backed by solid evidence and methods should be supported by an understanding of why they work, but that evidence and understanding need not have the rigor of a mathematical proof. At the risk of stern rebukes from my mathematics colleagues, I will say up front that I believe that students should focus on how we use mathematical results to solve problems. For this goal, we need to know why mathematical results are true, but we do not need to know how we prove them to be true. An example is the limit result needed to derive the formula for the derivative of the exponential function. The proof of this result appears in most calculus books and is indeed a beautiful piece of mathematics; however, understanding it does not help us compute derivatives or apply them to solve problems. Graphs and numerical computations strongly hint at the correct limit result. While not rigorous, these methods are more convincing to anyone but a professional mathematician and use problem solving skills that will be useful in other contexts. Similarly, the derivation of the Akaike information criterion (AIC) is very difficult, otherwise it would have been done prior to its actual discovery in the 1970s; nevertheless, it is not difficult to explain AIC in general terms. The mathematical error of presenting it without proof is far less serious in this book than would be the modeling error of omitting it.

Most of the sections are highly focused, often on one extended example. Mathematics experts know that we learn much more from a deep study of one problem than from many superficial examples. Many of my biological settings are in ecology, the area of biology I know best, but I have also tried to find settings of very broad interest such as environmental biology, conservation biology, physiology, and the biology of DNA. In particular, these areas are more likely to interest lower division undergraduates, many of whom are pre-medicine majors rather than biology majors, and most of whom have very little knowledge of biology.

I have attempted to be brief, in the hope that readers will work harder to read a short presentation than a long one. I use examples as contexts in which to present ideas rather than instances where a formula is used to obtain an answer. Hence, the number of examples is limited, but each example is treated with some depth. Similarly, I include only a small number of figures. Each figure is essential to the presentation, and the reader should work hard to understand each one. Being able to explain[2] a figure represents a high level of understanding.

Broad modeling problems require a variety of mathematical approaches. Hence, some topics are ideal for problems that are distributed among the relevant sections rather than being incorporated into a single project. I have indicated these connections within the problems themselves and also called attention to them in each chapter introduction. It is possible to combine all of the problems on a given model into one large project if desired.

---

[2] An *explanation* includes context and analysis in addition to mere *description*.

Technology

Some mathematical modeling work must be done by hand, while other work is greatly facilitated by the use of computers. I view both hand methods and computational methods as tools in my modeling toolbox. I try to identify the best tool for any particular task without a bias either for or against technology. I do have a bias against using computer algebra systems to do routine algebra and calculus. This stems from frequently encountering problems where valuable results can be found only with the careful use of algebraic substitution and simplification, which requires a human touch. I could not resist the temptation to point some of these out in the text.

There are a multitude of platforms for doing mathematical modeling tasks on computers. None of these is ideal, and the choice of which to use is a matter of taste. Rather than trying to find the very best tool for each individual task, my preference is to work with one tool that is reasonably good for any task (save symbolic computation) and is readily available. By these criteria, my choice is R, which runs smoothly in any standard operating system and is popular among biologists. Matlab is also an excellent choice. Both R and Matlab are programming environments, as opposed to packaged software or programming languages. Spreadsheets and other packaged software provide easy access to mathematics because of their intuitive graphical interface; however, programming is limited and the details of formulas are hidden from view, making it impossible to see the overall structure of a program at a glance. Reusability is limited, as anyone who has tried to modify a spreadsheet created by another author can attest. By comparison, one can see an entire R or Matlab program at a glance and adapt prior work to a similar context with minimal changes. High-level languages, such as Java and C++, offer sophisticated programming capabilities, but they are difficult to learn compared to the languages used in programming environments such as R or Matlab.

The choice between R and Matlab is a matter of personal taste. It is easier to get professional-quality graphics with Matlab, but R has a more intuitive syntax that facilitates programming. Matlab requires an add-on toolbox for probability and statistics, while R requires supplementation for dynamical systems. I use R because students can get it for free and install it seamlessly in any operating system. R lacks the excellent documentation that comes with Matlab; however, I maintain a collection of R scripts for various algorithms presented in the text, and these are readily available http://www.springer.com/978-1-4614-7275-9. These scripts are designed to be simple rather than robust; that is, compared to professionally written programs, they are easier to understand but less efficient and they lack error detection machinery. Their presence allows students to replace the difficulty of having to learn R from scratch with the much lesser difficulty of having to be able to read an R program and make minor modifications.

Topics

Of course, no book on mathematics for the life sciences can be complete. Some important areas do not appear here at all because, they do not fall into the broad categories of mathematical modeling, probability, and dynamical systems. Several concessions have been made in the interest of accessibility. Some topics are given only a partial treatment as compared to the treatment they would receive in a higher level course; for example, I do not find eigenvectors for complex eigenvalues, since they are not generally needed in biology. Others are presented in a roundabout way.

Finding the "correct" order of topics in this book was an insoluble problem. Mathematics is a hierarchical subject, but the hierarchy is not linear. Arguments can be given for significant rearrangements of the topics that are included here. Ultimately, the only reasonable solution was to group topics into related clusters. In particular, Parts II and III could easily have been reversed. Those who read this book for their own benefit or to design a course should be flexible in the way they structure their study. Each part introduction contains a graph with sections as nodes and arrows indicating which sections are necessary background for others. A syllabus that moves frequently between chapters is entirely possible, but for me to have written the book in that way would have excluded other topic orders.

One feature of mathematical models that causes difficulties for students is the appearance of parameters, which are constants whose values are not necessarily assigned. Without parameters, a function is merely an example to be used for routine calculations. With parameters, a function can be a model, which can serve as an environment for theoretical experimentation. Even the reader with a solid background in calculus should study Section 1.1.

The remainder of Chapter 1 can serve as a review of calculus or a conceptually oriented calculus primer. This chapter is not a complete treatment of calculus, which would require far more space than is available in one chapter. I present here only those aspects of calculus that provide the necessary background for the modeling, probability, and dynamical systems that make up the rest of the book. The reader who works through this chapter will be well equipped with the calculus background needed for the purpose at hand. The material in this book has been used successfully with life science graduate students who had no background in calculus. Anyone who requires a more complete understanding of calculus can consult any calculus book.

After the calculus primer comes a chapter on mathematical modeling, which is the necessary focus of any study of mathematics for those whose purpose is to use mathematics to better understand science. Even the most mathematical of topics, such as probability, are best seen by scientists from a viewpoint of mathematical modeling. Unfortunately for the science student, mathematical modeling has not been granted a place in the standard mathematics and science curricula. In mathematics books, we generally present mathematical ideas and then look for their applications to science. The result is a collection of idiosyncratic examples devoid of the analysis necessary for good mathematical modeling. In science books, the mathematics is usually presented as a collection of formulas, to be used as facts when required. Neither approach teaches modeling skills. If we are to use mathematics to improve our understanding of the natural and physical world, we must focus on the connections of mathematics to science.

Chapters 3 and 4 present the basic ideas and some applications of probability, including applications commonly classified as statistics. The treatment given here is organized differently from the treatment of this topic in statistics or probability books. Mathematicians generally use an axiomatic approach to introduce probability. My colleagues in biology helped me appreciate that the central topic of probability for scientists is that of the probability distribution, and this topic is best approached informally by thinking of a probability distribution as a mathematical model of a data set. My aim has been to get to probability distributions as quickly as possible while saving other topics, such as conditional probability, for later. The essentials of probability distributions form the subject of Chapter 3. Chapter 4 includes additional topics that build onto or supplement the basic material on probability distributions. The high point of these two chapters is Section 4.4, which looks at the question of how likely it is that a subpopulation used to provide sample data is distinct from some larger population.

The final three chapters introduce the mathematics of dynamical systems, which consist of one or more related quantities that change in time according to prescribed rules. These rules may be in the form of difference equations, where time is taken as discrete, or differential equations, where time is taken as continuous. It is usual to make this the primary distinction within the area of dynamical systems; however, there are valuable connections to be made between the two kinds, particularly for models with only one dynamic quantity. For this reason, I have chosen to treat all dynamic models of one variable together in Chapter 5 before presenting multivariable discrete systems in Chapter 6 and multivariable continuous systems in Chapter 7. For reasons presented in the modeling chapter, I believe that continuous models are almost always preferable to discrete models. Nevertheless, the analysis of continuous models requires an understanding of some discrete mathematics. Hence, Chapter 6 precedes Chapter 7. The reader whose primary interest is in continuous dynamical systems needs the tools developed in Section 6.3, and these tools are more easily acquired with Sections 6.1 and 6.2 as background. The high points of the three chapters on dynamical systems are the graphical and analytical tools used for continuous systems; these are the topics of Sections 7.3 and 7.5 respectively. The book contains three additional sections on discrete dynamical systems, presented in Appendix A.

Advice for the Reader

How one reads a book depends on what one wishes to get from the reading. I assume that my reader wants a working knowledge of mathematics that will enable him/her to read biological literature with quantitative content or to read a more advanced book on mathematical biology. At the same time, many readers will be interested in only a portion of the topics presented here. As noted above, each part begins with a schematic diagram showing the logical relationships among the topics of that part and any essential topics from earlier parts. In particular, the reader is cautioned not to skip Chapters 1 and 2 to get to some other topic more quickly. People who try to learn to play the organ without having already learned to play the piano are starting with an enormous handicap; the same is true for anyone who attempts to learn probability or dynamical systems without an adequate mastery of calculus and mathematical modeling. Not every section in Chapters 1 and 2 is essential for the remainder of the book; however, parts of these chapters are indispensable background and should be mastered before moving on. In particular, an understanding of parameters (Section 1.1) and the basic concepts of mathematical modeling (Section 2.2) is essential.

It is natural to try to work a large number of problems as quickly as possible. However, this is not the best way to learn mathematics. A mathematician learning something new will work through a relatively small number of examples carefully rather than a large number of examples superficially. At a talk I heard on mathematics pedagogy, the speaker asked the audience, "Why do we ask our students to work problems? Is it because we want to know the answer?" Usually we don't care about the answer; we work problems to learn mathematics. Keep this in mind when you are working a problem: your goal is to learn mathematics, not to get the answer to the problem. There are only a small number of routine problems in this book. Most problems are guided case studies and require quite a bit of time for a thorough understanding. Carefully working a small number of these will benefit the reader more than a cursory look at a larger number.

Course Designs

There is no standard curriculum of mathematics for biology. Mathematical biology can be incorporated into a calculus course, or calculus can be incorporated into a mathematical biology course for students who have not had calculus. There are also mathematical biology courses with a calculus prerequisite, and these can be offered for students with or without backgrounds in linear algebra and differential equations. Many institutions treat probability/statistics as being distinct from mathematics. However, the difficulty of finding room for either in the program of a biology major suggests the idea of incorporating some probability and some topics often included in a statistics course within the mathematical biology or calculus-for-biology course. I have tried to make this book suitable for a variety of plans.

Before listing possible course plans using the material in this book, it is important to start with a broad discussion of pacing. Books for some lower division courses are generally written under the assumption that each section will require 1 day of class. At this pace, it is not difficult to put more than 30 sections into a standard 3-credit course. This is typically what is done in a differential equations course, but not a calculus course. At the University of Nebraska, we cover something like 32 sections in our first-semester calculus course; however, we structure this course with a lecture-recitation format and offer it for 5 credits. This means that our actual rate of coverage is more like 6 sections per credit hour than 10 sections per credit hour. My own mathematics-for-biology course was originally a 5-credit course with a one-semester calculus prerequisite, and I covered approximately 32 sections of this book, which is again an average of only 6 sections per credit hour. The next time I use this book for a course, it will be a 5-credit calculus-for-biology course with 3 h of lecture and 2 h of recitation/laboratory. I expect to do only the 24 sections of Chapters 1, 2, 5, and 6. For students at the calculus level, I would certainly not try to do more than 16 sections for a 3-credit course. A slightly faster pace could be used with students who are more sophisticated or very highly motivated. I spend less than half of the total class time presenting lectures; in particular, I mark out days for laboratory-style activities such as collecting data from a virtual laboratory, writing a computer program to run a computer simulation, or working through one of the more difficult problems either in small groups or as a "committee of the whole." The standard requirement that mathematics courses cover as much material as possible sacrifices depth for breadth; a mathematics course for biology students should have some balance between the two, with some case studies being included at the expense of broad coverage.

A 2-Course Sequence of 4-Credit Courses

It should be possible to do almost the entire book with a total of 8 credit hours. I would do Chapters 1, 2, and 5 in a first-semester calculus-for-biology course and most of Chapters 3, 4, 6, and 7 in a second-semester probability and dynamical systems course.

A 2-Course Sequence of 3-Credit Calculus-for-Biology Courses

Given two courses for students with no calculus background, I would use Chapters 1 and 2 for the first semester and then make the second semester a dynamical systems course that would include Chapters 5–7, and possibly with parts of Appendix A. Both of these courses would be well focused, and the second course could be open to strong students with a background somewhat beyond one course in calculus.

A 3-Credit Calculus-for-Biology Course

In a 3-credit calculus-for-biology course, I would expect to complete all of Chapter 1 in about half of the semester or perhaps a little more. I would probably try to do some dynamical systems rather than a complete treatment of Chapter 2. It would be possible to do Sections 2.1, 2.2, 2.5, and 2.6 along with all of Chapter 5. I would present only a minimal version of Section 2.5, the point being to do just enough to set up Section 2.6.

A 3-Credit Empirical Modeling and Probability Course with a Calculus I Prerequisite

One could teach a course on empirical modeling and probability as an alternative to a standard statistics course. For such a course, I would do Sections 1.1, 2.1–2.4, 2.7, all of Chapter 3, and as much of Chapter 4 as could be done without rushing.

A 3-Credit Dynamical Systems Course with a Calculus I Prerequisite

A course on dynamical systems could not reasonably assume an adequate modeling background, so it would be necessary to start with Sections 1.1, 2.1, 2.2, 2.5, and 2.6, with 2.1 done in a cursory manner. It would then be possible to cover all of the material in Chapters 5–7. If any extra time is available, Section A.1 would round out the course.

Lincoln, NE, USA   Glenn Ledder

Acknowledgements A number of people contributed to this book in significant ways. The initial development of this material was funded by a grant from the National Science Foundation. The presentations of Section 1.1 and Chapter 6 grew out of suggestions from Lester Caudill and Sebastian Schreiber respectively. Lester Caudill, Renee Fister, Meredith Greer, Carole Hom, Eric Marland, Pam and Phil Ryan, Elsa Schaefer, and Robin Snyder read early drafts of the manuscript. Rebecca Ledder read the penultimate draft of the entire manuscript and made many suggestions for the improvement of clarity and concision. My editors, Achi Dosanjh and Donna Chernyk, worked with me to turn the manuscript into a book. All of these people helped improve the result; I alone am responsible for any flaws.

Contents

Part III Dynamical Systems

Part I
Calculus and Modeling

The first two chapters contain core material in mathematics: calculus and the basic elements of mathematical modeling. Section 1.1 is an essential preparatory section for the entire book. A significant part of the power of mathematics lies in its capacity for generalization. A single symbol can represent a range of numerical values, allowing the mathematical work to be done on a whole class of problems rather than an individual example. One cannot read any quantitative work in biology or any other science without an ability to understand how symbols are used in a particular context. This is a topic that most readers will find difficult, but one that is essential to master.

Aside from the opening section, Chapter 1 contains five sections on differential calculus and three on integral calculus. Each of these topics includes material on concepts, techniques, and applications. Contrary to the view of most students, calculus is largely a conceptual topic. Concepts are needed to understand the important applications of calculus; hence, the reader should spend enough time on the conceptual material for a thorough understanding. In particular, the derivative concept is essential to understanding dynamical systems, while the definite integral concept is essential for continuous probability distributions. The amount of effort to be expended on techniques is a matter of taste. Differentiation techniques are needed in the rest of the book, but all such cases are fairly elementary. Integration techniques are not needed for the rest of the book. Both of these can be done using computer algebra systems if desired. The applications that appear in Chapter 1 can generally be considered as ends in themselves, and can be accorded as much or as little interest as the reader desires. There are a few exceptions. Related rates (in Section 1.6) are vital background for nondimensionalization, which appears in Section 2.6, and linear approximation (in Section 1.4) is vital background for the nonlinear dynamics that appears in Chapter 5 and 7 and Appendix A.1.

In the preface, I described mathematical modeling as the tendons that connect the muscle of mathematics to the bones of science. Colleagues who teach science and engineering courses often say that their students do not seem to be able to do the mathematics necessary for their subject. The real problem is not so much an inability to do mathematics but an inability to harness the power of mathematics in a scientific context. More or different mathematics will not address this problem; it requires attention to mathematical modeling, which is largely absent from courses in either mathematics or science. This is the purpose of Chapter 2. The first two sections provide basic terminology and ideas. There does not seem to be a lot of material in these sections, and there are very few associated problems. The reader should plan to reread these sections several times while working through the rest of the book, as the ideas in them are hard to understand well without prior experience in mathematical modeling. The remainder of the chapter is divided into sections on mechanistic modeling, which starts with assumptions about the scientific setting, and empirical modeling, which starts with examination of data. The three sections on empirical modeling are of value to anyone who collects and analyzes data, but they are not essential background for the rest of the book. The basic ideas of mechanistic modeling (Section 2.5) are helpful to try to work through, but the reader does not need to be expert on this subject. Most biologists need to be able to read and understand mechanistic models but do not need to be able to create them. The reader of Part III of this book is often asked to interpret mathematical models mechanistically, but is never asked to construct one. In contrast, nondimensionalization (Section 2.6) is a vital skill for anyone who wants to do any work with dynamical systems. It would have been impossible to write a useful introduction to dynamical systems using complete versions of well-known models without relying on the power of nondimensionalization to simplify model analysis.

The accompanying sketch shows the interdependencies of the sections in Part I. Sections 1.1–1.3 and 2.2 are necessary background for the remainder of the book. Sections 1.7 and 1.8 are needed for Part II, while Section 2.5 and 2.6 are needed for Part III. Sections 1.4–1.6, and 1.9 are important topics in calculus, but are not necessary for the rest of the book; similarly, the empirical modeling topics in Section 2.3, 2.4, and 2.7 are not needed later.

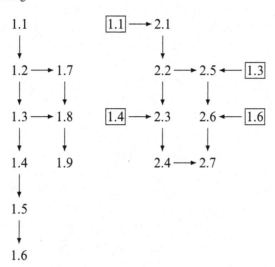

Chapter 1
A Brief Summary of Calculus

Calculus is one of the greatest intellectual achievements of humankind. It allows us to solve mathematical problems that cannot be solved with ordinary algebra, and that in turn allows us to make predictions about the behavior of real-world systems that could not be made with algebra alone. Beyond its usefulness, calculus has an elegant beauty that leads mathematicians to view it as a work of art.

A full presentation of calculus requires the equivalent of four standard-length American college courses, although this is usually achieved through three larger one-semester courses. Because many methods in probability and dynamical systems are calculus-based, biologists need to know some calculus in order to proceed to these other areas. However, much of the important material in probability and dynamical systems does not require a complete calculus background. The typical reader of this book has probably taken no more than the first course in calculus (corresponding to barely one-third of the full presentation), and this should be adequate. Other readers may not have had a good first course in calculus, or their prior calculus study may be part of the distant past. We therefore begin with a brief summary of the calculus background required for the mathematical work that follows. Most of this material corresponds to that found in standard calculus books, but there are some exceptions. First, standard calculus books have nothing comparable to the first section of this chapter. Second, the presentation in this chapter is more intuitive and less formal than what is typically found in calculus books. The reader who wants a more complete or rigorous treatment should still find the material of this chapter to be helpful background when reading a more standard presentation of calculus.

Many calculus books begin with a review of topics drawn from prior courses, particularly precalculus and trigonometry. In the interest of space, these topics are not presented here. The reader should have a reference available for use when needed. However, there is one precalculus topic that is always overlooked in mathematics books and which is therefore included here: parameters. Mathematicians are comfortable with abstraction and easily appreciate the role each mathematical symbol plays in a formula. However, this vital skill does not come easily to most people, and this is why a study of parameters is the place to start a study of calculus. The reader is therefore urged not to move beyond the opening section without having first acquired some facility with the material in it.

Section 1.2 discusses rates of change for functions defined only at discrete points, functions defined on a continuous interval but known only at discrete points, and functions defined and known on a continuous interval. Discrete rates of change and discrete approximations to a continuous rate of change are computed using algebra, but rates of change for continuous functions cannot be calculated exactly without using calculus. The problem of computing rates of change leads to the definition of the derivative.

G. Ledder, *Mathematics for the Life Sciences: Calculus, Modeling, Probability, and Dynamical Systems*, Springer Undergraduate Texts in Mathematics and Technology, DOI 10.1007/978-1-4614-7276-6_1, © Springer Science+Business Media, LLC 2013

The development of differential calculus continues in Section 1.3 with techniques for using a small number of predetermined derivative formulas and rules to obtain derivative formulas for all functions composed of simple elements. Complete mastery of differentiation techniques is unnecessary, but the reader should have enough experience to be able to compute the relatively simple derivative formulas used in subsequent application problems.

Sections 1.4–1.6 present additional topics in differential calculus, beginning in Section 1.4 with several mathematical ideas that can be grouped together under the broad heading of *local behavior*, an imprecise term that essentially means "what you see when you zoom in on a point on a graph." In some cases, this means identifying a line tangent to the graph of a function; in other cases it means identifying local maximum and minimum points on a graph. The following section treats the related subject of optimization. Here one wants to find the value of a variable or parameter that maximizes or minimizes some desired objective function. This is global behavior, rather than local behavior, but the topics are related because local extrema are points where global extrema are likely to occur. Optimization problems occur naturally in biology. Natural selection works by tuning an organism's genome to solve optimization problems; hence, we often know the solution of a biological optimization problem even though we don't fully understand the problem itself. It can be an interesting modeling task to try to construct a mathematical optimization problem whose solution corresponds to the biological solution we see in observation or experiment. Section 1.6 contains two mathematical topics that can be grouped together as related rates. The general idea is that an algebraic equation that relates more than one dependent variable can be differentiated to produce an algebraic equation that relates the derivatives of those variables. These topics feature in the development of many models that will be seen later.

Having completed a brief summary of differential calculus, we turn to integral calculus for the final three sections. Section 1.7 introduces the concept of the definite integral, which shares center stage with the derivative in the drama of calculus. We begin by examining ways to determine how much of a quantity accumulates over time using the rate of accumulation. This is a way of introducing the definite integral that profits most from the reader's everyday intuition about accumulation of changes. The definite integral is also tied to aggregation over space and area under a graph.

Section 1.8 presents the key result that the integral of a rate of change is the total change. This result allows us to compute definite integrals if we can recognize the integrand as the derivative of a known function. Section 1.9 elaborates on this theme by extending the method to integrals over infinite regions and integrals that can be converted to a more tractable form through the method of substitution. As with differentiation, the aim of this treatment of integration is not to make the reader a master of technique, but to enable the reader to do the more elementary computations of calculus that are needed in application examples.

After studying this chapter, you should be very comfortable with the concepts of the derivative and the definite integral and how they are applied to a variety of problem types, and you should have an intuitive feel for continuous mathematics. You should be able to compute simple derivatives and definite integrals, but you do not need to be an expert at calculus computations.

The problem sets include several case studies that are split over multiple sections:

Section	1.1	1.2	1.4, 1.5	1.6	1.7	1.8, 1.9	Chapter 2
Measles infection	1.1.8		1.4.1			1.8.9	
Weightlifting	1.1.13						2.4.9
Population growth	1.1.15				1.7.9	1.8.12	2.2.7
Demographics	1.1.16	1.2.11				1.8.12	
Swimming speed of fish			1.5.3				2.6.6
Dogs and calculus			1.5.5	1.6.7			2.6.7
Organism growth				1.6.5		1.9.11	2.4.7
Optimal organism size					1.7.10	1.9.12	

1.1 Working with Parameters

After studying this section, you should be able to:

- Perform algebraic manipulations on functions with parameters.
- Graph functions with parameters.
- Identify the mathematical significance of a parameter in a function.
- Interpret graphs of system properties in terms of a parameter.

We begin with an application problem typical of those found in a precalculus text.

Example 1.1.1. Q) Jan takes two acetaminophen tablets (650 mg) for her headache. The amount y of acetaminophen in Jan's system is given by the function

$$y = 650e^{-0.3t}, \tag{1.1.1}$$

where t is the time in hours after the dose and y is given in milligrams. At what time will there be only 130 mg of acetaminophen in Jan's system?

A) We solve this problem by setting $y = 130$ in (1.1.1) and solving for t:

$$e^{-0.3t} = \frac{130}{650} = 0.2.$$

We can take a natural logarithm now, but it is easier to take the reciprocal of the equation first:

$$e^{0.3t} = 5.$$

Now the natural logarithm yields

$$0.3t = \ln 5,$$

or

$$t = \frac{\ln 5}{0.3} \approx 5.36.$$

According to (1.1.1), the amount of acetaminophen in Jan's system will be 130 mg at about 5 h and 22 min (5.36 h) after the dose.   □

Example 1.1.1 is fine as far as it goes, but that isn't very far. There is an underlying mathematical model, which we can write as

$$y = Ae^{-kt}, \qquad A, k > 0, \tag{1.1.2}$$

where the quantities A and k are *parameters*.

Parameter: *a quantity in a mathematical model that can vary over some range, but takes a specific value in any instance of the model.*

Equation (1.1.1) is only an *instance* of the model (1.1.2).[1] The model has been *used*, but not fully *utilized*, because the results cannot be extended to related or broader questions.

---

[1] We'll consider more general versions of the model in Examples 1.1.3–1.1.5.

1. If we want to see what would happen with a different initial dose, or with aspirin instead of acetaminophen, we have to repeat the whole calculation.
2. We obtain quantitative results but no useful qualitative information.

In contrast, the same mathematical work can be done with the full model rather than an instance of the model, with advantages corresponding to the drawbacks of Example 1.1.1.

1. Work done in solving the full model need not be repeated when parameter values are changed.
2. Analysis of the full model can yield qualitative results that disclose model weaknesses or enhance scientific understanding.

Analysis of mathematical models requires more experience working with parameters than the reader of this book is likely to have—hence the need for this opening section. An understanding of parameters is also helpful for understanding the concepts and methods of calculus.

1.1.1 Scaling Parameters

Example 1.1.2. Q)   Consider the one-parameter family of functions defined by $f(x) = mx$, where $m \in \mathbb{R}$.[2] How are we to understand this family of functions?

A)   With modern technology, we can easily plot graphs and do computations with specific functions (that is, functions with no unspecified parameters). The presence of parameters makes the job more difficult. As a first try, we can graph several examples of the function by selecting a few values of the parameter. Figure 1.1.1 illustrates the family $y = mx$.[3] Graphically, the effect of the parameter m is to control the slope of the line $y = mx$. Algebraically, the effect of the parameter m is to modify all function values by the multiplicative constant m. This property of proportional change marks m as a *scaling* parameter.   □

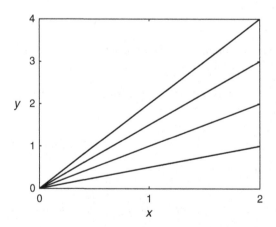

Fig. 1.1.1 The function $y = mx$, with $m = 0, 0.5, 1, 1.5, 2$

Scaling parameters have a very simple effect on the graph of a function. They magnify the function values by a fixed factor, so they cannot change the "shape" of the graph.

---

[2] This notation means that the parameter m can be any real number.

[3] We use the word "illustrate" rather than "plot" or "graph" here because we can't simultaneously plot all lines $y = mx$. Instead, we plot several representative examples together and infer the properties of the whole family.

Example 1.1.3. Consider the family of functions

$$y = Ae^{-0.3t}.$$

The parameter A in this family, like the parameter m in the family $y = mx$, is a scaling parameter. In Example 1.1.1, where $A = 650$, the function value is reduced from 650 to 130 in 5.36 h. In the general case, the function value after 5.36 h is

$$y = Ae^{-(0.3)(5.36)} = 0.2A.$$

A time of 5.36 h reduces any dose to 20% of its initial value. Figure 1.1.2 illustrates the family of functions, using $A = 650$ and $A = 325$, corresponding to doses of one and two standard acetaminophen tablets. The scaling parameter no longer represents the slope of the graph, but it remains true that two instances with values of A differing by a factor of 2 have results at any given time that also differ by a factor of 2. $\qquad\square$

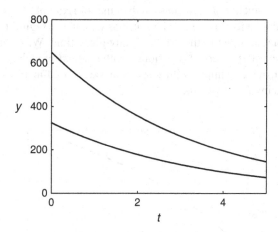

Fig. 1.1.2 The function $y = Ae^{-0.3t}$, with $A = 325, 650$

We are now ready to rework Example 1.1.1 using the family $y = Ae^{-0.3t}$.

Example 1.1.4. Q) Jan takes acetaminophen of initial dose A mg. The amount y of acetaminophen in Jan's system is given by the function

$$y = Ae^{-0.3t}, \tag{1.1.3}$$

where t is the time in hours after the dose and y is given in milligrams. At what time will there be only 130 mg of acetaminophen in Jan's system?
A) We must solve

$$130 = Ae^{-0.3t}$$

for t, which we can do following the same steps as in Example 1.1.1. Dividing by A yields

$$e^{-0.3t} = \frac{130}{A}.$$

Taking the reciprocal of the equation gives us

$$e^{0.3t} = \frac{A}{130},$$

and then the natural logarithm yields

$$0.3t = \ln\left(\frac{A}{130}\right) = \ln A - \ln 130.$$

Thus, we arrive at the answer

$$t = \frac{\ln A - \ln 130}{0.3}. \tag{1.1.4}$$

Because no numerical value has been specified for A, the answer retains A as a parameter. We can understand this answer by plotting a graph of time to 130 mg against A, shown in Figure 1.1.3. Doses smaller than 650 mg take less time to be reduced to 130 mg than a 650-mg dose, but the curvature of the graph means that a dose of only half the size takes more than half as long.   □

Notice that A is *not* a scaling parameter in (1.1.4). It magnifies $y(t)$ by a fixed factor, but its effect on answers to more complicated questions is more subtle. Consequently, the result of Example 1.1.4 contains much more information than that of Example 1.1.1. We can now easily determine the required time for any size of dose, just as we can use Figure 1.1.2 to determine the amount of medication remaining in the system at any given time. While application problems generally consist of simple numerical questions, mathematical models permit a large variety of questions, some answered simply with numerical values or functions of time and others answered by formulas involving parameters.

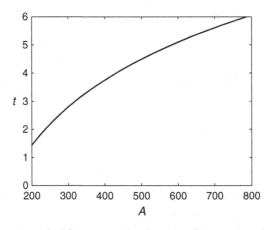

Fig. 1.1.3 The time (in hours) required for an acetaminophen dose of A mg to be reduced to 130 mg

Scaling parameters simplify mathematical analysis. In Section 2.4, we will see a method of analysis that works for $f(x) = qx/(a+x)$, but does not work for $g(x) = sx/(1+shx)$; the method relies on identification of q as a scaling parameter. The simple properties of scaling parameters will also be utilized in Section 2.6.

1.1.2 Nonlinear Parameters

Most parameters are not scaling parameters. We can still study their effects by using thought experiments similar to those of Examples 1.1.3 and 1.1.4.

Example 1.1.5. Q) Determine the key features of the one-parameter function defined by $f(t) = e^{-kt}$, where $k > 0$, on the interval $t > 0$.

A) All of these functions have $f(0) = 1$, all are decreasing, and all are positive. One way to see the effect of k is to find the times required for the function to decrease to one-half of its initial value. Let T be the desired time. Then

$$e^{-kT} = \frac{1}{2},$$

or

$$e^{kT} = 2.$$

Taking logarithms yields $kT = \ln 2$. Thus, e^{-kt} is reduced to one-half of its initial value at time

$$T = \frac{\ln 2}{k}. \tag{1.1.5}$$

The time T defined here is called the *half-life* for the decaying exponential function $y = e^{-kt}$. The half-life decreases as k increases, so larger k values make the function decrease faster. Figure 1.1.4 illustrates the function e^{-kt}.   □

Decaying exponential functions of the form $y_0 e^{-kt}$ are used to model radioactive decay, drug clearance, and many other phenomena. The parameter y_0, which represents the initial value, is entirely dependent on context. The parameter k, which represents the decay rate, is a property of the specific quantity undergoing decay. Published sources generally provide the half-life T rather than the rate parameter k, but the relationship in (1.1.3) makes it easy to connect the desired parameter value to the known half-life.

Example 1.1.6. Suppose the half-life of a drug in the human body is 4 h. Then the rate constant k is

$$k = \frac{\ln 2}{4} \approx 0.173 \, \text{hr}^{-1}.$$

□

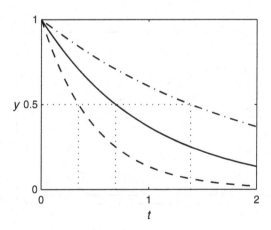

Fig. 1.1.4 The function $y = e^{-kt}$, with $k = 0.5$ (*dash-dotted line*), $k = 1$ (*solid line*), and $k = 2$ (*dashed line*), showing the points $(T, 0.5)$

1.1.3 Bifurcations

In each of our examples so far, the parameter has had only a quantitative effect on the function. Nonlinear parameters can have a qualitative effect as well.

Example 1.1.7. Q) Determine the nonnegative solutions of the equation $f(x) = x^3 - 2x^2 + bx = 0$, where b is any real number.

A) We can start by factoring out an x:

$$f(x) = x^3 - 2x^2 + bx = x(x^2 - 2x + b).$$

Thus, the function f always has a root at $x = 0$ and may have additional roots at points where $x^2 - 2x + b = 0$. Applying the quadratic formula to the latter equation, we have

$$x = \frac{2 \pm \sqrt{4 - 4b}}{2} = 1 \pm \sqrt{1 - b}.$$

There are two special cases. At $b = 1$, the quadratic formula yields only one root, $x = 1$; hence, the roots are $x = 0$ and $x = 1$. At $b = 0$, one of the roots from the quadratic formula is 0 and the other is 2, so the roots are $x = 0$ and $x = 2$. These special cases divide the range of b values into three intervals where the numbers of roots are different.

- For $b > 1$, there are no roots from the quadratic formula.
- For $0 < b < 1$, the square root quantity is between 0 and 1; hence the quadratic formula yields two roots, one in the interval $0 < x < 1$ and one in the interval $1 < x < 2$.
- For $b < 0$, the square root yields a number larger than 1, so there is one positive root in the interval $x > 2$.

Figure 1.1.5 is a plot of the roots as a function of b. The horizontal line indicates that $x = 0$ is a root for all values of b. The curve indicates roots that come from the quadratic formula, showing that there is one such root if $b < 0$, two such roots if $0 < b < 1$, and no roots (other than 0) for $b > 1$. □

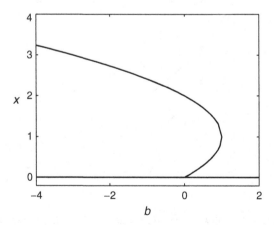

Fig. 1.1.5 The nonnegative roots of $x^3 - 2x^2 + bx$ as a function of b

The function $f(x) = x^3 - 2x^2 + bx$ of Example 1.1.7 is a case where the qualitative properties change as a parameter crosses a specific value. This kind of behavior is called a *bifurcation*.[4] Bifurcations are common in biological models and are often of critical importance. For example, conservation models generally include a parameter representing the ratio of populations in successive years. If this parameter is greater than or equal to one, the population will be successful, while a value of 0.999 represents a population headed for extinction.

Problems

1.1.1.* Rework Example 1.1.1 for an initial dose of 390 mg. Check your result for the time needed for the drug amount to drop to 130 mg against the appropriate figure in the text.

1.1.2.* For the data of Example 1.1.1, find a formula for the time t_z at which the amount of acetaminophen in Jan's system is some unspecified amount z. Plot a graph of t_z versus the parameter z. Interpret the graph in terms of the drug clearance process.

1.1.3. The clearance rate constant k in (1.1.2) is slightly different for different individuals, as well as being considerably different for different drugs. Suppose k is the clearance rate for acetaminophen of a patient who takes two acetaminophen tablets (650 mg). Determine the time t_1 at which a patient's system has only 130 mg of acetaminophen, in terms of the parameter k. Plot a graph of t_1 versus k. Interpret the graph in terms of the drug clearance process.

1.1.4. Do an Internet search to find the half-life of acetaminophen in a healthy body. Look at a variety of sites so that you will find a range of values. Determine the constant k for the largest and smallest half-lives and plot the functions $f(t) = Ae^{-kt}$ using these values together on the same axes. Use a standard dose for A.

1.1.5. Do an Internet search to find the half-life of naproxen sodium in a healthy body. Look at a variety of sites so that you will find a range of values. Determine the constant k for the largest and smallest half-lives and plot the functions $f(t) = Ae^{-kt}$ using these values together on the same axes. Use a standard dose for A.

1.1.6. Determine the nonnegative roots of the polynomial $x^3 - bx^2 + x$ as a function of b. Plot these roots as in Figure 1.1.5.

1.1.7.*

(a) The one-parameter function $f(x) = x^2 - bx$, where b is any real number, has a graph that is a parabola. Find the vertex of the parabola in terms of the parameter b. [Hint: All quadratic equations can be written in the form $f = a(x - h)^2 + k$, where the vertex is at (h, k) and a is a parameter that represents the "broadness" of the parabola. By setting the formula given for the function equal to the desired form and requiring the polynomials to be identical, determine three equations that relate the desired parameters h, k, and a to the given parameter b.]

(b) Plot f with $b = -2, 0, 2, 4$. Verify that the vertices are in the location determined in part (a).

1.1.8. Data for the number of infected cells during the course of a measles infection show roughly 1,470 infected cells at 14 days after exposure, which is approximately the peak of the infection [5].

---

[4] It is pronounced BYE-fur-ca-shun.

(a) What features of the function

$$f(t) = t(21 - t)(t + k), \qquad k > 0$$

make it a plausible model for the data?
(b) What value for k is necessary to get $f(14) = 1,470$?
(c) Plot f with the value of k from part (b). Your plot should only include the portion of the graph that is biologically relevant. Confirm that the peak is in fact close to 14 days after infection.

(This problem is continued in Problem 1.4.1.)

1.1.9. Influenza is an example of a disease that exists in multiple strains with frequent mutations. This means that individuals who recover from the disease are immune to the current strain, but not to the next strain to reach the population. Now suppose a new strain reaches a closed population. Under these circumstances, the standard model for such diseases[5] can be used to predict the fraction d of individuals who contract the disease in terms of the basic reproductive number R_0, a parameter that represents the average number of secondary infections due to one infected individual in a population of susceptibles. Values of R_0 can vary from 0 to about 20,[6] with measles estimated at 12–18 and smallpox at 5–7 [4]. If $R_0 < 1$, then the disease can only infect a few individuals before it dies out. If $R_0 > 1$, then a disease epidemic occurs, and the model predicts that the fraction who get the disease satisfies the equation

$$R_0 d + \ln(1 - d) = 0.$$

(a) Solve the equation for R_0 and use a "guess-and-check" method to estimate the value of d for $R_0 = 5$. Independent of disease mortality, what does this suggest about the impact of smallpox on a Native American population newly exposed by explorers in the late fifteenth century?
(b) Plot d as a function of R_0 for the range $0 < R_0 < 5$. (Keep in mind that the given formula is not correct for all values of R_0. Note that tabular data for two related variables can be plotted with either variable on the horizontal axis.)

1.1.10. A resource x is consumed at the rate

$$f(x) = \frac{Ax}{B + x},$$

where A and B are positive parameters.

(a) What is the maximum consumption rate in terms of A and/or B, achieved in the limit as $x \to \infty$?
(b) Suppose the actual consumption rate is half of the maximum from part (a). What is the corresponding resource level x?

1.1.11. In this problem, we consider the additional difficulties of modeling ibuprofen amounts in a patient, given that ibuprofen is slowly absorbed from the digestive system and relatively quickly eliminated from the body.

---

[5] See Problem 7.1.3.

[6] The mathematical epidemiologists Valerie Tweedle and Robert J. Smith? (The question mark in "Smith?" is part of the spelling of his name. I do not know whether this is to distinguish him from all the other Robert J. Smiths in the world or to disprove the popular belief that mathematicians are the most uninteresting people alive. I prefer not to ask.) claim that the most infectious "disease" is the social disease they call "Bieber fever," whose basic reproductive number they estimate to be about 27 [2].

(a) Plot a concentration curve for ibuprofen using the model $y = Ae^{-kt}$. Assume a dose of 200 mg and a rate constant of $k = 0.35$.

(b) How long does it take for one-half of the drug to be eliminated? (Do this algebraically, not by trial and error.)

(c) A more sophisticated drug clearance model takes account of the need for an oral dose to be absorbed by the digestive system. This model is

$$y = \frac{Ab}{b-k}(e^{-kt} - e^{-bt}),$$

where A is the amount of the dose, k is the clearance rate for the bloodstream, and b is the clearance rate for the digestive system. Normally, $b > k$. Use this model to plot a drug versus time curve for a 200-mg dose of ibuprofen, assuming $k = 0.35$ and $b = 0.46$.

(d) Describe the important differences between the curves of part (a) and part (c). Consider qualitative features of the graphs as well as quantitative properties such as the peak value and the time at which the peak occurs.

1.1.12.* Medication regimens for multiple doses are designed so that the minimum level (just before a dose) is high enough to be therapeutic and the maximum level (just after a dose) is not high enough to be toxic. We explore this idea in this problem.

(a) Suppose a person has been taking 650-mg doses of acetaminophen every 4 h for several days. This will result in a drug versus time curve that repeats over a period of 4 h. Let A be the unknown concentration immediately after taking a dose. Use the model for acetaminophen drug clearance, with $k = 0.3$, to obtain a formula for the amount of drug present after 4 h, immediately before the next dose.

(b) Use the result from part (a) to obtain a formula for the amount present immediately after the next dose.

(c) Given that the concentration curve is periodic, the amount present immediately after that next dose (from part (b)) must be A. Use this fact to calculate the value of A and to determine the minimum amount present during the 4-h period.

(d) Repeat parts (a)–(c) with doses of B mg of a drug with rate constant k taken every T hours. [Keep in mind that we are using A to represent the amount in the system at time 0, which includes both the new dose and whatever is remaining of previous doses.]

(e) Prepare a drug versus time curve showing the amount of acetaminophen from parts (a)–(c) over a 24-h period.[7]

1.1.13. In Olympic-style weightlifting, competitors are scored in two events, called the snatch and the clean and jerk; each competitor's final score is the sum of his/her best lift in each event. As with many other sports in which body size is important, a weightlifting competition takes place in several classes. Using mathematical models, it is possible to compare the totals across the classes to calculate a theoretical all-class champion. The International Weightlifting Federation recognizes the use of the Sinclair coefficient for this purpose, but no Olympic all-class medals are awarded. The Sinclair coefficient $C(m)$ is calculated using the model

$$C(m) = 10^{A[\log_{10}(m/b)]^2},$$

where m is the mass of a weightlifter, b is the mass of the heavyweight world record holder, and A is a constant determined by some modeling procedure. The parameters A and b are reset

---

[7] Note that we are assuming the patient has been taking the drug for several days already. The results would be more complicated if our study began with the very first dose.

after each Olympic competition and then fixed for the next 4 years.[8] In the period from the 2008 Olympics to the 2012 Olympics, the parameter values for men were $A = 0.78478$ and $b = 173.961$. The Sinclair-adjusted total score for a lifter is obtained by multiplying his/her actual score by the Sinclair coefficient for that lifter's mass. Table 1.1.1 includes the masses and actual scores for the gold medalists in each event of the men's division, along with the three silver medalists who very nearly won their event.

(a) Show that the 342.5 kg score achieved at the 1988 Olympic games by Naim Suleymanoglu in the 60 kg class corresponds to a Sinclair-adjusted total score of 504.0 using the 2008–2012 parameter values. At the time of this writing, Suleymanoglu is the only weightlifter to have earned a Sinclair-adjusted score over 490.
(b) Look up the values of the Sinclair parameters for the period between the 2012 and 2016 Olympics. What is Suleymanoglu's Sinclair-adjusted total score with these new parameter values?
(c) Calculate the Sinclair-adjusted total scores for each of the lifters in the table, using the 2008–2012 parameter values. If there had been all-class medals based on Sinclair-adjusted total scores, who would have won the three medals?
(d) Recalculate the Sinclair-adjusted total scores for the 2008 Olympics using the parameter values adopted in 2012. With these values, are there any changes in the top three places?

(This problem is continued in Problem 2.4.9.)

1.1.14. R. McNeill Alexander presents a simple model for the optimal amount of territory for a bird to try to maintain [1]. Let A be the area of the territory and let p be the daily food availability per unit area, so that the total amount of food available per day is pA. If k is the amount of time (in days) per unit area needed to defend the territory, then $(1 - kA)$ is the amount of time (in days) available for collecting food. Assuming that the bird can collect food at the rate of q units per day, then the amount that can be collected in the time not spent defending the territory is $q(1 - kA)$. The optimal territory is that which has just enough food available so that the bird can collect all of it.

(a) Determine the optimal territory size in terms of the various parameters in the model.
(b) Let $Q = q/p$. Use this new quantity to obtain a formula for optimal territory size that depends on just two parameters.
(c) What is the biological meaning of the parameter Q? [Hint: The units of Q are the units of q divided by the units of p.]
(d) Suppose $k = 1$. This means that $A = 1$ represents the largest territory that can be defended. We can then interpret a specific value of A as the fraction of defendable territory that the bird should choose. Plot the optimal territory size as a function of the parameter Q.
(e) Interpret the graph biologically.

---

[8] This seems very strange. The ordering of two weightlifters who competed in the 2004 Olympics might have changed after the 2008 Olympics, even though the two performances being compared were already historical. This is not the only strange thing about Olympic weightlifting. The International Weightlifting Federation changed the weight classes in 1993 and again in 1998, so currently recognized world records are all for performances after 1997, in spite of obvious problems. For example, the pre-1993 world record for the 60 kg class is a full 5% larger than the currently accepted world record for the 62 kg class. An additional quirk deals with the rules for breaking ties, under which the world record in the 85 kg class was claimed by the silver medalist in the 2008 Olympics. Andrei Rybakou posted the score of 394 kg a few minutes before fellow competitor Lu Yong posted the same score. Rybakou got the world record because he was the first to get the score, but Yong was awarded the gold medal by virtue of being 0.28 kg smaller (about 0.5 pounds) than Rybakou at the official weigh-in. For the sake of completeness, we note that a "world record" score was also achieved in the 105 kg weight class by gold medalist Andrei Aramnau; the quotation marks are used because Aramnau's score at 105 kg was less than the pre-1993 world record for the 100 kg class.

Table 1.1.1 Selected medalists from the 2012 Summer Olympics Men's Weightlifting competition

Name	Class	Place	Mass	Total Lifted
Long Qingquan	56 kg	First	55.37	292
Zhang Xiangxiang	62 kg	First	61.91	319
Liao Hui	69 kg	First	68.97	348
Sa Jae-Hyouk	77 kg	First	76.46	366
Li Hongli	77 kg	Second	76.91	366
Lu Yong	85 kg	First	84.41	394
Andrei Rybakou	85 kg	Second	84.69	394
Ilya Ilin	94 kg	First	93.64	406
Andrei Aramnau	105 kg	First	104.76	436
Matthias Steiner	> 105 kg	First	145.93	461
Evgeny Chigishev	> 105 kg	Second	124.13	460

1.1.15. Under a given set of assumptions (see Problem 2.2.7), we can obtain the model

$$r = \frac{\ln n}{a}$$

for the fractional yearly population growth (i.e., $r = 0.01$ corresponds to a growth rate of 1% per year) as a function of the average number of children per adult (n) and the average age of the mother when a child is born (a).

(a) Plot r as a function of n over the reasonable range $0.5 \leq n \leq 2$, with a values of 25, 30, and 35. These three curves should be plotted together on one set of axes. Use this figure to discuss the dependence of growth rate on number of children, focusing on how this dependence changes for different average reproduction ages.

(b) Plot r as a function of a over the reasonable range $25 \leq a \leq 35$, with n values of 1, 1.5, and 2. These three curves should be plotted together on one set of axes. Use this figure to discuss the dependence of growth rate on average reproductive age, focusing on how this dependence changes for different average numbers of children.

1.1.16.* Consider a species that has a constant death rate of m individuals per capita per time unit.

(a) Given a constant per capita death rate, survival of individuals in a cohort born at the same time is mathematically identical to maintenance of a drug in a patient who gets a single dose at time 0. Use this fact to find a formula for the fraction $y(t)$ of individuals alive at age t. In particular, find the fraction s of individuals alive at age 1 as a function of m.

(b) Suppose individuals of our hypothetical species mature at age 1 and reproduce at constant rate n until age 2. It can be shown (Problem 1.8.12) that the birth rate necessary for the overall population to stay at the same size is

$$n = \frac{m}{e^{-m} - e^{-2m}}.$$

Combine this formula with the result of part (a) to obtain an equation that determines the birth rate necessary for population maintenance as a function of the probability of surviving to maturity.

(c) Plot the resulting function $n(s)$. Choose a biologically reasonable range of s values for the plot. Discuss the biological significance of the result by comparing the birth rate requirements for animal species with different probabilities of survival to maturity.

(This problem is continued in Problem 1.2.11.)

1.2 Rates of Change and the Derivative

After studying this section, you should be able to:

- Compute rates of change for a function defined at discrete points.
- Approximate the instantaneous rate of change for a function defined on a continuous interval.
- Discuss rates of change, tangent slopes, and spatial gradients in terms of the derivative concept.

The ancient Greek philosopher Heraclitus taught that the world is in a perpetual state of change, a view shared by modern science. Some change is obvious and unidirectional, such as the changes observed in evolution and aging. Some change is obvious, without necessarily appearing unidirectional, such as seasonal temperature changes. Some change is not obvious, but is nevertheless present on a small scale, as in the exchange of gases in the lungs when you breathe. The primacy of change has consequences for mathematical modeling in biology. An approximate description of physical processes often consists of rules that predict rates of change in terms of the current state of the system. Any serious attempt at mathematical modeling must begin with development of the mathematics of rates of change.

1.2.1 Rate of Change for a Function of Discrete Time

A function of discrete time is any quantity that has a clearly defined value at regularly spaced times. We need not know how to compute the values, as long as we can measure them.

Example 1.2.1. Table 1.2.1 shows a portion of the dataset for the mean annual atmospheric carbon dioxide concentration at Mauna Loa observatory in Hawaii [8]. The full data set is shown in Figure 1.2.1. Because this is real data, there is no formula that allows us to calculate the function values.   □

Table 1.2.1 Average annual CO_2 concentration at Mauna Loa, in ppm, for selected years

Year	1960	1965	1970	1975	1980	1985	1990
Average CO_2	316.91	320.04	325.68	331.08	338.68	345.87	354.16

Year	1995	2000	2005	2007	2008	2009	2010
Average CO_2	360.63	369.40	379.78	383.72	385.57	387.36	389.78

Given data for a discrete function of time, we can use the difference in function values to calculate an *average* rate of change.

Example 1.2.2. From Table 1.2.1, the average rate of change of CO_2 concentration for the period 1960–1970 is

$$\frac{\Delta y}{\Delta t} = \frac{325.68 - 316.91}{1970 - 1960} = 0.877.$$

Similarly, the average rate of change for the period 2005–2010 is

$$\frac{\Delta y}{\Delta t} = \frac{389.78 - 379.78}{2010 - 2005} = 2.000.$$

□

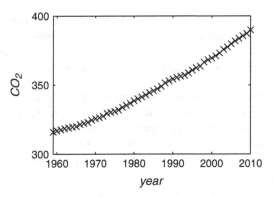

Fig. 1.2.1 Average annual CO_2 concentration at Mauna Loa, in parts per million, from 1960 to 2010

Check Your Understanding 1.2.1:
Find the average rate of change of CO_2 concentration at Mauna Loa for the period 1975–1980.

How rapidly is the carbon dioxide concentration of the air at Mauna Loa changing right now? Unlike Example 1.2.2, this question is about the *instantaneous* rate of change. This is a much more difficult question. In principle, we could imagine a device that would actually measure the rate of change of carbon dioxide concentration. In practice, we do not have such a device, so the best we can do is to calculate an approximate instantaneous rate of change from the discrete Mauna Loa data. This is problematic. For example, the average rate of change from 2008 to 2009 is 1.79, while the average rate of change from 2009 to 2010 is 2.42. Each of these values has equal claim to be *the* rate of change at 2009. Perhaps the best estimate is the average, which is 2.105. However, this is still not an *instantaneous* rate of change, since there is a seasonal pattern that would only be apparent through examination of monthly data. Although the carbon dioxide concentration is generally increasing over a period of years, it has annual peaks every May and falls through the Northern Hemisphere summer.[9] Hence, the instantaneous rate of change is negative for a few months of each year.

1.2.2 Rate of Change for a Function of Continuous Time

As with atmospheric carbon dioxide, most quantities in the natural and physical world change over continuous time rather than discrete time. For example, population changes are themselves discrete, but they can occur at arbitrary times. Discrete change over continuous time poses mathematical difficulties, so we confine ourselves at present to continuous quantities that change over continuous time.[10]

Example 1.2.3. While a quantity of some chemical includes a discrete number of individual molecules, there are so many of these that we can reasonably assume that the mass of a chemical can take on any positive value, and chemical reactions occur continuously. Likewise, changes

---

[9] This is the pattern as measured at Mauna Loa.

[10] Discrete changes occurring in continuous time are best dealt with in *mathematical modeling*, rather than mathematics per se. This topic is discussed in some detail in Section 2.2.

in the amount of a drug in a patient's system are distributed over a continuous interval of time rather than concentrated at a discrete set of times. These assumptions were implicit in Example 1.1.1, where we used the function $y = 650e^{-0.3t}$ to model the amount (in milligrams) of acetaminophen in a patient's system at time t hours after a dose of 650 mg.   □

Given the importance of rates of change, we must find a way to understand the rate of change of continuous quantities over continuous time. This issue was a major theme of mathematical thought in the sixteenth and seventeenth centuries, and it was resolved by the invention of calculus. The problem is conceptually difficult because we tend to focus our thinking on exact quantities. To understand calculus, one has to think about approximations. Our plan of attack is actually rather simple:

1. Find a way to approximate the rate of change so that the error is arbitrarily small (but not zero, as that would indicate an exact answer).
2. Calculate what would happen to the approximation if the error actually *could* be reduced to zero. This requires the concept of the *limit*, the central notion on which all of calculus is based.[11]

Because the rate of change of a discrete function $y(t)$ is denoted by $\frac{\Delta y}{\Delta t}$, it is both suggestive and convenient to use the notation $\frac{dy}{dt}(t)$ to represent the instantaneous rate of change of a continuous function $y(t)$ at the time t.[12] Of course we do not yet know how to calculate the rate of change, or even whether there is any reasonable definition for it.

Table 1.2.2 Acetaminophen data for Example 1.2.4   .

Time (h)	0	0.5	1.0	1.5	2.0	2.5	3.0	3.5	4.0
Acetaminophen (mg)	650.0	559.5	481.5	414.5	356.7	307.0	264.3	227.5	195.8

Example 1.2.4. Table 1.2.2 shows some data for acetaminophen levels in the hypothetical patient of Example 1.1.1. Suppose we want to know the rate of change of the acetaminophen amount immediately after the dose at time 0; that is, $\frac{dy}{dt}(0)$. With the data in Table 1.2.2, the best we can do is calculate an average rate of change using the first two data points:

$$\frac{dy}{dt}(0) \approx \frac{\Delta y}{\Delta t}(0, 0.5) \approx \frac{559.5 - 650}{0.5} \approx -181.$$

□

In our hypothetical drug clearance scenario, we have a formula, $y = 650e^{-0.3t}$, which can be used to calculate more data points than those in Table 1.2.2. We can approximate the instantaneous rate of change at time 0 more accurately than we did in Example 1.2.4 by using a smaller time interval for the average rate of change. We could even use a parameter to represent the duration of the time interval, and then we would be able to see how the average rate of change depends on that duration.[13]

---

[11] The limit is a very subtle mathematical concept that was not rigorously defined until many years after the original development of calculus. The creators of calculus were able to manage without a rigorous definition of the limit, and so can we.

[12] The actual meaning of the distinct symbols dy and dt will be considered later. For now, the reader should think of $\frac{dy}{dt}$ as a single symbol rather than a quotient.

[13] This is the reason for the assertion in Section 1.1 that understanding parameters is very helpful for understanding calculus.

Example 1.2.5. In Example 1.2.4, we calculated an approximate rate of change at time 0 for the function $y = 650e^{-0.3t}$ using data at time 0 and time 0.5. Assuming that we can compute y for any positive time h, we have the approximation

$$\frac{dy}{dt}(0) \approx \frac{\Delta y}{\Delta t}(0, h) \approx \frac{650e^{-0.3h} - 650}{h} = 650\frac{e^{-0.3h} - 1}{h}.$$

The approximation of Example 1.2.4 can be obtained from this formula using $h = 0.5$, and better approximations can be obtained using smaller values.[14] Table 1.2.3 shows the approximations for several values of h. Each is one-tenth of the previous value, which allows us to get to very small values of h with only a few calculations. From the data, it seems likely that the rate of change is exactly

$$\frac{dy}{dt}(0) = -195.$$

Note that we cannot obtain this value by using the approximation formula with $h = 0$. Both $e^{-0.3h} - 1$ and h evaluate to 0 for $h = 0$. Of course, this had to happen, because these quantities are the changes in y and t, both of which are 0 if the end of the time interval used for the approximation is the same as the beginning. □

Table 1.2.3 Approximations for $dy/dt(0)$ in Example 1.2.5

h (hours)	1.0	0.1	0.01	0.001	0.0001	0.00001
$\Delta y/\Delta t$	-168	-192.1	-194.7	-194.97	-195.00	-195.00

There was nothing special about the choices of the function $y = 650e^{-0.3t}$ and the evaluation point $t = 0$. A similar calculation could be accomplished using most functions and most evaluation points. We arrived at what appears to be the correct instantaneous rate of change by taking h to be progressively smaller and looking for the logical conclusion of the trend. We had to identify this logical conclusion without allowing the time interval to shrink all the way to 0.

Check Your Understanding 1.2.2:
Determine a formula for the average rate of change for the function $y = 650e^{-0.3t}$ in an interval of duration h that begins at $t = 3$. Use this formula to generate a table of approximations for dy/dt at $t = 3$, similar to Table 1.2.3. From the table, conjecture the value of dy/dt at $t = 3$ to two decimal places.

1.2.3 The Derivative

So far, we have explored rates of change using algebra. In Example 1.2.5, we succeeded in calculating an approximate rate of change good to at least two decimal places, but only for one value of t. This is not a satisfactory method, because the calculation would have to be repeated for many other times to get a complete picture of the rate of change. A better answer requires the mathematics of calculus.

---

[14] In practice, too small a value of h causes difficulties because of the way real numbers must be stored in computers. This is not a serious practical difficulty; we just have to avoid ridiculously small values of h.

In general, suppose we have a function $y(t)$ and a time t for which we want to know the rate of change dy/dt. Using a time interval of arbitrary positive length h, we obtain the approximation

$$\frac{dy}{dt}(t) \approx \frac{\Delta y}{\Delta t}(t) = \frac{y(t+h) - y(t)}{h}. \tag{1.2.1}$$

This approximation formula never works for the one value we care about ($h = 0$), but in most cases we can infer the correct rate of change using calculations with $h \neq 0$.

Think of our problem in this way: Given a function $y(t)$ and a time t, find a numerical value for the rate of change that "makes sense" for $h = 0$, where "makes sense" means that the value is consistent with the values that we can compute for arbitrarily small h. To determine such a value, if there is one, we can compute the approximation for a sequence of h values that get progressively smaller. If this procedure yields the same result no matter what we actually pick for the decreasing sequence of h values, we call this result the *limit* of the approximation as h approaches 0.

With the introduction of the limit concept, we are now prepared to define the *derivative* to be the mathematical solution of the rate of change problem.

Derivative of a function y(t): *the function defined by the formula*

$$\frac{dy}{dt}(t) = \lim_{h \to 0} \frac{\Delta y}{\Delta t}(t, h), \tag{1.2.2}$$

where

$$\frac{\Delta y}{\Delta t}(t, h) = \frac{y(t+h) - y(t)}{h}, \tag{1.2.3}$$

provided the limit exists.

There are two difficulties with this definition of the derivative.

1. From a theoretical perspective, the definition rests on the limit, which we haven't properly defined. This is not mathematically satisfying, but it will not prevent us from getting the right answers. The interested reader can consult any calculus book.
2. From a practical perspective, it is a significant disadvantage to have to calculate a quantity using a complicated procedure rather than a single formula. This difficulty will be eliminated in Section 1.3.

1.2.4 Slope of a Tangent to a Graph

Any data indicating the values of a quantity at different times can be plotted on a graph. If the data represents a continuous process, the graph consists of infinitely many points on a smooth curve. Often this curve is computed using a formula that gives a result for any time within some given interval. The average rate of change of a continuous function on any interval $(t, t + h)$ is given by (1.2.3) and is represented on the graph by the slope of the straight line connecting the points $(t, y(t))$ and $(t + h, y(t + h))$. As the time interval duration h decreases toward 0, the straight line connecting the approximation points becomes closer and closer to the tangent line at the fixed point $(t, y(t))$. Thus:

The rate of change of a function of continuous time corresponds to the slope of the tangent to the graph of the function.

The interpretation of the rate of change as the slope of a smooth curve on a graph holds for any graph, including graphs where the independent variable is something other than time.

Example 1.2.6. A possible model for the rate of change of a consumed resource x is given by the differential equation

$$\frac{dx}{dt} = -\frac{x}{1+x},$$

given appropriate units for the resource and the time. We can think of the quantity $x/(1+x)$ as a function $y = f(x)$ that represents the consumption rate in terms of the current amount of resource. This function is illustrated in Figure 1.2.2, along with the tangent line at the point $(1, 0.5)$. The slope of the tangent line at this point, using the definition of the derivative, is

$$\frac{dy}{dx}(1) = \lim_{h \to 0} \frac{y(1+h) - y(1)}{h} = \lim_{h \to 0} \frac{\frac{1+h}{2+h} - 0.5}{h}.$$

Using numerical approximations, we obtain the result

$$\frac{dy}{dx}(1) \approx 0.25.$$

We will show in Section 1.3 that this result is exactly correct.   □

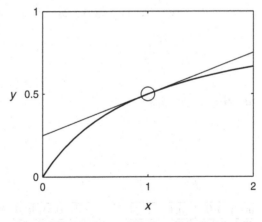

y

x

Fig. 1.2.2 The resource consumption function from Example 1.2.6, along with the tangent line at the point $(1, 0.5)$

Example 1.2.7. Suppose the temperature in a laboratory environment is controlled by heating a flat metal plate to a fixed temperature that is $20°$ higher than normal room temperature. Using a mathematical model (Newton's law of cooling), we can derive the formula $T(x) = 20e^{-kx}$ for the difference between the temperature a distance x from the hot plate and room temperature. Several members of this one-parameter family are plotted in Figure 1.2.3 along with the tangent lines at position $x = 1$. Using the definition of the derivative, we can write the slope as

$$\frac{dT}{dx}(1) = \lim_{h \to 0} \frac{T(1+h) - T(1)}{h} = \lim_{h \to 0} \frac{20e^{-k-kh} - 20e^{-k}}{h} = 20e^{-k} \lim_{h \to 0} \frac{e^{-h} - 1}{h}.$$

Numerical approximations suggest the limit result

$$\lim_{h \to 0} \frac{e^{-h} - 1}{h} = -1,$$

with the result

$$\frac{dT}{dx}(1) = -20e^{-k}.$$

Indeed, this formula was used to calculate the slopes of the tangents plotted in the figure. □

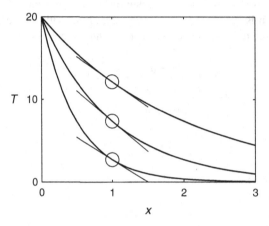

Fig. 1.2.3 The temperature profiles of Example 1.2.7, with $k = 0.5, 1, 2$ (*top* to *bottom*), along with the tangent lines at the point $x = 1$

Check Your Understanding Answers

1. 1.500
2.

$$\frac{dy}{dt}(3) \approx 650e^{-0.9} \frac{e^{-0.3h} - 1}{h}.$$

h (hours)	1.0	0.1	0.01	0.001	0.0001	0.00001
$\Delta y / \Delta t$	−68.5	−78.1	−79.2	−79.27	−79.28	−79.28

Problems

1.2.1.* Use the data in Table 1.2.1 to determine the average rates of change for the 10-year intervals beginning in 1960, 1970, 1980, 1990, and 2000. Plot these average rates of change versus time. Is there an apparent trend?

1.2.2. Use the data in Table 1.2.1 to determine the average rates of change for the 5-year intervals beginning in 1970, 1975, 1980, 1985, 1990, 1995, 2000, and 2005. Plot these average rates of change versus time. Is there an apparent trend?

1.2.3. Use the Mauna Loa CO_2 annual mean data [8] to determine average rates of change for the 2-year intervals beginning in 1970, 1972, ..., and 2008. Plot these average rates of change versus time. Notice that it is difficult to give a quantitative summary of these values, because of the high variability of the data. We will address the issue of extracting summary information from variable data in Section 2.3 and 2.4. Qualitatively, it is certainly clear that carbon dioxide concentration in the atmosphere has been increasing for a long time.

1.2.4. Use the Mauna Loa CO_2 monthly mean data [8] to determine average rates of change for the 1-month intervals beginning in January 2011 and continuing to the end of the data. Plot these average rates of change versus time. How predictable is the seasonal variation from 1 year to the next?

1.2.5.*

(a) Use the data in Table 1.2.2 to approximate the average rates of change on the intervals $[0.5, 1]$, $[1, 1.5]$, and $[0.5, 1.5]$.
(b) Approximate the rate of change of the function $y = 650e^{-0.3t}$ at time $t = 1$ using forward differences (Equation (1.2.1)) and the time intervals $h = 0.1, 0.01, 0.001, 0.0001$, and 0.00001. Use these results to infer the value of $dy/dt(1)$.
(c) Which of the estimates in part (a) comes closest to the answer determined in part (b)?

1.2.6. Repeat Problem 1.2.5 with the time $t = 2$.

1.2.7. Use the results of Example 1.2.5, Problem 1.2.5b, and Problem 1.2.6b to plot three points on a graph in which the horizontal coordinate is y and the vertical coordinate is dy/dt. Conjecture the quantitative relationship between these quantities for this function.

1.2.8.* Approximate the rate of change of the function $y = \sqrt{x}$ at time $x = 0$ using forward differences (Equation (1.2.1)) and the time intervals $h = 0.1, 0.01, 0.001$, and 0.0001. What happens when you try to use these results to infer the value of $dy/dx(0)$? Conjecture the slope of the tangent line at the point $(0, 0)$.

1.2.9.* Use the approximation method to determine the slopes of the tangent lines to the graph of

$$f(x) = \frac{x}{2+x}$$

at the points $x = 1, 2, 3$. Does the trend of these calculated tangent line slopes reflect the shape of the graph of f?

1.2.10. Repeat Problem 1.2.9 with the function $f(x) = xe^x$ at the points $x = 0, 1, 2$.

1.2.11. (Continued from Problem 1.1.16.)
In Problem 1.1.16, we derived the formula

$$n(s) = \frac{-\ln s}{s - s^2}$$

for the average number of offspring per individual in a population needed to maintain the total population size, where s is the probability of living long enough to reach maturity. (The formula includes other assumptions about death rates and maximum age for reproduction.) Use a sequence of s values that rapidly approaches 1 to conjecture the value of

$$\lim_{s \to 1^-} n(s).$$

[Note: In general, limits can be defined using points to the right or left of the limiting value. In the definition of the derivative, for example, it does not matter whether one looks at small positive values or small negative values as $h \to 0$. In this case, the superscript in $s \to 1^-$ indicates that this limit is to be considered specifically for values of s that are less than 1. In the biological context, $s > 1$ is not meaningful.]

1.3 Computing Derivatives

After studying this section, you should be able to:

- Compute derivatives for power functions, exponential functions, logarithmic functions, sine functions, cosine functions, and more complicated functions composed from these elementary ones.
- Compute partial derivatives.
- Use different notations for derivatives.

In Section 1.2, we defined the derivative dy/dt by the formula

$$\frac{dy}{dt}(t) = \lim_{h \to 0} \frac{\Delta y}{\Delta t}(t, h), \tag{1.3.1}$$

where

$$\frac{\Delta y}{\Delta t}(t, h) = \frac{y(t + h) - y(t)}{h}. \tag{1.3.2}$$

The derivative is defined in this way so that it will represent the rate of change of a quantity y with respect to the independent variable t. While this definition is useful for understanding what derivatives are, it is not practical as a method for calculating actual values. A more convenient strategy is to use the mathematical definition to determine:

- A small number of elementary derivative formulas and
- Some very general rules for reducing complex derivative problems to elementary ones.

1.3.1 Two Notations

In the formula $y = 650e^{-0.3t}$, it is natural to think of y as a dependent variable, which is what led to the notation dy/dt for the rate of change of y with respect to t. However, sometimes there is an advantage to using the mathematical concept of a *function*, and the notation for this concept leads to an alternative notation for the derivative.

Function: *a rule that associates a single output value to each allowable value of an independent variable.*

The formula $650e^{-0.3t}$ is a function of t in this mathematical sense. All real numbers are allowed for the independent variable t, although negative values for t lack a physical meaning in the biological context of the example. Suppose we use the name f to indicate the function that produces the output $650e^{-0.3t}$ for any input t. Then our mathematical model for acetaminophen clearance can be written as

$$y = f(t) = 650e^{-0.3t}.$$

The distinction between y and $f(t)$ is subtle, but it is worth understanding. The symbol f represents the rule from which the quantity represented by the symbol y is calculated, given t as the input variable. Thus, the meaning of y depends on a nonmathematical context, whereas f is a strictly mathematical quantity. The symbol used for the independent variable in a function is not relevant. For example, the same function f could be defined by the formula

$$f(v) = 650e^{-0.3v}.$$

Regardless of whether the input quantity is represented by the symbol t or the symbol v, the function f defines a unique output quantity by raising e to the power of -0.3 times the input quantity and multiplying the result by 650. In similar fashion, there is no calculation required to determine $f(x^2)$ for this function f; it is:

$$f(x^2) = 650e^{-0.3x^2}.$$

We do not need to know anything about x to write this result; indeed, we do not even need to know whether x^2 means x times x or is an independent symbol like x_2.

When computing derivatives, it is usually convenient to think of formulas as representing functions rather than quantities. The mathematical notation f' is used in such cases to represent the derivative of the function f. Thus, if the acetaminophen concentration is given by

$$y = f(t) = 650e^{-0.3t},$$

then the rate of change can be denoted either as dy/dt or $f'(t)$. In practice, we often use notation that ignores the distinction between quantities and functions, so there is no harm in using the mixed notations $y'(t)$ or df/dt, as long as the meaning is clear from the context. As a general rule, this text will use whichever notation seems better suited to the context. For derivative computation, the prime notation seems to be the better choice.

1.3.2 Elementary Derivative Formulas

Table 1.3.1 summarizes the elementary derivative formulas that we need to be able to find derivatives for all polynomial, rational, exponential, logarithmic, and trigonometric functions, as well as more complicated functions composed of these elements. The derivatives of arctangent and arcsine are included for reference.

Table 1.3.1 Elementary derivative formulas

$f(x)$	$x^p \ [p \neq 0]$	e^{ax}	$\ln x$	$\sin ax$	$\cos ax$	$\arctan \dfrac{x}{a}$	$\arcsin \dfrac{x}{a}$
$f'(x)$	px^{p-1}	ae^{ax}	$\dfrac{1}{x}$	$a\cos ax$	$-a\sin ax$	$\dfrac{a}{a^2+x^2}$	$\dfrac{1}{\sqrt{a^2-x^2}}$

These formulas can be obtained from the definition of the derivative with varying amounts of difficulty and use of the following limit results, which we present without proof. These values can be approximately confirmed with graphs and/or numerical experiments.

$$\lim_{x\to 0}\frac{e^x-1}{x}=1,\quad \lim_{x\to 0}\frac{\ln(1+x)}{x}=1,\quad \lim_{x\to 0}\frac{\sin x}{x}=1,\quad \lim_{x\to 0}\frac{1-\cos x}{x}=0. \quad (1.3.3)$$

Example 1.3.1. Let $f(x)=e^{ax}$. Then the average rate of change from x to $x+h$ is

$$\frac{\Delta f}{\Delta x}(x,h)=\frac{e^{ax+ah}-e^{ax}}{h}=\frac{e^{ax}e^{ah}-e^{ax}}{h}=e^{ax}\left(\frac{e^{ah}-1}{h}\right)=ae^{ax}\left(\frac{e^{ah}-1}{ah}\right).$$

Thus,[15]

$$f'(x)=\lim_{h\to 0}\left[ae^{ax}\left(\frac{e^{ah}-1}{ah}\right)\right].$$

This calculation looks uninviting, but there are two things we can do to simplify it. First, we observe that ae^{ax} is independent of h, so it can be factored out of the limit expression. We now have

$$f'(x)=\left(\lim_{h\to 0}\frac{e^{ah}-1}{ah}\right)ae^{ax}.$$

Next, suppose we let $y=ah$. Having h go to 0 is equivalent to having y go to 0, so we can simplify the limit expression and obtain

$$f'(x)=\left(\lim_{y\to 0}\frac{e^y-1}{y}\right)ae^{ax}.$$

This is the closest we can get to computing the derivative using algebra alone. However, the limit in this last formula is exactly the same as the first limit result of (1.3.3) (since the symbol chosen for the independent variable does not matter), so we have the final result

$$f'(x)=ae^{ax}.$$

$\square$

1.3.3 General Derivative Rules

The task of symbolic differentiation is greatly facilitated by the use of general rules that allow us to reduce differentiation of various algebraic structures to differentiation of elementary functions. The derivation of the product rule is left as a problem.[16] A mathematical justification for the chain rule is deferred to Section 1.6, at which point we will be able to use linear approximation.[17] The quotient rule can be derived by combining the product rule with the chain rule.[18]

Let f and g be differentiable functions and let a and b be real constants.

---

[15] It seems silly to have the extra factor a in the denominator of the fraction, since it must be balanced by an extra factor a in the front. The point will be clear in the calculation that follows. Mathematics contains many instances where something that appears to make extra work turns out to be helpful by the end.

[16] Problem 1.3.31.

[17] Section 1.4.

[18] See Problem 1.3.32.

Linearity Rule:

$$[af(x) + bg(x)]' = af'(x) + bg'(x). \tag{1.3.4}$$

Example 1.3.2. Let $f(x) = 3x^2 + 4\sqrt{x}$. Using the linearity rule, we can reduce the task of calculating f' to application of the power rule from Table 1.3.1:

$$f' = 3\left(x^2\right)' + 4\left(x^{1/2}\right)' = 3 \times 2x + 4 \times \frac{1}{2}x^{-1/2} = 6x + \frac{2}{\sqrt{x}}. \qquad \square$$

Example 1.3.3. Let $f(x) = 2^x + \ln(3x)$. This derivative can also be computed using the linearity rule and basic rules. However, some algebraic manipulation of the exponential and logarithmic functions is necessary. We have

$$2^x = e^{x\ln 2}, \qquad \ln(3x) = \ln 3 + \ln x.$$

Thus,

$$f(x) = e^{x\ln 2} + \ln 3 + \ln x.$$

The first term is of the form e^{ax}, with $a = \ln 2$. The second term is a constant; hence, its derivative is 0. We therefore have

$$f' = \left(e^{x\ln 2}\right)' + (\ln 3)' + (\ln x)' = (\ln 2)e^{x\ln 2} + 0 + \frac{1}{x} = (\ln 2)2^x + \frac{1}{x}. \qquad \square$$

Product Rule:

$$[f(x)g(x)]' = f'(x)g(x) + f(x)g'(x). \tag{1.3.5}$$

Example 1.3.4. Let $f(x) = x^2\cos 3x$. This function is a product of two elementary functions listed in Table 1.3.1. Hence, we apply the product rule and obtain

$$\left[x^2\cos 3x\right]' = (x^2)'\cos 3x + x^2(\cos 3x)' = 2x\cos 3x + x^2(-3\sin 3x)$$

$$= 2x\cos 3x - 3x^2\sin 3x. \qquad \square$$

Check Your Understanding 1.3.1:
Find the derivative of $f(x) = (x+1)\sin 2x$.

Quotient Rule:

$$\left[\frac{f(x)}{g(x)}\right]' = \frac{f'(x)g(x) - f(x)g'(x)}{[g(x)]^2}, \quad g \neq 0. \tag{1.3.6}$$

Example 1.3.5. Let

$$f(x) = \frac{x}{a+x}.$$

This function is a quotient, with a parameter a. Using the quotient rule, we have

$$\left[\frac{x}{a+x}\right]' = \frac{(x)'(a+x) - x(a+x)'}{(a+x)^2} = \frac{(a+x) - x \times 1}{(a+x)^2} = \frac{a}{(a+x)^2}. \qquad \Box$$

Compare Example 1.3.5 with Example 1.2.6. In the earlier example, we had the special case $a = 1$, and we found the derivative value $f'(1) \approx 0.25$. It took quite a bit of effort to accomplish this result, which is specific to one value of a and one value of x. The result of Example 1.3.5 holds for all a and all x, and it yields the exact result 1/4 for $f'(1)$ with $a = 1$ rather than the approximate result 0.25 of the earlier example.

> **Check Your Understanding 1.3.2:**
> Find the derivative of $\dfrac{e^{2x}}{1+e^x}$.

Example 1.3.6. Let

$$f(x) = \frac{x}{y(x)+x}.$$

This function is similar to that in Example 1.3.5, except that it has a function $y(x)$ in place of the parameter a. The quotient rule still applies:

$$\left[\frac{x}{y(x)+x}\right]' = \frac{(x)'(y+x) - x(y(x)+x)'}{(y+x)^2} = \frac{(y+x) - x(y'+1)}{(y+x)^2}$$

$$= \frac{y+x-xy'-x}{(y+x)^2} = \frac{y-xy'}{(y+x)^2}.$$

Note that it is helpful to use the full notation $y(x)$ wherever it is important to remember that y is a function of x rather than a parameter, while using the abbreviated notation y wherever it does not matter that y is a function. Note also that we cannot compute y', since y is not given. In such cases, it is acceptable to leave y' in the answer. $\qquad \Box$

> **Chain Rule:**
> $$[f(g(x))]' = f'(g(x))g'(x). \qquad (1.3.7)$$

The chain rule is often confusing to students. It is helpful to think of it in a verbal form rather than the abstract mathematical form:

> The derivative of $f(g(x))$ is f' evaluated at $g(x)$ multiplied by the derivative of $g(x)$.

As the examples show, it is also helpful to focus on the specific function f, while continuing to think of g in general terms.

Example 1.3.7. Let $f(x) = \sin(e^{2x})$. This function is a composition of the form $\sin(g(x))$, so we apply the chain rule. This rule says, "The derivative of $\sin(g(x))$ is $\cos(g(x))$ multiplied by the derivative of $g(x)$." Hence,

$$[\sin(e^{2x})]' = \cos(e^{2x}) \times (e^{2x})' = \cos(e^{2x}) \times 2e^{2x} = 2e^{2x}\cos(e^{2x}). \qquad \square$$

Check Your Understanding 1.3.3:
Find the derivative of $f(x) = e^{\sin 2x}$.

Our small list of elementary formulas and general rules allows us to compute derivatives for any function that is built up from sums, products, quotients, and compositions of elementary functions. The mechanics of the calculation can become messy for functions whose structure is complicated; the important thing is to reduce the structure "from the outside in."

Example 1.3.8. Let $f(x) = \sin(x^2 e^{3x})$. This function is a composition of the form $\sin(g(x))$, just like the function in Example 1.3.7. The only difference is that $g(x)$ is now a product. Hence,

$$[\sin(x^2 e^{3x})]' = \cos(x^2 e^{3x}) \times (x^2 e^{3x})' = \cos(x^2 e^{3x}) \times (2xe^{3x} + x^2 \times 3e^{3x})$$
$$= (2x + 3x^2)e^{3x}\cos(x^2 e^{3x}).$$

$$\square$$

1.3.4 Partial Derivatives

We can define derivatives for functions of more than one variable by treating all but one of the variables as parameters. We also modify the notation.[19]

Partial derivative of $f(x,y)$ with respect to x: *the formula obtained by treating y as a parameter and differentiating with respect to x. This quantity is denoted by $\dfrac{\partial f}{\partial x}$. Thus,*

$$\frac{\partial f}{\partial x}(x,y) = \lim_{h \to 0} \frac{f(x+h,y) - f(x,y)}{h}.$$

Example 1.3.9. Let

$$f(x,y) = \frac{sxy}{a+x}, \qquad s,a > 0.$$

To compute the partial derivative of f with respect to x, we treat y (as well as s and a) as a parameter:

$$\frac{\partial f}{\partial x}(x,y) = sy\frac{\partial}{\partial x}\left(\frac{x}{a+x}\right) = sy\frac{a}{(a+x)^2} = \frac{asy}{(a+x)^2},$$

---

[19] In some contexts, it is important to have a partial derivative notation that is distinguishable from ordinary derivative notation.

where we have used the result of Example 1.3.5. Similarly,

$$\frac{\partial f}{\partial y}(x,y) = \frac{sx}{a+x}\frac{\partial}{\partial y}(y) = \frac{sx}{a+x}.$$

□

Partial derivatives are important in models of quantities that depend on more than one independent variable; for example, a population in a region near a river might depend on the distance x from the river as well as the time t. The partial derivative with respect to t represents the instantaneous rate of change at any particular location x. The partial derivative with respect to x represents the spatial gradient at any particular time t. Most applications of partial derivatives are beyond the scope of this book; however, we will use them in Chapter 7 to analyze dynamical system models.

Check Your Understanding Answers

1. $f'(x) = \sin 2x + 2(x+1)\cos 2x.$
2. $f'(x) = \dfrac{2e^{2x} + e^{3x}}{(1+e^x)^2}.$
3. $f'(x) = 2(\cos 2x)e^{\sin 2x}.$

Problems

1.3.1.* Use the definition of the derivative to compute the derivative of $f(x) = x^2$.

1.3.2. Use the definition of the derivative to compute the derivative of $f(x) = x^3$.

1.3.3.* Use the definition of the derivative and the appropriate limit result(s) from (1.3.3) to compute the derivative of $f(x) = \sin x$. [Hint: You will need a trigonometric identity for $\sin(x+h)$.]

1.3.4. Use the definition of the derivative and the appropriate limit result(s) from (1.3.3) to compute the derivative of $f(x) = \cos x$. [Hint: You will need a trigonometric identity for $\cos(x+h)$.]

1.3.5. Use the definition of the derivative and the appropriate limit result(s) from (1.3.3) to compute the derivative of $f(x) = \ln x$. [Hint: Write $\ln(x+h)$ as $\ln[x(1+h/x)]$.]

In Problems 1.3.6–1.3.30, compute the derivative of the indicated function. Simplify the resulting formulas.

1.3.6. $f(x) = 3x^4 - \dfrac{1}{x}$

1.3.7.* $f(x) = e^{-5x} + \sin \pi x$

1.3.8. $g(x) = 3^x + \ln \pi x$

1.3.9. $f(t) = t^3 e^{3t}$

1.3.10. $f(t) = t^2 \cos 4t$

1.3.11.* $g(x) = x^2 \ln x$

1.3.12. $g(x) = e^{ax} \sin kx, \quad a, k \neq 0$

1.3.13.* $f(x) = \tan x$ [Hint: Write $\tan x$ in terms of $\sin x$ and $\cos x$.]

1.3.14. $f(x) = \dfrac{\cos x}{1 + x}$

1.3.15. $f(v) = \dfrac{e^v}{e^v + v}$

1.3.16.* $g(t) = \dfrac{e^{v(t)}}{e^{v(t)} + v(t)}$

1.3.17.* $g(t) = \dfrac{t^2}{t^3 + 1}$

1.3.18. $g(t) = \dfrac{\sqrt{x}}{x^2 + 1}$

1.3.19. $f(x) = e^{x^2}$

1.3.20. $g(x) = xe^{x^2}$

1.3.21. $y(u) = \ln(2 + \sin u)$

1.3.22.* $g(x) = \sin(e^{2x} + \ln x)$

1.3.23. $f(t) = (t^2 + 1)e^{t^2}$

1.3.24. $f(t) = \dfrac{e^{t^2}}{t^2 + 1}$

1.3.25.* $f(x) = e^{\pi \sin x}$

1.3.26. $g(x) = e^{x \sin x}$

1.3.27.* $g(t) = \cos\left(\dfrac{\pi}{t^2 + 1}\right)$

1.3.28. $g(y) = \ln(e^{x^2} + x^2)$

1.3.29. $f(x) = \sin e^{x^2}$

1.3.30. $g(x) = e^{\sin x^2}$

1.3.31. Use the definition of the derivative to derive the product rule.

1.3.32. Use the product rule with the chain rule to derive the quotient rule.

1.3.33.* In Problem 1.1.15, we derived the formula

$$n = \frac{-\ln s}{s - s^2}$$

for the average number of offspring per individual in a population needed to maintain the total population size, where s is the probability of living long enough to reach maturity. (The formula includes other assumptions about death rates and maximum age for reproduction.) Use the appropriate formula of (1.3.3) to compute the limiting fertility rate as the survival probability approaches 1. [Note: You can factor limits and you can convert one limit to another by a substitution, as in Example 1.3.1.]

1.4 Local Behavior and Linear Approximation

After studying this section, you should be able to:

- Use the derivative to determine local behavior of functions near points on the graph.
- Use the derivative for linear approximation.

Interesting features of a graph can be classified into two types. *Global behavior* refers to general features that are best seen over a large interval, such as the overall maximum and minimum values and trends of increase and decrease. In contrast, the term *local behavior* refers to the limiting behavior one sees when zooming in on a point on a graph. Local behavior includes slope, relative maxima and minima (peaks and valleys), and concavity (whether the graph curves up or down); calculus is necessary to determine these features.

1.4.1 Tangent Lines

Figure 1.4.1 shows the graph of $y = x/(1+x)$ along with the line tangent to the graph at the point $(1, 1/2)$. Figure 1.4.1a shows the *global* behavior of the function because the domain of x values for the graph is large. Figure 1.4.1b shows only a small portion of the graph, corresponding to the box in Figure 1.4.1a and magnified so that the plot uses the same amount of space on the paper. As we zoom in on a particular point on a graph, the graph looks more and more like the tangent line; hence, the tangent line can reasonably be said to show the local behavior of the function.

We can use the derivative to find the equation of the tangent line.

Example 1.4.1. Given

$$y = f(x) = \frac{x}{1+x},$$

the slope of the tangent at any point x is (see Example 1.3.5)

$$f'(x) = \frac{(x)'(1+x) - x(1+x)'}{(1+x)^2} = \frac{1+x-x}{(1+x)^2} = \frac{1}{(1+x)^2};$$

in particular, the slope at the point $\left(1, \frac{1}{2}\right)$ is $f'(1) = \frac{1}{4}$. Now suppose (x, y) is any point on the tangent line other than the point $\left(1, \frac{1}{2}\right)$. The slope of the line connecting these two points is

$$m = \frac{\Delta y}{\Delta x} = \frac{y - \frac{1}{2}}{x - 1}.$$

This slope must be the value obtained from the derivative; hence, we have the equation

$$\frac{y - \frac{1}{2}}{x - 1} = \frac{1}{4}.$$

Solving for y yields the formula

$$y = \frac{1}{2} + \frac{1}{4}(x - 1).$$

This is the tangent line shown in Figure 1.4.1.   □

The tangent line to a graph $y = f(x)$ at a point (x_0, y_0) is given by

$$y = x_0 + f'(x_0)(x - x_0). \qquad (1.4.1)$$

1.4.2 Local Extrema

Sometimes the tangent line alone does not give an adequate picture of local behavior.

Example 1.4.2. The function $f(x) = x^3 - 3x$ has derivative $f'(x) = 3x^2 - 3$. Note that $f'(-1) = f'(1) = 0$. This means that the tangent lines at the points $x = -1$ and $x = 1$ are horizontal. However, as the graph in Figure 1.4.2 shows, there is a very significant difference in the local behavior of the function at these two points. The function shows a *local maximum* at $x = -1$ and a *local minimum* at $x = 1$. A horizontal tangent line is *suggestive* of a local extremum, but is of no help in distinguishing the two. As we will see, a horizontal tangent is not even a guarantee of a local extremum. $\qquad\qquad\square$

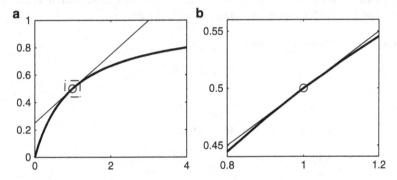

Fig. 1.4.1 The function $x/(1+x)$, along with the tangent line at the point $(1, 0.5)$. The small box in (**a**) is the entire viewing window in (**b**)

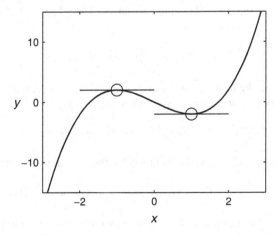

Fig. 1.4.2 The function $f(x) = x^3 - 3x$ from Example 1.4.2, along with the tangent lines at the points $(-1, 2)$ and $(1, 2)$

The graph in Figure 1.4.2 shows that the function $f(x) = x^3 - 3x$ is increasing for $x < -1$, decreasing for $-1 < x < 1$, and increasing again for $x > 1$. As we zoom in on the point $(-1, 2)$, the essential fact (other than the horizontal tangent slope) is that the function is increasing to the left of the point and decreasing to the right, marking the point as a local maximum. Mathematically, f' changes from positive to negative at a local maximum. Similarly, f' changes from negative to positive at the local minimum point $(1, -2)$.

Local maximum (minimum) of a continuous function f: *a point x_0 where f' changes from positive to negative (negative to positive).*

If $f' = 0$ at a point x_0, then the tangent line at x_0 is horizontal. If f' does not exist at a point, then there is no tangent line. These are the only two cases where a local extremum can occur. It is convenient to have a term that takes these two cases together.

Critical point of a continuous function f: *a point where $f' = 0$ or f' does not exist.*

A systematic way to identify local extrema by calculation is to first find all the critical points and then use some method for classifying them. The definition of a local maximum or minimum classifies critical points of a continuous function using the first derivative.

Example 1.4.3. Consider the function

$$f(x) = x^4 - 8x^3 + 18x^2 - 16x,$$

for which

$$f'(x) = 4x^3 - 24x^2 + 36x - 16 = 4(x^3 - 6x^2 + 9x - 4).$$

Critical points are the solutions of

$$x^3 - 6x^2 + 9x - 4 = 0.$$

Normally we must resort to numerical approximation to solve a cubic equation, but observe that $x = 1$ is a solution of this equation. This means that $x - 1$ is a factor of $x^3 - 6x^2 + 9x - 4$. The factorization must have the form

$$x^3 - 6x^2 + 9x - 4 = (x - 1)(x^2 + ax + 4),$$

for some value of a, with the terms x^2 and 4 in the second factor necessary to obtain x^3 and -4 in the product. Multiplying out the product in terms of a yields

$$x^3 - 6x^2 + 9x - 4 = x^3 + (a - 1)x^2 + (4 - a)x - 4,$$

from which we obtain $a = -5$. Thus, the equation for the critical points becomes

$$0 = x^3 - 6x^2 + 9x - 4 = (x - 1)(x^2 - 5x + 4) = (x - 1)[(x - 4)(x - 1)] = (x - 1)^2(x - 4).$$

The critical points are $x = 1$ and $x = 4$. Moreover, we can use the result of the factoring to rewrite f' as

$$f'(x) = 4(x^3 - 6x^2 + 9x - 4) = 4(x - 1)^2(x - 4).$$

From this form, we can see that f' is positive for $x > 4$ and negative for $x < 4$ (except for the critical point $x = 1$). The function changes from decreasing to increasing at $x = 4$, so $x = 4$ is a local minimum. The point $x = 1$ is not a local extremum because the function is decreasing on both sides of the point. A graph of this function appears in Figure 1.4.3.   □

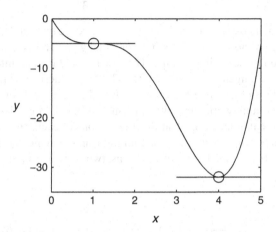

Fig. 1.4.3 The graph of $y = x^4 - 8x^3 + 18x^2 - 16x$, showing the points with horizontal tangents

The first derivative indicates whether a function is increasing or decreasing. Because the second derivative is the rate of change of the first derivative, it can be used to determine if the slope is increasing or decreasing. This information, when applied to a point where the slope is zero, can often be used to determine whether such a point is a local extremum.

Example 1.4.4. The function $f(x) = x^3 - 3x$ has derivatives $f'(x) = 3x^2 - 3$ and $f''(x) = 6x$. Specifically, $f'(-1) = 0$ and $f''(-1) < 0$. A negative value for the second derivative means that the slope of the graph is decreasing at $x = -1$. Since the slope is 0 at $x = -1$, this decreasing trend implies that the slope is positive to the left of $x = -1$ and negative to the right of $x = -1$. These facts establish that the point $(-1, 2)$ meets the requirements for a local maximum. Similarly, the facts $f'(1) = 0$ and $f''(1) > 0$ establish that the point $(1, -2)$ is a local minimum. These results are consistent with the graphical evidence in Figure 1.4.2 and the first derivative analysis in Example 1.4.2.   □

The argument of Example 1.4.4 applies at any point where $f' = 0$, as long as f'' exists and is not 0. We present this result as a theorem.[20]

Theorem 1.4.1 (Second Derivative Test). *If $f'(x_0) = 0$ and $f''(x_0) < 0$, then x_0 is a local maximum; similarly, if $f'(x_0) = 0$ and $f''(x_0) > 0$, then x_0 is a local minimum.*

Check Your Understanding 1.4.1:
Apply the second derivative test to the problem of Example 1.4.3.

---

[20] Don't be intimidated by the word "theorem." Mathematicians use the word to indicate a significant mathematical result that follows logically from a relatively simple set of mathematical assumptions. The student should think of a theorem as a statement that combines some quantitative or qualitative result with a set of conditions that are sufficient to ensure that the result is correct.

Example 1.4.5. Suppose we want to analyze the family of functions $f(x) = x^3 - 2x^2 + bx$, where b is any real number. We start by computing the derivative, $f'(x) = 3x^2 - 4x + b$. The function has a horizontal tangent line if the derivative is 0, which occurs at the points

$$x = \frac{4 \pm \sqrt{16 - 12b}}{6} = \frac{2 \pm \sqrt{4 - 3b}}{3}.$$

The existence of any such points requires $3b < 4$; hence there are two principle cases to consider. If $b > 4/3$, then there are no points with a horizontal tangent. The derivative must always be positive, so the function is increasing for all x. The case $b < 4/3$ is more interesting. Here, there are two points where the tangent line is horizontal. The second derivative is $f'' = 6x - 4$, which is positive for $x > 2/3$ and negative for $x < 2/3$. By Theorem 1.4.1, any critical point with $x > 2/3$ is a local minimum, while any critical point with $x < 2/3$ is a local maximum. From the formula for the critical points, it is clear that there is one of each. Hence, as x increases, the graph increases to a maximum, decreases to a minimum, and then increases again. Figure 1.4.4 illustrates several members of the family of functions, two that have a pair of local extrema and two that do not.

For the special case $b = 4/3$, we have

$$f(x) = x^3 - 2x^2 + \frac{4}{3}x, \quad f'(x) = 3x^2 - 4x + \frac{4}{3}, \quad f''(x) = 6x - 4.$$

From $f' = 0$, we obtain a single critical point:

$$x = \frac{2 \pm \sqrt{0}}{3} = \frac{2}{3}.$$

Theorem 1.4.1 does not apply, because $f''(2/3) = 0$.   □

To see what to do when $f'' = 0$ at a critical point, it helps to review what we have learned so far about local behavior. For any point where $f' \neq 0$, the tangent line serves as an adequate representation of local behavior. For the special case $f' = 0$, we need to examine f'' to fully determine local behavior. There is a general rule at work here:

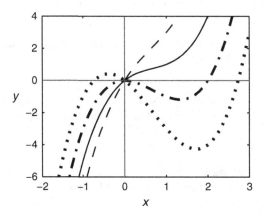

Fig. 1.4.4 The function $f(x) = x^3 - 2x^2 + bx$, with $b = -2$ (*dotted*), $b = 0$ (*dash-dotted*), $b = 2$ (*solid*) and $b = 4$ (*dashed*)

> **The local behavior of a function f near a point x_0 is determined by the lowest order derivative whose value is not zero, provided such a nonzero derivative exists.**

Example 1.4.6. Let $f(x) = x^3 - 2x^2 + \frac{4}{3}x$, which is the function from Example 1.4.5 with $b = 4/3$. We have already seen that $f'(2/3) = f''(2/3) = 0$. From $f''(x) = 6x - 4$, we have $f'''(x) = 6$. Although f''' is not the slope of the function or the rate of change of the slope, it nevertheless indicates the local behavior at points where $f' = f'' = 0$. Just as a positive value of f' means that the function is increasing on both sides of the point, a positive value of f''', given $f' = f'' = 0$, indicates that the function is increasing on both sides. As shown in Figure 1.4.5, the point $x = 2/3$ is not a local extremum.   □

1.4.3 Linear Approximation

The formula for the line tangent to the graph of $y = f(x)$ at a point can be used as the *linear approximation* for the function. For x near x_0, we have

$$f(x) \approx f(x_0) + f'(x_0)(x - x_0). \tag{1.4.2}$$

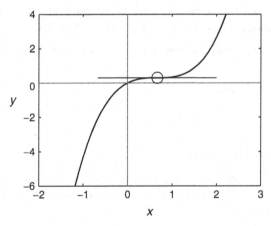

Fig. 1.4.5 The function $f(x) = x^3 - 2x^2 + (4/3)x$

Linear approximation can be used to estimate the change in a function value due to a change in the independent variable. It is convenient to rearrange the formula. Defining Δx and Δf to be the changes in the variable values from base values x_0 and $f(x_0)$, we have

$$\Delta x = x - x_0, \qquad \Delta f = f(x) - f(x_0),$$

and the linear approximation formula becomes

$$\Delta f \approx f'(x_0)\Delta x. \tag{1.4.3}$$

This simple formula says that a small change in the independent variable x results in a change in the function f that is proportional to the change in x; the proportionality factor is the constant $f'(x_0)$. Thus, the derivative can be interpreted as the *marginal value* of the function.

In the simple context of a function $f(x)$, it seems silly to use a linear approximation, since better approximate values can be obtained using a calculator or computer.[21] This does not mean that linear approximation has no value; it simply means that the value of linear approximation can only be seen in more complex contexts. For example, much of the analysis of dynamical systems in Chapters 5 and 7 and Appendix A.1 will be done with the use of linear approximation.

Check Your Understanding Answers

1. We have $f'' = 12x^2 - 48x + 36$, so $f''(1) = 0$ and $f''(4) = 36$. The test indicates that there is a local minimum at $x = 4$, but it does not apply at $x = 1$.

Problems

1.4.1. (Continued from Problem 1.1.8.)
Find the local maximum of the function $f(t) = t(21 - t)(t + 1)$ that has been used to model the number of infected cells on day t of a measles infection [5]. Verify that your result is consistent with the graph of the function.
  (This problem is continued in Problem 1.8.9.)

1.4.2. Determine equations for the tangent lines to the graph of $f(x) = x^3 - 4x$ at the points $x = 0, 1, 2$. Plot these tangent lines on the graph of the function, using the interval $-3 \le x \le 3$.

1.4.3. Repeat Problem 1.4.2 with the function $f(x) = xe^x$ at the points $x = 0, 1, 2, 3$ and the interval $0 \le x \le 5$.

1.4.4. Repeat Problem 1.4.2 with the function

$$f(x) = \frac{x}{2 + x}$$

at the points $x = 0, 2, 4$ and the interval $0 \le x \le 5$.

1.4.5. In Example 1.4.6, we classified the critical point of the polynomial $f(x) = x^3 - 2x^2 + \frac{4}{3}x$ by using the third derivative. Obtain the same result by considering the sign of the first derivative on intervals, as in Example 1.4.3.

1.4.6.* Use the first derivative to find any local extrema for the function $f(x) = \dfrac{x^2 - 1}{x^2 + 3}$.

1.4.7. Use the first derivative to find any local extrema for the function $f(x) = \dfrac{x}{x^2 + 1}$.

  In Problems 1.4.8–1.4.11, identify local extrema and confirm your results by plotting the function. Note that you can use a variety of methods, based on the first derivative, as in Example 1.4.3, the second derivative, as in Example 1.4.4, and higher order derivatives, as in Example 1.4.6. Also note that polynomial equations for critical points can be factored if you can find a root, as in Example 1.4.3. Sometimes you can guess a root by trying small integers; you can also use a graph to guess a root.

---

[21] In most cases where a decimal value is obtained by a calculator or computer from a formula that includes functions other than polynomials, the computed value is a numerical approximation. The point here is that the numerical approximation obtained by a computer is almost always more accurate than the linear approximation.

1.4.8.* $f(x) = x^4 - 8x^2$.

1.4.9. $f(x) = x^4 - 8x^3$.

1.4.10.* $f(x) = x^4 - 4x^3 + 16x + 8$.

1.4.11. $f(x) = x^3 + 6x^2 + 6x + 1$.

1.4.12. Plot the function $f(x) = 0.1x^5 - 0.19x^3 + 0.07x$ on the interval $-3 \le x \le 3$. Can you identify local extrema from this graph? Now use the first derivative to find and classify critical points. Confirm your results by plotting the function on the interval $-1.5 \le x \le 1.5$.

1.4.13. Consider the family of functions $f(x) = x + ax^2 - x^3$, where a is any real number.

(a) Find and simplify the formula for the critical points of f. Are there any requirements that a must satisfy for these critical points to exist?
(b) Use the second derivative to show that the classification of a critical point $x^\star$ depends only on whether it is larger or smaller than $a/3$.
(c) Use the results of parts (a) and (b) to classify the two critical points. Note that you do not need to compute f'' at the critical points.

1.4.14.* Consider the family of functions $f(x) = x^4 - 4x^3 - 2bx^2$, where $b > 0$.

(a) This function has one critical point that does not depend on b. Find and classify this critical point.
(b) Factor f' to obtain a quadratic polynomial $q(x)$, with a coefficient that depends on b, whose roots are critical points of f. Are there any requirements that b must satisfy for these critical points to exist?
(c) Suppose $x^\star$ is one of the critical points satisfying $q(x^\star) = 0$. Combine this with the formula for f'' to obtain a formula for $f''(x^\star)$ that does not contain b.
(d) Use the formula for $f''(x^\star)$ from part (c) to design a test for determining if $x^\star$ is a maximum or minimum directly from the value of $x^\star$ (so that you won't have to calculate f'' at $x^\star$).
(e) Use the test from part (d) to classify the critical points that satisfy $q(x) = 0$.

Biological parameters can seldom be measured exactly. One important use of linear approximation is to estimate the error in a calculated quantity that results from an error in an input quantity. Typically, one is interested in the relative error (error divided by value) rather than the absolute error. For example, suppose the value of an input quantity x is estimated to be 5, but it is expected that this amount could have as much as a 4% error, and that we would like to estimate the error in the value x^2. It requires three steps to connect percentage error in x to percentage error in x^2. First, we need to determine the absolute error in x, which is 4% of 5, or $\Delta x = \pm 0.2$. Note that the error could be either positive or negative. Second, we need to use linear approximation to estimate the absolute error in the calculated quantity. Since the derivative of x^2 is $2x$, and $x \approx 5$, we have $\Delta x^2 \approx 10\Delta x \approx \pm 2$. Finally, we compute the relative error in the calculated quantity. In the example, the calculated value is 25, so an absolute error of ± 2 corresponds to a relative error of $\pm 8\%$. Note that the actual value of 5.2^2 is 27.04, which is an error of 8.16%. Using calculus to obtain information at a point slightly removed ($x = 5.2$ rather than $x = 5$) almost never gives the exact answer, but it usually gives a useful approximation. Real data is not good enough to warrant anything better than an approximation. (If we can make as much as a 4% error in a measurement, how do we know that the maximum error is *exactly* 4% rather than 3.9%?)

1.4.15. (a) In the above example, the estimated relative error in x^2 corresponding to a measurement error of $\pm 4\%$ in the measurement $x = 5$ is $\pm 8\%$. Show that the estimated relative error in x^2 corresponding to a measurement error of $\pm p\%$ in the measurement $x = x_0$ is $\pm 2p\%$.

(b) Use calculus to find the estimated relative error in the volume of a sphere if the radius is r_0 and there is a measurement error of $\pm p\%$.

(c) In general, what is the estimated relative error in a function Ax^n if the error in the measurement x_0 is $\pm p\%$?

1.4.16.* Use calculus to estimate the maximum relative error in e^{2x} and e^{-2x} if there is a maximum error of $\pm 1\%$ in the measurement $x = x_0 > 0$.

1.4.17. Suppose $f(x)$ and $g(x)$ are positive and increasing. Assume the measurement $x = x_0$ has an absolute error of Δx. Show that the estimated relative error in the product fg is the sum of the estimated relative errors in the individual factors f and g.

1.5 Optimization

After studying this section, you should be able to:

- Use the derivative to determine extreme values of a function on a given domain.
- Explain the marginal value theorem with reference to lines on a graph of the function that represents the net resource collection in one patch.

Optimization problems have the goal of finding the best value of a variable for achieving some purpose. In medicine, we may want to find the best dosage to benefit a patient. In biology, we may want to understand natural selection as a process by which a genome maximizes fitness. In both of these examples, we can employ a mathematical approach, provided we can construct a function that quantifies the desired outcome in terms of the parameter we want to control. We get a *global extremum* problem rather than a local extremum problem. In other words, we are interested in finding that value of the parameter that yields the largest or smallest outcome *over a broad range of parameter values*. This difference changes the nature of the problem in a small but important way. Usually, we would expect the global maximum or minimum on an interval $a \leq x \leq b$ to be a local maximum or minimum, and therefore also a critical point. However, the function could be always increasing on the interval, in which case the maximum is at $x^* = b$ and the minimum is at $x^* = a$, or always decreasing, with the opposite conclusions. Other combinations are also possible.

> **Theorem 1.5.1 (Global Extrema).** *Suppose x^* yields the largest or smallest value of a continuous function $f(x)$ on an interval $a \leq x \leq b$. Then either x^* is a critical point, $x^* = a$, or $x^* = b$.*

Example 1.5.1. Q) Suppose an organism takes in resources at a rate proportional to its surface area and uses resources (for maintenance of its tissues) at a rate proportional to its mass. How large should the organism be at maturity if natural selection works to maximize the availability of resources for reproduction?

A) To answer this question, we need a mathematical model for the resource balance of the organism. Let x represent the organism's length and assume that the organism maintains the same shape as it grows. We assume that the intake rate is ax^2 and the maintenance rate is bx^3, where a and b are positive parameters.[22] The function we want to maximize is the surplus of resource intake rate minus resource expenditure rate, which is

$$f(x) = ax^2 - bx^3.$$

First we need to consider the possible values of x. Certainly we must have $x > 0$. We cannot have $f < 0$, as that would mean that the organism spends more resources on maintenance than it is able to collect; hence we must also have $x < a/b$. In this example, the endpoints $x = 0$ and $x = a/b$ both yield the minimum value $f = 0$. Hence, biological considerations in conjunction with Theorem 1.5.1 guarantee that the optimal size is a critical point of f. Differentiating f and setting the result equal to 0 yields

$$0 = 2ax - 3bx^2 = x(2a - 3bx).$$

There are two critical points, but we have already ruled out $x = 0$. Hence the optimal size must be

$$x = \frac{2a}{3b}.$$

We have found the correct answer to the mathematical problem; however, we have not yet found a meaningful biological interpretation. Consider that the maximum possible size for our organism is a/b. Thus, the optimal size is 2/3 of the maximum size. This is a very robust result, because it does not depend on the specific parameters a and b. Note that our perception of the result depends on how we state it. Two-thirds of the maximum length sounds large, but it corresponds to less than one-third of the maximum mass (8/27).

Figure 1.5.1 shows the function $f(x) = x^2 - x^3$ and indicates the maximum at $x = 2/3$ that we have calculated. At present, this is merely an example that illustrates our work with one set of parameter values. We shall see in Section 2.6 that a slight change in the formulation of the model makes it possible to illustrate all sets of parameter values with a single figure. □

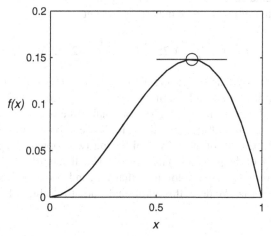

Fig. 1.5.1 The function $f(x) = x^2 - x^3$

Example 1.5.2. The gene responsible for sickle cell disease has two possible alleles, **A** and **a**. Individuals with the **aa** genotype have sickle cell anemia. Under natural conditions, people with sickle cell anemia did not live long enough to have children. Since the gene reduces fitness, one

---

[22] See Example 1.6.3 for justification of these assumptions.

would expect it to have been eliminated by natural selection. Indeed, people whose ancestors come from cold climates do not have the **a** allele, but this allele is not uncommon in parts of Africa. Individuals with the **Aa** genotype have a higher natural resistance to malaria than individuals with the **AA** genotype; where malaria is endemic, individuals with the **Aa** genotype have a longer life expectancy, which means they are likely to have more surviving children. Thus, the **a** allele has an indirect benefit along with its direct liability.

To model the population dynamics of the sickle cell gene, let p be the fraction of **A** genes in the gene pool. We assume that the fitness of the **aa** genotype is 0, the fitness of the **Aa** genotype is 1, and the fitness of the **AA** genotype is a parameter w. According to the theory of genetics, the overall fitness of the gene pool is

$$F(p) = wp^2 + 2p(1-p) = wp^2 + 2p - 2p^2 = 2p + (w-2)p^2.$$

If conditions remain fixed for many generations, natural selection works to achieve optimum fitness. Mathematically, this is a global extremum problem in which we seek the value of p that produces maximum fitness, with $0 \leq p \leq 1$. The problem is somewhat complicated by the parameter w. Theorem 1.5.1 tells us that global extrema for $F(p)$ can only occur at $p = 0$, $p = 1$, and critical points of F. These critical points, and the global extrema, will depend on the parameter w. As examples, we consider two specific values: 0.75 and 1.2.

If $w = 1.2$, then

$$F(p) = 2p - 0.8p^2, \qquad \frac{dF}{dp} = 2 - 1.6p.$$

Solving $dF/dp = 0$ for p yields the critical point $p = 1.25$. This critical point lies outside the interval $0 \leq p \leq 1$, so it is not relevant. With no critical points in the interval of interest, the optimal value must occur at one of the endpoints, $p = 0$ and $p = 1$. These values yield $F(0) = 0$ and $F(1) = 0.4$. Hence, the global maximum fitness occurs at $p = 1$, as shown in Figure 1.5.2. Thus, the sickle cell trait should disappear from the population. This makes sense biologically, because with $w = 1.2$, the **AA** genotype is more fit than the **Aa** genotype. Unfortunately, this is not the case for the sickle cell trait. In reality, $w = 1$ in an environment that is free of malaria and $0 < w < 1$ in an environment where malaria is present.

If $w = 0.75$, then

$$F(p) = 2p - 1.25p^2, \qquad \frac{dF}{dp} = 2 - 2.5p.$$

Solving $dF/dp = 0$ for p yields the critical point $p = 0.8$. This critical point lies inside the interval $0 \leq p \leq 1$, so it is a possible optimal value, along with $p = 0$ and $p = 1$. We have $F(0) = 0$, $F(0.8) = 0.8$, and $F(1) = 0.75$. The optimal value occurs at $p = 0.8$, as shown in Figure 1.5.2. This means that 20% of the alleles in the gene pool will be of type **a**, which in turn means that the probability of an individual having two copies of the sickle cell allele is $0.2^2 = 0.04$. Hence, 4% of the population inherits sickle cell anemia.

As we saw in Section 1.1, there is a lot more that we can learn about the model if we do not choose specific values for w. We defer this additional study to Problem 1.5.4.   □

1.5.1 The Marginal Value Theorem

Suppose you are visiting an apple orchard, and you are allowed to take as many apples as you can pick in one hour. For the sake of a good thought experiment, we assume that the trees are far enough apart that it takes about a minute to walk from one tree to the next. What strategy do you follow? At first thought, it makes sense to start at the nearest tree and pick all of its apples before moving on. You begin with the apples that are easily within reach, but it gets

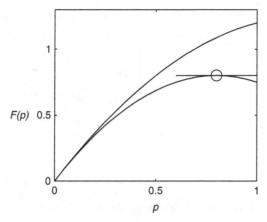

Fig. 1.5.2 The fitness function of Example 1.5.2 with $w = 1.2$ (*upper*) and $w = 0.75$ (*lower*)

progressively more difficult to pick the next apple. At some point, you have to jump to get an apple, and eventually you will have to climb the tree to get more. So you change your strategy. You decide to move on to another tree, even though you could still have picked more apples at the first tree. There is clearly a trade-off at work. It costs some time to move from one tree to the next, which suggests that you should stay longer at the first tree. However, the apples get harder to pick as time goes on, which suggests that you should move to a new tree sooner.

For animals that forage, getting close to the right time to move from one patch to another is critical. Suppose some individuals are good at picking the right time to move, while others tend to linger too long and others are too quick to move. Suppose further that this "perseverance" trait is genetically controlled. In each generation, individuals with the optimal amount of perseverance will be more successful at foraging; consequently, they will produce more offspring, each one inheriting its perseverance from its parents. Over time, the population will be dominated by individuals who have just the right amount of perseverance.

Our scenario has two phases for each patch, one phase for changing patches and one phase for collecting food. Alternatively, we could just look at the collection phase using a function that starts out negative to account for the resources used up in the process of changing patches. This formulation is easier. To complete it, we need to prescribe a function to represent the net resource gain as a function of time. This function should be negative at time 0, always increasing, and curved downward. In mathematical notation the requirements are

$$f(0) < 0, \qquad f'(t) > 0, \qquad f''(t) < 0.$$

We consider an example and then look at the general case.

Example 1.5.3. Suppose the net resource gain for a forager is given by the function $f(t) = \sqrt{t} - 1$. Note that this function has the properties indicated in the previous discussion—it is negative at time 0, increasing, and curved down.

The challenge for us now is to determine the appropriate function to maximize. Clearly, it is not the function f, which we have already said is always increasing. The maximum of f is achieved only in the limit $t \to \infty$, but it is clear from our thought experiment that it is bad to stay too long in the same patch.

The key to finding the correct function to optimize is to recognize that the forager will move on to the next patch when that optimal time is reached. Hence, the total amount of resources collected over a long period of time will include resources from multiple patches. In effect, an optimal forager will collect resources at the largest average rate. Since each patch in our model is identical, the average rate over multiple foraging steps is the same as the average rate for each step, and that is the net resource gain divided by the time spent at the patch:

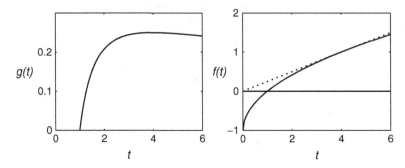

Fig. 1.5.3 The average resource gain function g and the resource gain function f from Example 1.5.3. The *dotted line* on the graph of f connects the origin to the point of maximum average resource gain; at this point, the secant slope (average resource gain) is equal to the tangent slope (marginal resource gain)

$$g(t) = \frac{\sqrt{t} - 1}{t} = t^{-1/2} - t^{-1}.$$

We also need to determine the time interval to be considered. The average gain of resources for our given f is negative up to the time $t = 1$, which is the amount of time required for the forager to collect enough food to make up for the initial deficit. Of course, the optimal amount of time spent in the patch must be at least 1 time unit. While there is no strict maximum time, it is clear that the average gain of resources is 0 in the limit as $t \to \infty$, so we may consider the interval $1 \le t < \infty$. Our function is continuous on this interval, zero at both endpoints, and positive in the middle; clearly, there is an optimal time somewhere in the interval, and it must be a time for which $g' = 0$. If we find just one critical point, that point must be the optimal time.

Using the formula for g, we have

$$g' = -\frac{1}{2}t^{-3/2} + t^{-2}.$$

This quantity is zero when

$$t^{-2} = \frac{1}{2}t^{-3/2}.$$

Taking reciprocals on both sides yields the equation

$$t^2 = 2t^{3/2},$$

or $\sqrt{t} = 2$. Thus, the forager should stay in a patch for 4 time units before moving on. Figure 1.5.3 shows the functions $f(t)$ and $g(t)$, along with a dotted line connecting the point $(0,0)$ and the point $(4, f(4))$.   □

In Example 1.5.3, the optimal point on the graph of f appears to be the point for which the tangent passes through the origin. This is no accident. We can derive this result mathematically,[23] but it can also be explained graphically. For any time t, the quantity $g(t) = f(t)/t$ represents the slope of the line connecting the point $(0,0)$ and the point $(t, f(t))$. Thus, we can visualize the function g on the graph of f as the slope of a line. The optimum point is the one for which this slope is the largest; graphically, it is clear that this is achieved when the line whose slope is $g(t)$ is tangent to the graph of f. This principle is summarized by the marginal value theorem.

[23] See Problem 1.5.8.

Theorem 1.5.2 (Marginal Value Theorem). *Suppose $f(t)$ satisfies the properties $f(0) <$ 0, $f' > 0$, and $f'' < 0$. Then the maximum of the average value $f(t)/t$ occurs at the point where that average is equal to the marginal value $f'(t)$.*

Problems

1.5.1.* Find the global maximum of the function

$$f(r) = -r \ln r$$

on the interval $(0, 1]$. You can use the first derivative test to rule out the possibility of a global maximum as $r \to 0$. Plot a graph of the function to confirm your result.

1.5.2.* Find the global maximum and minimum of the function

$$f(r) = \frac{q}{r} + \ln r$$

on the interval $r > 0$, where $q > 0$ is fixed. To confirm your result, you need to determine the behavior of the function as $r \to 0$ and $r \to \infty$. The easiest way to do this is to examine f' on the intervals $(0, r^\star)$ and $(r^\star, \infty)$.

1.5.3. Consider a fish that swims at speed v relative to the water against a current of speed $u < v$.

(a) How fast is the fish actually moving relative to a fixed reference point?
(b) The energy expended by the fish per unit time is equal to the speed relative to the water multiplied by the force required to move through the water. Assuming that force to be kv^2, where $k > 0$ is a parameter that depends on the shape of the fish, find a formula for the energy expenditure per unit time.
(c) Combine the answers to parts (a) and (b) to determine the energy expenditure per unit distance actually traveled.
(d)* Find the speed the fish should swim relative to the water to minimize the energy expenditure per unit distance traveled. Experiments have shown that fish generally do swim at roughly this speed when not trying to catch prey or avoid being caught by a predator.

(This problem is continued in Problem 2.6.6.)

1.5.4.* In the sickle cell problem of Example 1.5.3, we derived a formula for the gene pool fitness F as a function of the fraction p of non-sickle alleles in the gene pool. We used this formula to predict the optimal value $p = p^\star$ for $w = 0.75$ and $w = 1.2$. Here we consider a more complete analysis of the model.

(a) Find the value of w for which $F'(1) = 0$.
(b) Suppose w is larger than the value found in part (a). Determine the optimal p for this case. What should happen to the sickle cell trait in this case?
(c) Suppose w is smaller than the value found in part (a). Determine the optimal p for this case.
(d) Plot a graph of the optimal p as a function of w. Explain the graph in biological terms.
(e) Determine the fraction of individuals who inherit sickle cell disease (individuals of the **aa** genotype). Plot this fraction as a function of w.

1.5.5.* In what is arguably the best known modern mathematics paper on elementary calculus, Hope College mathematics professor Tim Pennings reported on experiments he conducted with his dog Elvis [9]. In each experiment, Pennings threw a tennis ball into a lake at an angle to the shoreline. Elvis ran along the shore toward the point closest to the ball, but he jumped into the water for the final swim before reaching that closest point. Pennings measured his dog's running and swimming speeds and used calculus to determine the optimal point for Elvis to switch from running to swimming so as to minimize the total time required to reach the ball. Repeated experiments showed that Elvis consistently chose a point near the optimal point, even though Pennings varied the distance and angle of the throw. In this problem, we consider Pennings's optimization model along with an improved model that shows how a lot of careful thought put into the modeling can result in more meaningful answers.

(a) A dog owner standing near a lake with a straight shoreline throws a ball into the water. (See Figure 1.5.4.) The ball lands at a point Y meters (m) away from the shore, as in the figure. The dog begins to run with speed r along the shore toward the point P that is directly across from the ball's location. At a distance x from the point P, the dog jumps into the water and swims the remaining distance to the ball at speed $s < r$. Determine the total time $t(x)$ required for the dog to get to the ball. Assume that the distance between the start and the point P is X. [Note: Use the equation "distance $=$ rate $\times$ time" to determine the times spent running and swimming.]

(b) The function $t(x)$ from part (a) contains parameters X, Y, r, s. Without loss of generality, we can assume $s < r = 1$, which removes the parameter r from the problem. The derivative dt/dx then contains only two parameters. Find the critical point x^* in terms of these two parameters.

(c) Under what circumstances is this critical point the optimal solution? [Hint: Think about the range of possible x values.]

(d) The distance x is only one of the variables that could be used to indicate the optimal jump point. One could also use either of the angles marked in the figure. While we would prefer the answer to be in terms of ϕ rather than θ, the calculations work better in terms of θ. Begin by writing down formulas for the swimming distance and the running distance in terms of θ rather than x, using the simplification $r = 1$. Then find the total time as a function $t(\theta)$.

(e) The function $t(\theta)$ from part (c) contains the same parameters as the original function $t(x)$, and the same parameter disappears in the computation of $dt/d\theta$. Find a simple trigonometric equation for the critical point θ^* in terms of the remaining parameters. This critical point is the optimal solution whenever the condition from part (c) is met.

(f) Note that θ and ϕ are complementary angles, meaning that the triangle containing them is a right triangle. What does this say about the relationship between the cosine of one of the angles and the sine of the other? Use this information to write the answer to part (e) in the form $\cos\phi^* = C$, where C depends on one or more of the parameters.

(g) The model of parts (a) and (b) yields an answer for the optimal x, and the model of parts (d)–(f) yields an answer for the optimal angle, either as θ or ϕ. Suppose you were going to design a robot dog to participate in a race from the starting point to the finishing point. Each of your answers carries with it an implication of the capabilities the robot would require. For example, the robot for part (b) must be able to measure x as it moves, while the robot for part (f) must be able to measure ϕ. We can assume that the speed s is built into the robot's program, so that it does not need to be measured. What must the robot for part (b) be able to measure that the robot for part (f) does not need? The answer is of biological importance, because the capability of optimal behavior must be acquired by natural selection. The better model is the one that gives the simplest plausible explanation of behavior. Which of the two models is better in this sense?

(This problem is continued in Problems 1.6.7 and 2.6.7.)

1.5.6. In a well-known optimal foraging paper, R. Zach reported data on the foraging behavior of crows [12]. The crows break whelk shells by carrying them above a rocky area and then dropping them onto the rocks. Zach measured the probability of a shell breaking in a fall from a height of x m and observed that the crows consistently dropped the whelks from a height of approximately 5.25 m. He explained these results as the solution of an optimal foraging problem. Zach did not use calculus to solve the problem, but we will do so in this problem.

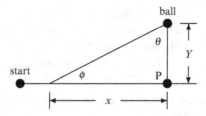

Fig. 1.5.4 The geometry for Problem 1.5.5

It seems reasonable that the optimal strategy for the crows would be to minimize the product of the height and the average number of drops required for the shell to break. This assumption is based on the fact that the energy the crow expends on an upward flight is proportional to the total height. The product of the height per attempt and the number of attempts therefore represents the total energy cost of breaking a shell. Note, however, that the average number of drops required should be a function of the height of the drop. Thus, the function to be minimized is

$$f(x) = xn(x),$$

where we need a model for the function $n(x)$.

(a) What properties should we expect a graph of $n(x)$ to have? Think in terms of slopes and limiting values.
(b) Joe Mahaffy has proposed the model

$$n(x) = 1 + \frac{a}{x - b}$$

for the average number of drops needed to break a shell falling from a height of x [6]. Mahaffy found the estimates $a \approx 16$ and $b \approx 1.2$ using Zach's data for shell breaking experiments.[24] Plot the functions $n(x)$ and $f(x)$ with these parameter values. Your plot should only cover the range of x values that make sense (What happens if $x = 1$, for example?). What happens physically if x is outside of the reasonable range? Does Mahaffy's model have the right properties?
(c) Determine the optimal dropping height for Mahaffy's model with the parameter values he obtained from Zach's data. How does the answer compare with what Zach observed? Does this answer support the model?
(d) Determine the optimal dropping height for unspecified parameter values.
(e) Suppose $b = 1.2$. Plot the optimal dropping height as a function of the parameter a.
(f) Does a larger value of a indicate a harder surface or a softer one? How do you know?

1.5.7. Let $f(t) = \frac{t - 1}{t + 1}$.

(a) Show that f has the properties required of a function that can be the net resources in the marginal value theorem.

---

[24] These experiments were simply to determine data points for the function $n(x)$; as such they did not have anything to do with the independent measurement of the crows' behavior.

(b) Use the marginal value theorem to compute the time at which the forager should move to a new patch.

(c) Plot $f(t)$ along with the line segment connecting the origin and the point found in part (b). Confirm that you have found the correct optimal solution.

1.5.8. Derive the result stated in the marginal value theorem.

1.5.9.* The marginal value theorem determines the optimal time at which to move to a new patch. A more useful result would be the optimal level of remaining resources. We explore this issue in this problem. Let $f(t)$ be the net amount of resources collected at time t and let $x(t)$ be the size of the patch at time t. Assume that the patch is initially of unit size and that the initial resource deficit is an amount D; that is, $f(0) = -D$, with $D < 1$.

(a) Suppose the rate at which the forager obtains resources from a patch is proportional to the resource level of the patch; that is,

$$\frac{dx}{dt} = -kx, \qquad k > 0,$$

where the parameter k represents the feeding rate of the forager. Use this equation, along with the information about the initial size, to determine the resource amount $x(t)$. [See the discussion of (1.6.2) in the text. Note that your answer will contain the unspecified parameter k.]

(b) Explain why the quantity $f(t) + x(t)$ should be constant. Use this fact, the initial data, and the answer to part (a) to obtain a formula for $f(t)$.

(c) Apply the marginal value theorem to the function f from part (b) to determine an equation relating the optimal time $t^\star$ and the parameters k and D. You will not be able to solve this equation by hand.

(d) Use the result of part (a) to eliminate $t^\star$ from the equation of part (c). The result should be an equation that has only $x^\star$ and D.

(e) Plot the optimal residual resource amount $x^\star$ against D, using the result of part (d). Note that the equation does not need to be solved for $x^\star$. Given any $x^\star$ between 0 and 1, you can calculate the corresponding value of D and plot the point in the $Dx^\star$ plane.

(f) Interpret the graph of part (e), noting that D is the cost of travel between patches and $x^\star$ is the optimal amount of resource left at the time the forager moves to a new patch. Your discussion should note whether $x^\star$ increases or decreases with D and explain this result. It should also address the issues of how large D can be, what happens in the limit as $D \to 0$, and what resource deficit amounts correspond to the limiting values $x^\star = 0$ and $x^\star = 1$.

(g) The final result does not include the rate parameter k. Explain why it is reasonable that the relationship of $x^\star$ and D should not depend on k.

1.5.10. In this problem, we develop a different version of the marginal value theorem. Suppose the total resources collected in time t, starting when the forager reaches the patch, is $m(t)$ and the time required to change patches is T. As in the text version, our goal is to maximize the average rate of resource collection.

(a) We require m to satisfy the properties $m(0) = 0$, $m'(0) > 0$, and $m''(0) < 0$. Explain the biological justification for these properties.

(b) Write down a formula that gives the average rate of resource gain from a patch, given that the resource gain is $m(t)$ for the time in the patch and 0 for the time traveling to the next patch.

(c) Let $t^\star$ be the optimal time to switch patches. Find a formula that relates $m'(t^\star)$ and $m(t^\star)$.

(d) Suppose $m(t) = t/(1+t)$. Use the result of part (c) to determine the optimal switching time in terms of the parameter T.

(e) Sketch a graph of the resource accumulation, given $T = 1$. Note that this should be $m(t)$ up to the time $t^\star$ and remain constant for an additional time of duration T. Add the line segment that connects the beginning of the curve to the end and add a line segment that is tangent to the curve $m(t)$ at the point $t^\star$. What is the geometric relationship between these lines?

1.5.11. Many optimization problems have multiple independent variables but only a linear objective function and constraints. These *linear programming* problems have mathematical properties and techniques that are quite different from the problems found in calculus books. Here we consider a particularly simple example, adapted from work originally done by G.E. Belovsky in 1978 [3] and subsequently described by R.M. Alexander [1]. Suppose a moose eats x kg of aquatic plants and y kg of terrestrial plants each day. Experiments have shown that a moose can extract four times as much energy from a given amount of terrestrial plants than from the corresponding amount of aquatic plants. Thus, the total energy the moose gets from its food is proportional to the function $z = x + 4y$. This is the function to be maximized. Two constraints control the possibilities for the moose diet. Salt is available primarily from aquatic plants, and the moose can meet its salt requirement by consuming 17 kg of these per day; assume that it would require 51 kg of terrestrial plants to meet the salt requirement without any aquatic plants. Also, the digestive capacity of the moose limits its total consumption to 33 kg per day.

(a) Write down a pair of inequalities to represent the total consumption constraint and the salt requirement. For the salt requirement, note that 3 kg of terrestrial plants and 1 kg of aquatic plants produce the same amount of salt, in the same way that 4 kg of aquatic plants have the same energy value as 1 kg of terrestrial plants. The problem is to maximize z while satisfying the two constraints.
(b) Plot the (triangular) region in the xy plane that satisfies both constraints. This is the *feasible region* for the problem.
(c) Certainly one possibility is that the moose could simply eat 33 kg of aquatic plants, yielding a total value of $z = 33$. Add the graph of the line $x + 4y = 33$ to the graph of the feasible region. All of the points on this line produce the same value of the objective function.
(d) The line from part (c) divides the feasible region into two portions, one that produces $z > 33$ and one that produces $z < 33$. The optimal solution must be in the $z > 33$ portion. Now plot the line $x + 4y = 70$ on the same graph. This confines the optimal solution to a smaller subregion.
(e) Notice the simple geometric relationship between the lines from parts (c) and (d). Any line of constant z has the same geometric relationship, which means that such lines can be drawn without any calculations. Sketch several more lines of constant z, each with a progressively larger value. Stop with the line that is just tangent to the feasible region.
(f) Find the optimal point in the feasible region, which is the point of tangency found in part (e).

1.6 Related Rates

After studying this section, you should be able to:

- Differentiate equations to obtain relationships between derivatives.
- Apply the chain rule to find rate relationships among variables.

In Section 1.2, we defined the derivative for a function of continuous time. We used the definition in Section 1.3 to devise a calculation scheme for obtaining a derivative formula from a formula for the function. This is useful whenever we have a known formula for a function, but that is not always the case. Sometimes we have an equation that defines the relationship between two functions, even though we don't know either of the functions explicitly.

Example 1.6.1. Many organisms, such as fish, grow without changing geometric proportions. A model for the growth of such an organism can use length, volume, or mass as the primary measure of the organism's size, since these are all related by simple formulas. Let $x(t)$ be the length, $V(t)$ the volume, and $m(t)$ the mass for an organism of fixed proportions. Then the relationship between volume and length can be modeled as $V(t) = c[x(t)]^3$ for some constant shape factor c. For example, we would have $V = x^3$ for a cubic shape. In the case of a sphere, the volume is given by $V = (4/3)\pi r^3$. Since the radius would be half of the length, this gives us

$$V = \frac{4}{3}\pi \left(\frac{x}{2}\right)^3 = \frac{\pi}{6}x^3.$$

In general, the factor c indicates the volume of an organism of unit length. If the measurement chosen for x is the largest dimension, then of course $c < 1$; the greater the deviation from a cubical shape, the smaller the value of c.

Similarly, mass and volume are related by the formula $m = \rho V$, where ρ is the density of the material of which the organism is made. Living organisms are mostly water, so $\rho \approx 1$.

Thus, the functions $x(t)$, $V(t)$, and $m(t)$ are related by the equations

$$V(t) = c[x(t)]^3, \qquad m(t) = \rho V(t), \qquad m(t) = \rho c[x(t)]^3, \qquad (1.6.1)$$

regardless of the actual growth history of the organism.   □

Whenever we have an equation relating functions, as in Example 1.6.1, we can obtain an explicit relationship between the derivatives by differentiating that equation.

Example 1.6.2. Suppose an organism grows with length $x(t)$ and mass $m(t)$ in a way that maintains the geometric proportions. Then the length and mass are related by the third formula of (1.6.1), which we can write as

$$m(t) = k[x(t)]^3, \qquad k > 0,$$

by defining a new parameter $k = c\rho$ for convenience to replace the two individual parameters. Differentiating the left side of the equation with respect to t simply results in the derivative $m'(t)$. Differentiating the right side of the equation requires the chain rule, because we have an explicit function of x but are differentiating with respect to t. Application of the chain rule results in the equation

$$m'(t) = 3k[x(t)]^2 x'(t).$$

The result of differentiating an equation is an algebraic equation that relates the rates of change.   □

1.6.1 Differential Equations

Frequently, functions are defined by the properties of their derivatives. For example, we might have a mathematical model that defines a quantity y as a function of t by the equation

$$\frac{dy}{dt} = -ky, \tag{1.6.2}$$

where k is some positive parameter. Equation (1.6.2) is called a *differential equation* because the equation includes the derivative of y as well as y itself. Differential equation models arise in many contexts, and they constitute a portion of the content of Chapter 5 and all of Chapter 7. Most differential equations are difficult to solve; however, (1.6.2) can easily be solved by inspection. A review of the basic derivative formulas in Table 1.3.1 shows that the derivative of e^{ax} with respect to x is ae^{ax}. Hence, the derivative of e^{-kt} with respect to t is $-ke^{-kt}$, which is $-k$ times the original quantity. The function defined by $y = e^{-kt}$ satisfies (1.6.2) because both sides of the equation evaluate to $-ke^{-kt}$. More generally, any function $y = Ae^{-kt}$ satisfies the differential equation.

Mathematically, (1.6.2) represents the same model as the explicit formula $y = Ae^{-kt}$. It may seem to be less useful, but it has the significant advantage of being directly equivalent to a conceptual statement about a process of change. A literal translation of (1.6.2) is "The rate of change of the quantity y is equal to $-k$ times the quantity y." This is correct as far as it goes, but the equation actually makes a much stronger statement. The parameter k is arbitrary; not only is the choice of symbol unimportant, but the actual numerical value only matters in a specific context. What is really important about (1.6.2) is that values of the rate of change of the quantity y are simply a multiple of the quantity present at any time. A mathematical modeler would read the equation as "The rate of change of the quantity y is proportional to the amount of y present." This is a mechanistic assumption about the physical process of change.[25]

In many mathematical models, the rate of change of one quantity is a function of a related quantity.

Example 1.6.3. We can only eat as fast as we can digest our food. If an organism grows proportionally, as in Examples 1.6.1 and 1.6.2, then the length and diameter of its digestive tract are proportional to its overall length. This means that the area of the digestive tract is proportional to the square of the organism's length. Therefore, it is reasonable to guess that the rate of change of an organism's mass is also proportional to the square of its length:

$$\frac{dm}{dt} = a[x(t)]^2,$$

where a is a parameter that represents the food consumption capability of an individual of unit length. Our model contains the unknown length along with the derivative of the unknown mass, but the mass and length are related by the formula $m = kx^3$ from Example 1.6.2. Differentiating $m = kx^3$ with respect to t as in Example 1.6.2 yields the formula $m' = 3kx^2x'$, which we can use to replace m' by x' in our growth model. We then have

$$3kx^2\frac{dx}{dt} = ax^2,$$

which we can rewrite as

$$\frac{dx}{dt} = \frac{a}{3k} \equiv r,$$

using the single parameter r for convenience in place of the combined expression $a/(3k)$. This differential equation for the length of the organism is one that we can solve by inspection. The derivative of a function ax with respect to x is a, so the derivative of a function rt with respect to t is r. Thus, the function $x = rt$ satisfies the differential equation, as does the more general formula $x = C + rt$ where C is an arbitrary constant. □

---

[25] See Section 2.5 on mechanistic modeling.

Example 1.6.3 seems to show that organisms grow equal increments of length in equal amounts of time. However, extensive data on growing organisms shows that this is not what happens. Instead, subsequent intervals in time produce progressively smaller changes in length as an organism grows. What went wrong?

As we shall explore in more detail in Chapter 2, a mathematical model is only as good as its underlying assumptions. Sometimes assumptions that seem plausible lead to predictions that are contradicted by experimental data. Occasionally it is the data that is wrong, but usually the problem is a flaw in the model. This does not mean that the model has no value at all. Finding and correcting flaws in models can lead to biological insight. The flaw in the model of Example 1.6.3 is discussed in Problem 1.6.5.

1.6.2 The Chain Rule

The chain rule is of vital importance in many applications of mathematics, not just the computation of derivatives. In particular, it is often used to change variables in a differential equation model, an essential technique in mathematical modeling. When we introduced the chain rule in Section 1.3, we merely presented it without any justification. Now we can use linear approximation for this purpose. To make the argument easier to follow, we provide a simple context.

Consider a long rod placed so that one end is in a campfire and the other end is resting on the cold ground outside of the fire. An insect is walking along the rod in the direction away from the fire. We define the following quantities:

- t is the time;
- x is the position of any point along the rod, measured from the hot end;
- $s(t)$ is a function that indicates the location of the insect on the x axis;
- $T(x)$ is a function that indicates the temperature at any point x on the rod;
- $u(t)$ is a function that indicates the temperature at the insect's location at any time.

Assume that we know the temperature function $T(x)$, the location function $s(t)$, and the derivatives of both functions. How are we to calculate $u(t)$ and du/dt?

Let t_1 be the current time. Then the insect is at the point $x_1 = s(t_1)$. In a small amount of time, the insect's position changes by an amount Δx, which we can relate to the change in time using linear approximation:

$$\Delta x \approx \frac{ds}{dt}(t_1)\Delta t.$$

This change in position creates a corresponding change in the insect's temperature, which we can approximate as

$$\Delta u \approx \frac{dT}{dx}(s(t_1))\Delta x \approx \frac{dT}{dx}(s(t_1))\frac{ds}{dt}(t_1)\Delta t.$$

Dividing by Δt gives us the ratio of temperature change to change in time:

$$\frac{\Delta u}{\Delta t} \approx \frac{dT}{dx}(s(t_1))\frac{ds}{dt}(t_1).$$

This result becomes exact in the limit as Δt goes to 0:

$$\frac{du}{dt}(t_1) = \lim_{\Delta t \to 0}\frac{\Delta u}{\Delta t} = \frac{dT}{dx}(s(t_1))\frac{ds}{dt}(t_1).$$

The time t_1 is not specified, so the result holds for all time. This is the chain rule.

Chain Rule in terms of functions

If

$$u(t) = T(s(t)),$$ (1.6.3)

then

$$\frac{du}{dt}(t) = \frac{dT}{dx}(s(t))\frac{ds}{dt}(t).$$ (1.6.4)

In the context of the moving insect scenario, du/dt is the rate of change of the insect's temperature with respect to time, dT/dx is the rate of change of temperature with respect to distance *at any point* on the rod, $\frac{dT}{dx}(s(t))$ is the rate of change of temperature with respect to distance *at the insect's location*, and ds/dt is the velocity of the insect's motion.

Example 1.6.4. Suppose $s(t) = 3t$ and $T(x) = e^{-2x}$ in the moving insect scenario. Then $\frac{ds}{dt} = 3$, $\frac{dT}{dx} = -2e^{-2x}$, and $\frac{dT}{dx}(s(t)) = -2e^{-2(3t)} = -2e^{-6t}$. Using (1.6.4),

$$\frac{du}{dt}(t) = \frac{dT}{dx}(s(t))\frac{ds}{dt}(t) = -2e^{-6t} \times 3 = -6e^{-6t}.$$

We can check the result in this simple example by calculating and differentiating u directly. We have

$$u = T(s(t)) = e^{-2(3t)} = e^{-6t},$$

so

$$\frac{du}{dt} = -6e^{-6t}.$$

$\square$

The notation of (1.6.4) is precise, but somewhat difficult to interpret. It is common in mathematical modeling to address this difficulty by abandoning the standard function notation of mathematics. If you carefully reread the thought experiment, you will see that the quantities T, x, s, t, and u are related by the equations $x = s(t)$, which gives the insect's location, and $u = T(s(t))$, which gives the insect's temperature. In a sense, t, x, and u are variables, while T and s are functions that are used to indicate relationships between pairs of variables. The temperature of the insect depends on the location of the insect, which in turn depends on the time.

Given the dependencies of position on time and temperature on position, there are three meaningful derivatives in terms of the variables: du/dt, dx/dt, and du/dx. The first of these is clearly the rate of change of the insect's temperature with respect to time. The second is equivalent to ds/dt, since the insect's position x is given by the function $s(t)$. The spatial derivative du/dx represents the rate of change of the insect's temperature with respect to its position. In general, the rate of change of temperature with respect to position is dT/dx; this is the same as the rate of change of insect temperature to insect position if it is evaluated specifically at $x = s(t)$. Thus, du/dx is a simpler notation for $\frac{dT}{dx}(s(t))$.

Substituting dx/dt for ds/dt and du/dx for $\frac{dT}{dx}(s(t))$ in (1.6.4) yields the equation

$$\frac{du}{dt} = \frac{du}{dx}\frac{dx}{dt}.$$

While less mathematically precise than (1.6.4) because it fails to indicate that du/dx must be evaluated at $x = s(t)$, this version of the chain rule is much easier to interpret. It looks like an equation about multiplication of fractions, an interpretation that, while not precisely correct, makes the formula easy to remember. To summarize, we have the following rule:

Chain Rule in terms of variables
If t, x, and u are variables such that u depends on x and x depends on t, then

$$\frac{du}{dt} = \frac{du}{dx}\frac{dx}{dt}.$$ (1.6.5)

In applying the chain rule as written in (1.6.5), one must be careful to remember that the derivative $\frac{du}{dx}$ is to be evaluated specifically at $x = s(t)$, and not some other value of x.

Problems

1.6.1.* Sunflowers make seeds at a constant rate, which means that the area of the flower head increases at a constant rate. Suppose that each seed occupies $3\,\text{mm}^2$ of head area and the head generates 20 seeds per day. How rapidly is the flower head diameter changing when the current diameter is D mm?

1.6.2. Suppose the number of prey animals eaten per predator per day is given by

$$y = \frac{x}{1+hx}, \qquad h > 0,$$

where x is the number of prey animals per unit area.

(a) Determine the rate at which predation is changing with respect to time in terms of the quantities $x(t)$ and $x'(t)$.
(b) Suppose the prey population is growing at the rate rx, where $r > 0$. Use this additional information to determine the rate at which predation is changing with respect to time in terms of $x(t)$ alone.

1.6.3. The male fiddler crab *Uca pugnax* grows one oversize claw whose only purpose appears to be attracting a mate. The relationship between the claw mass C and the body mass B (not counting the claw mass), both given in milligrams, has been reported to be

$$C = 0.036B^{1.356}.$$

(a) Find a formula to relate the growth rates of the claw mass and body mass.
(b) Plot the ratio of the claw mass growth rate to the body mass growth rate, as a function of the body mass size in grams. Assume the body mass can be as large as $40\,\text{g}$.
(c) For what body mass size in grams is the growth rate of the claw as large as the growth rate of the body mass?
(d) At what body mass size in grams is the mass of the claw equal to the mass of the body?
(e) Assume that the density of body tissues is roughly 1 mg per mm^3 of volume, that the largest dimension of the claw is about 44 mm, and that the body mass is 30 g. Use this information to estimate the shape factor in the length-volume relationship ($V = cx^3$) for a large male claw.

1.6.4. The focal length of the lens in the human eye adapts to the distance of the object being viewed. The necessary length is given approximately by the thin lens formula

$$\frac{1}{f} = \frac{1}{x} + \frac{1}{v},$$

where f is the focal length of the lens, x is the distance of the object, and v is a positive constant. [The lens has a maximum focal length f_0; muscles in the eye automatically decrease the focal length when viewing nearby objects.]

(a) What is the physical significance of the parameter v? [Hint: Find a value of x for which the relationship between f and v is particularly simple.][26]

(b) Suppose an object is approaching you at a speed of 10 miles per hour (16.1 km/h, or 4.47 m/s). How rapidly must your eye's focal length change when the distance of the object is 30 m? What about 10 m? You may assume $v = 17$ mm.

1.6.5. The von Bertalanffy equation, which is often used to model the growth of an organism, is

$$\frac{dx}{dt} = r(x_\infty - x),$$

where $x(t)$ is the linear size of the organism, x_∞ is a theoretical maximum organism size, and r is a rate constant. In this problem, we derive this differential equation by correcting the model of Example 1.6.3. In that model, we made two quantitative assumptions: that mass is proportional to linear size cubed ($m = kx^3$) and that mass increases at a rate proportional to linear size squared ($dm/dt = ax^2$). The differential equation resulting from these assumptions incorporates a model for growth due to feeding, but it fails to account for the cost of staying alive. This is clearly important, as an organism needs to eat enough to maintain its weight before it can use additional food for growth. It is reasonable to assume that this maintenance need is proportional to the mass.

(a) Write down a differential equation model that relates changes in mass to the combination of an increase due to feeding and a decrease (proportional to the mass) due to maintenance requirements.

(b) Differentiate the equation $m = kx^3$ with respect to time and eliminate m from the differential equation of part (a).

(c) Rearrange the resulting equation so that it is equivalent to the von Bertalanffy equation. You will need to define the von Bertalanffy parameters x_∞ and r in terms of the parameters a, k, and the parameter you introduced in part (a).

(This problem is continued in Problem 1.9.11.)

1.6.6.* Suppose the biomass $x(t)$ of a plant changes according to the differential equation

$$\frac{dx}{dt} = rx - \frac{qx}{a+x}, \qquad a, q, r > 0,$$

where the first term represents plant growth and the second represents herbivory. Define new variables X and T and parameter q by

$$X = \frac{x}{a}, \qquad T = rt, \qquad Q = \frac{q}{ra}.$$

(a) Use the chain rule to obtain a differential equation for $X(T)$. [Hint: You can work with more than two factors, which is necessary here since you have a derivative formula for dx/dt along with algebraic relations between x and X and between t and T. Note that you must also use the algebraic formula to replace x by X; your result should have no lowercase symbols.]

---

[26] The parameter v is determined by the focal power of the cornea and the distance from the lens to the retina. If your focal length f_0 is smaller than v, then you need glasses.

(b) Note that X and T represent the plant biomass and time. If we want to study the effect of growth and herbivory on plant biomass, why might we prefer to study $X(T)$ rather than $x(t)$?

1.6.7. (Continued from Problem 1.5.5.)
In Problem 1.5.5, we worked through the optimization model used by Hope College mathematics professor Tim Pennings to determine the appropriate point for his dog to switch from running along the shore of a lake to swimming through the water, given the goal of reaching a ball in the lake as quickly as possible [9]. Pennings's paper has generated quite a bit of commentary,[27] particularly with regard to attempts to account for the dog's success at solving the problem.[28] In Problem 1.5.5, we found a simpler version of the original optimization problem that made the solution more intuitive. It is still difficult to see a connection between the mathematical solution and a mechanism for animal behavior. More recently, two French animal behavior specialists, Pierre Perruchet and Jorge Gallego, repeated Pennings's experiments, concluding that European dogs are as good as American dogs at consistently determining the optimal launch point. Their paper offers a more plausible explanation for the dogs' success that is based on related rates rather than optimization [10]. This problem explores their model.

(a) A dog owner standing near a lake with a straight shoreline throws a ball into the water. (See Figure 1.6.1.) The ball lands at a point Y meters away from the shore and X meters along the shore, as in the figure. The dog begins to run with speed r along the shore toward the point P that is directly across from the ball's location. Let $z(t)$ be the distance between the dog and the ball, and compute z' for the simplified case $r = 1$. [Hint: Let $x(t)$ be the distance along the shore between the dog and the point P. This distance is decreasing at the same rate that the distance of the dog from its initial location is increasing; however, this distance is easier to use in writing an equation to relate the distance z to known functions.]
(b) Plot the rate of change of the distance between the dog and the ball as a function of the distance between the dog and the point P. Assuming that the dog wants to get the ball as quickly as possible, why is it not a good idea for the dog to run all the way to the point P? [Keep in mind that $x(t)$ is decreasing.]
(c) Suppose the dog swims with speed $s < r = 1$. Determine the point x at which the dog should switch from running along the shore to swimming directly toward the ball, assuming that the immediate goal is to always progress as rapidly as possible. [Hint: What is the rate of change of z while the dog is swimming?]
(d) Plot the optimal jump distance x as a function of the dog's swimming speed s for the case $Y = 1$. Examine the behavior of the graph near the extreme values $s = 0$ and $s = 1$ to verify that the result makes sense.
(e) What must a dog be able to measure for the strategy suggested by the related rate solution to work? Is this a more or less plausible explanation of dog behavior than the strategy used in Problem 1.5.5?

---

[27] As of this writing, some of this discussion can be found in the Wikipedia entry for "Fetch (game)."

[28] While any discussion of dogs solving calculus problems is obviously tongue in cheek, it should not be at all surprising that animals can choose near-optimal strategies for finding objects quickly. The wild ancestors of today's dogs needed to hunt efficiently to survive, so natural selection developed whatever behavior strategies today's dogs now use to fetch balls thrown into water.

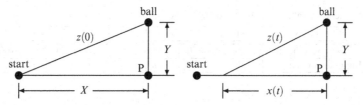

Fig. 1.6.1 The geometry of Problem 1.6.7

1.7 Accumulation and the Definite Integral

After studying this section, you should be able to:

- Approximate the total accumulation of a quantity distributed over time.
- Approximate the total aggregation of a quantity distributed over space.
- Discuss area, total accumulation, and total aggregation in terms of the definite integral concept.

On April 20, 2010, a drilling rig in the Gulf of Mexico called the Deepwater Horizon suffered a catastrophic explosion. By April 22, a large oil slick could be observed at the site. The spill continued to increase until the well was successfully capped on July 15. It has been estimated that the oil was leaking at a rate of $8400\,\mathrm{m}^3$/day at the time it was capped. Since the pressure in the oil reservoir was slowly decreasing as a result of the spill, the leakage rate must have started at a higher value. Scientists eventually estimated the initial leakage rate at $9900\,\mathrm{m}^3$/day, with an estimated reliability of $\pm10\%$ [7]. Given that the oil spill lasted for a total of 84 days, these rates allow us to estimate the total amount of oil that leaked from the well. At the lower rate, the total amount would be the product of $8400\,\mathrm{m}^3$/day and 84 days, which is about $705600\,\mathrm{m}^3$. At the higher rate, the total would be about $831600\,\mathrm{m}^3$. Without any additional data, our best guess is the average of these values, or $768600\,\mathrm{m}^3$.[29]

1.7.1 Estimating Total Volume from Flow Rate Data

The calculations in the above paragraph are estimates whose accuracy is limited by the lack of moment-to-moment exact data for the spill rate. A better approximation could be made if more data were available. We illustrate how this could be done with a hypothetical data set.

Table 1.7.1 Hypothetical flow rate data for the Deepwater Horizon oil spill

Day	0	14	28	42	56	70	84
Flow rate (m³/day)	9900	9620	9350	9100	8850	8620	8400

Example 1.7.1. Table 1.7.1 shows some hypothetical flow rate data for the Deepwater Horizon oil spill. For each of the six 2-week intervals of the spill, we can now make separate lower and

---

[29] This is enough to cover the entire state of Delaware to a depth of 1.2 cm. On the other hand, if the spill were spread across the entire Gulf of Mexico, it would only reach a depth of 0.05 mm. The enormity of the spill is a matter of perspective, and the environmental damage from the spill was highly dependent on concentration at any particular location.

Table 1.7.2 Hypothetical total flow approximations for the Deepwater Horizon oil spill

Interval (days)	0–14	14–28	28–42	42–56	56–70	70–84	0–84
Maximum volume (m³)	138600	134680	130900	127400	123900	120680	776160
Minimum volume (m³)	134680	130900	127400	123900	120680	117600	755160

upper estimates, just as we did for the entire spill in the first paragraph. These lead to a maximum volume estimate of $776160 \, \text{m}^3$ and a minimum of $755160 \, \text{m}^3$, as shown in Table 1.7.2. Our best estimate from these results is the average of $765660 \, \text{m}^3$. $\qquad\qquad\square$

> **Check Your Understanding 1.7.1:**
> Verify several of the entries in Table 1.7.2.

The mathematical value of these calculations lies not in the estimates themselves, which depend on uncertain data, but in the differences between the maximum and minimum estimates. For our original estimate, the difference is $831600 - 705600 = 126000$, whereas for the estimate of Example 1.7.1, it is only $776160 - 755160 = 21000$. The new estimates are significantly better than the old ones. Notice that these two values differ by a factor of 6; it is no accident that this is the same as the number of subintervals in the second calculation. Notice in Table 1.7.2 that the maximum estimate for the second time interval is the same as the minimum estimate for the first time interval. The same thing is true for the maximum estimate for the third time interval and the minimum estimate for the second, and so on. In fact, the lists of maximum and minimum estimates contain the same numbers, except for the maximum on the first interval and the minimum on the last interval. Thus, the entire difference between the sum of the maxima and the sum of the minima comes down to these two values: $138600 - 117600 = 21000$. The first of these values is the product of the flow rate 9900 and the time interval 14; similarly, the second is the product of 8400 and 14. Hence, if we took the flow rate difference of $9900 - 8400 = 1500$ and multiplied it by the time interval, we would once again get the same answer of 21000. The actual estimates of flow rates in Table 1.7.1 don't affect this result as long as the flow rates don't increase over any of the intervals.

The verbal argument of the preceding paragraph can be put in a more precise mathematical form. Let Q_0 be the flow rate on day 0, Q_1 the flow rate on day 14, and so on, so that Q_6 is the flow rate on day 84. We have $Q_0 = 9900$ and $Q_6 = 8400$. Assume that $Q_0 \geq Q_1 \geq \cdots \geq Q_6$. Since each Q_i underestimates the flow rate on interval i and each Q_{i-1} overestimates the same flow rate, we have

$$14(Q_1 + Q_2 + Q_3 + Q_4 + Q_5 + Q_6) \leq V \leq 14(Q_0 + Q_1 + Q_2 + Q_3 + Q_4 + Q_5).$$

The difference between the maximum and minimum estimates is

$$14(Q_0 - Q_6) = 14(9900 - 8400) = 21000.$$

Example 1.7.2. Assume that the flow rate for the Deepwater Horizon oil spill was 9900 at the beginning and 8400 at the end and was slowly decreasing throughout the interval. If we divide the time interval of 84 days into 42 equal subintervals of 2 days each, then the difference between the maximum and minimum estimates of total flow will be $1500 \times 2 = 3000$. Similarly, if we divide the time interval into n equal subintervals, then the time interval is $84/n$ days and the difference between the maximum and minimum estimates will be $1500 \times (84/n) = 126000/n$; thus, the difference becomes progressively smaller as the time interval decreases. If we had enough data to divide the 84 days into 126000 subintervals, each approximately 1 min in length, then the difference between the maximum and minimum estimates would only be $1\,\text{m}^3$. If we had a formula for computing the flow rates at all times, we could come arbitrarily close to an exact answer in the limit as the time interval drops to zero.   □

So how much oil actually did enter the Gulf of Mexico from the Deepwater Horizon spill? We can't answer the question without more data. With a formula for the flow rate as a function of time, we could construct maximum and minimum estimates for large numbers of small subintervals, as in Example 1.7.2. Theoretically, the difference between the maximum and minimum estimates approaches 0 in the limit as the number of subintervals becomes infinite. Thus, the maximum and minimum estimates converge to the same answer, which must be the exact amount of the spill.

1.7.2 The Definite Integral

In general, assume we have a formula $Q(t)$ at which some quantity V is produced over an interval of $a \le t \le b$. As a thought experiment, suppose we divide the interval $[a,b]$ into an arbitrarily large number of subintervals, each starting at a time t and of duration dt. If we use $Q(t)$ to approximate the rate for each of these time intervals, the total amount produced in that interval is approximately $Q(t)\,dt$. With a finite number of subintervals, the total is given approximately as $V \approx \sum Q(t)\,dt$.[30] As the number of subintervals increases to infinity, the interval duration dt approaches 0 and the approximation becomes exact. The summation notation ceases to be appropriate, and our new notation must specifically indicate the time interval, which the sum lacks. This new notation is that of the *definite integral*. In the case of the Deepwater Horizon oil spill, the volume is given in terms of the flow rate by

$$V = \int_0^{84} Q(t)\,dt.$$

There is simple visual interpretation of the definite integral. We can think of the differential dt that appears in the integral as representing an arbitrarily small amount of time. The product of $Q(t)$ (which in this case has the dimension of volume per time) and dt then represents an arbitrarily small amount of volume collected at time t. The integral from a to b indicates the sum of these arbitrarily small amounts collected at all times in the interval. Suppose we have a graph of $Q(t)$ versus t, and assume $Q \ge 0$. At any particular time in an interval $a \le t \le b$, we can draw a line segment that goes up from the t axis to the graph of $Q(t)$. Think of this line segment as having an arbitrarily small width dt. The height of the line segment is $Q(t)$, so the quantity $Q(t)\,dt$ represents the arbitrarily small area of the line segment. Adding these up gives us the whole area, so the quantity $\int_a^b Q(t)\,dt$ is the area under the graph of Q between $t = a$ and $t = b$. Figure 1.7.1a shows the area that represents the total volume of the oil spill, given a made-up function $Q(t)$ that matches the known data. The highlighted line segment represents the arbitrarily small volume spilled during the interval from time 30 to time $30 + dt$.

Example 1.7.3. Figure 1.7.1b illustrates $\displaystyle\int_0^1 x\,dx$. We do not need to approximate this integral because we can calculate it using ordinary geometry. The triangular region has a base of 1 unit and a height of 1 unit, so the area is 1/2. Hence,

---

[30] Indices have been omitted from the sum because we are thinking of an arbitrarily large number of subintervals rather than a fixed number.

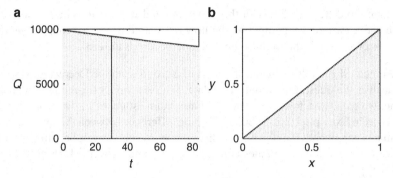

Fig. 1.7.1 (a) A function $Q(t)$ and the area represented by $\int_0^8 4Q(t)\,dt$—the *dark line* represents the area corresponding to the total volume for one specific t; (b) the function $f(x) = x$ and the area represented by $\int_0^1 x\,dx$

$$\int_0^1 x\,dx = \frac{1}{2}.$$

□

Example 1.7.4. Let A be the area between the graph of $y = 1/(1+x^2)$ and the x axis, for $x \geq 0$. Figure 1.7.2 shows the portion of this area that is on the interval $[0,8]$; however, we are interested in the area on the interval $[0,\infty)$. It is not clear that this area is finite, because the horizontal extent of the region is infinite. If the area is finite anyway, then we can write it as

$$A = \int_0^\infty \frac{1}{1+x^2}\,dx.$$

Note that we cannot approximate this integral using finite sums as we have in other examples. Nevertheless, we'll compute the value of the integral in Section 1.9.   □

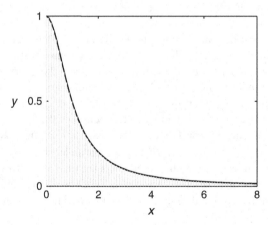

Fig. 1.7.2 The area represented by the integral $\int_0^8 1/(1+x^2)\,dx$

A more formal definition of the definite integral appears in Appendix B. It is not presented here because definite integrals are much more difficult to calculate from the definition than derivatives. In the next section, we obtain a formula that can sometimes be used to calculate definite integrals exactly; for now, we can use the method of Examples 1.7.1 and 1.7.2 to approximate definite integrals to any desired accuracy.

Example 1.7.5. Suppose we want to estimate $\int_0^1 x^2 \, dx$. With two subintervals, we have

$$\int_0^1 x^2 \, dx \approx 0.5(Q_1 + Q_2),$$

where the factor 0.5 is the interval width h and the numbers Q_1 and Q_2 are function values drawn from the subintervals $[0, 0.5]$ and $[0.5, 1]$. It seems reasonable that the best approximation would use the function values at the middle of the subintervals, yielding the approximation

$$\int_0^1 x^2 \, dx \approx 0.5(0.25^2 + 0.75^2) = 0.3125.$$

A better approximation, using the midpoint of each of 10 subintervals, is

$$\int_0^1 x^2 \, dx \approx 0.1(0.05^2 + 0.15^2 + \cdots + 0.95^2) = 0.3325.$$

Using 100 subintervals yields the approximation 0.333325. The fact that increasing the number of subintervals from 10 to 100 makes very little change in the approximation suggests that we are close to the correct answer. Given the values obtained so far, we might guess that $\int_0^1 x^2 \, dx = \frac{1}{3}$. This is the correct answer, as we will see in Section 1.8. □

1.7.3 Applications of the Definite Integral

The oil spill problem is merely one example of a large class of problems. We want to determine the total amount of some quantity—volume in the case of the oil spill—that accumulates over a time interval $a \leq t \leq b$. It is natural to estimate this result using constant approximated rates over small subintervals of time. In the case of the oil spill, the volume over a time interval is the product of the rate, in volume per time, with the length of the interval. Hence, we obtain an accumulation of volume by integrating the volume per unit time with respect to time. The same principle applies to any quantity that is accumulated over time. We can represent this as

$$\text{total stuff} = \int_a^b (\text{stuff per unit time}) \, dt. \tag{1.7.1}$$

Example 1.7.6. Suppose we want to connect the birth rate of a population with survival and fecundity[31] data. For simplicity, we consider only the females. Let $B(t)$ be the birth rate at time t. Each newborn individual has exactly one mother, so we could calculate the birth rate at time t by adding up the births to mothers of various ages. Consider the cohort of individuals who are of age a to $a + da$ at time t. These individuals were all born over an interval of duration da at time $t - a$. The birth rate at that time was $B(t - a)$; hence, the initial size of this cohort was $B(t-a) \, da$. If $\ell(a)$ is the fraction of individuals who survive to age a or greater, then the size of our cohort at time t is $\ell(a)B(t-a) \, da$. Now suppose the rate of births per capita for individuals of age a is $m(a)$. Then the total rate of births for mothers of age a is $m(a)\ell(a)B(t-a) \, da$. The overall birth rate is obtained by summing the birth rates for cohorts of mothers of all possible ages. Thus, the birth rate satisfies the equation

$$B(t) = \int_{a_1}^{a_2} B(t - a)\ell(a)m(a) \, da, \tag{1.7.2}$$

---

[31] Fecundity is the expected reproductive capacity, usually given as a function of the age of the mother.

where a_1 and a_2 are the minimum and maximum ages of reproduction.

In practice, several assumptions can be used to simplify (1.7.2). Suppose the population is growing exponentially with rate constant r. Then the birth rate is also growing exponentially with rate constant r, and we may assume $B(t) = B_0 e^{rt}$. Substituting this formula into (1.7.2), we obtain the equation

$$\int_{a_1}^{a_2} e^{-ra} \ell(a) m(a) \, da = 1. \tag{1.7.3}$$

This is the Euler–Lotka[32] equation, which can be used to determine the unknown growth rate r from survival and fecundity data.[33]

Under some circumstances, the Euler–Lotka equation can be further simplified by the assumption that the risk of death is independent of age. In this case, the survival function is $\ell(a) = e^{-\mu a}$, with $\mu > 0$, so we have

$$\int_{a_1}^{a_2} e^{-(r+\mu)a} m(a) \, da = 1. \tag{1.7.4}$$

It can be shown through probability analysis that the mortality parameter μ is the reciprocal of the life expectancy.   □

Some quantities are distributed over space rather than accumulated over time. In such cases, definite integrals can be used to determine the total amount of "stuff" aggregated over a spatial region. Here we focus on situations where it is possible to identify the density of "stuff" per unit length.[34] In this case, we have the generic formula

$$\text{total stuff} = \int_a^b (\text{stuff per unit length}) \, dx. \tag{1.7.5}$$

Example 1.7.7. Suppose a population of microorganisms is distributed along a stream of length 10, with linear density $1000e^{-0.1x}$ individuals per unit length. (Note that the "stuff" has the dimension "individuals," so the quantity we need to integrate must be given in individuals per unit length. Note also that it doesn't matter how the individuals are distributed—they could be lined up on the surface of the stream along a line running from one bank to the other, or they could be distributed beneath the surface as well as on the surface.) Then the number of individuals associated with a point along the line is $1000e^{-0.1x} \, dx$ and the total number of individuals in the stream is

$$p = \int_0^{10} 1000 e^{-0.1x} \, dx.$$

                                                                                            □

---

[32] The name Euler is pronounced "oiler."

[33] Leonhard Euler (1707–1783) was one of the most prolific mathematicians of all time. He remained productive even after becoming blind in 1766. Alfred J. Lotka (1880–1949) did a lot of pioneering work in mathematical population dynamics, including (1.7.3).

[34] In general, definite integrals can be used to determine the total amount of stuff aggregated over two-dimensional and three-dimensional regions, even when the density of stuff needs to be given per unit area or per unit volume. These multiple integrals are beyond the scope of our treatment. Details can be found in any calculus book that includes multivariable calculus.

Problems

1.7.1.* Consider the integral $y = \int_0^3 x^2\, dx$.

(a) Use three subintervals to obtain estimates y_L and y_R for the integral. Also find the difference between the two estimates.

(b) Repeat part (a) with six subintervals.

(c) Find a formula for the difference between the estimates with n subintervals.

(d) Using the formula from part (c), determine the number of subintervals necessary so that the difference between the estimates is not greater than 0.1.

(e) Use the number of subintervals you found in part (d) to obtain y_L and y_R.

(f) What is the best estimate you can give for the integral using the estimates of part (e)?

(g) Estimate the integral using the same number of subintervals as in part (e), but with the midpoint rather than the endpoints, as in Example 1.7.3.

(h) The methods used in parts (f) and (g) are called the trapezoidal rule and midpoint rule, respectively. A better rule, called Simpson's rule, is obtained by using the weighted average

$$S_n = \frac{2M_n + T_n}{3},$$

where S_n, M_n, and T_n are the Simpson, midpoint, and trapezoidal estimates using n subintervals. Determine the Simpson's rule approximations for $n = 3$ and $n = 6$.

(i) Based on the results from part (h), what do you think is the exact value of the integral?

1.7.2. Repeat Problem 1.7.1 for the integral $y = \int_0^1 x^3\, dx$, but with 2 and 4 subintervals instead of 3 and 6 subintervals in parts (a), (b), and (h) and a difference between estimates of 0.01 in parts (d)–(g).

1.7.3. A runner's respiration rate during a 30-min training run is $r(t)$ breaths per minute. Construct an integral to calculate the total number of breaths over the course of the run.

1.7.4. The development rate of many organisms shows an approximately linear dependence on temperature. Biologists measure the development of such organisms in terms of "degree-days." These accumulate over time at a rate equal to the difference between the current temperature and a threshold temperature T_m. This temperature varies among species, as does the total number of degree-days required for full development.

(a) Construct an integral to represent the total number of degree-days in t_1 days. This integral represents a function $F(t_1)$ that contains the parameter T_m and the function $T(t)$ in addition to the independent variable t_1.

(b) In their study of the insect species *Thrips major*, David Stacey and Mark Fellowes determined that a total of 230 degree-days are required for development, with the threshold value $T_m = 7\,°C$ [11]. Calculate the total amount of time needed for development of this species if the temperature is a constant $27\,°C$.

(c) Suppose the temperature is a constant value T. Calculate the corresponding development time. Plot the development time as a function of T.[35]

(d) Find data for mean daily temperature for the month of July in your area. Use this data set to estimate the number of days required for development of thrips born on June 30.

---

[35] Among the more subtle effects of global warming are changes in the development times for plants and animals. Some species have evolved to develop in conjunction with a host or food species. Different degree-day requirements and threshold temperatures mean that synchronized development of two species may be lost when the mean temperature changes.

(e) How should the definition of the function $F(t_1)$ be modified if the function $T(t)$ can be less than T_m? [Hint: Although the term "degree-day" reflects the connection with temperature data, we should think of degree-days as a measure of development rate rather than a record of temperature. Hence, the accumulation of degree-days should take into account the fact that development is not negative during periods when $T < T_m$.]

(f) Use the modified formula from part (e) to calculate the number of degree-days that accumulate in 1 day if the temperature is $T(t) = 14 + 14\sin(2\pi t)$. [Hint: You can get a reasonably good estimate of the integral value by using either a right-hand or left-hand sum with 24 subdivisions, which corresponds to using hourly average temperatures rather than exact temperatures.

(g) Use the result from part (f) to calculate the development time for the temperature $T(t) = 14 + 14\sin(2\pi t)$. Compare this result with the development time for a constant temperature of $14\,°C$.

1.7.5. Suppose $t = \int_0^Y f(y)\,dy$ represents the time required for Y grams of a toxin to be eliminated from an organism's bloodstream. What does $f(y)$ represent? In particular, what does $f(Y/2)$ represent?

1.7.6.* The population density of a plant species is $f(x)$ individuals per square meter, where x is the distance from a river, with $f(x) \geq 0$ for $x \leq 100$ and $f(x) = 0$ for $x \geq 100$. Construct an integral to calculate the number of plants along a section of river of length L. [Hint: You have to be careful about the integrand; pay attention to (1.7.5).]

1.7.7. The population density of a plant species is $f(r)$ individuals per square meter, where r is the distance from the center of a circular lake of radius $100\,$m. Assume the density falls to 0 at a distance of $100\,$m from the lake. Construct an integral to calculate the total number of plants. [Hint: You have to be careful about the integrand; pay attention to (1.7.5).]

1.7.8. Let $m(t)$ be the mass of a plant. The mass changes over time because of photosynthesis and root uptake, which increase the mass, and respiration, which decreases the mass. Let $p(t)$, $u(t)$, and $r(t)$ be the rates per unit time for mass changes due to photosynthesis, root uptake, and respiration, respectively. Construct a definite integral formula for the change in mass from time a to time b.[36]

1.7.9.* In Problem 2.2.7 we use the Euler–Lotka equation to investigate the dependence of population growth rates on average number of offspring and average age of reproduction. Part of the analysis depends on evaluating the integral

$$\int_0^1 (1 - x)(e^{0.1x} + e^{-0.1x})\,dx.$$

(a) Estimate how many subdivisions are necessary so that the approximation using left-hand sums and the approximation using right-hand sums differ by no more than 0.001.

(b) Approximate the integral using 5 subdivisions, with sampling points x_j at the middle of each interval.

(c) Repeat part (b) with 10 subdivisions.

(d) Compute the value of the integral using mathematical software (a computer algebra system or a calculator that can compute definite integrals).

---

[36] The formula from this problem could be used as part of a model, but it is not complete as it is. In any practical setting, the rates of photosynthesis, root uptake, and respiration would not be known functions of time; instead, they would be functions of the plant mass. The model equation would then have m on both sides and would be a problem to be solved rather than a quantity to be calculated.

(e) Use the known integral value to compute the error in the midpoint approximations from parts (b) and (c).

(f) Explain the discrepancy between the accuracy achieved with the midpoint approximation and accuracy predicted in part (a). [Hint: Examine the graph of the integrand and think about how the midpoint rule works.]

1.7.10. In Example 1.5.1, we proposed the model $f(x) = ax^2 - bx^3$ for the net rate at which an organism of size x can accumulate resources, where a and b are positive parameters. We then associated this function with the maximum reproduction rate of an organism of size x and used it as a way to tie the fitness of the organism to its mature size. Associating fitness with maximum reproduction rate makes sense if an organism is immortal; however, it seems clear that an organism that cannot expect to live forever should stop growing earlier than one that will never die. For real organisms, we should associate fitness with the total expected lifetime reproduction rather than the maximum rate of reproduction.

(a) Given that an organism so large that $f(x) = 0$ cannot grow, we can use the resource accumulation model to determine a maximum possible size, which will be a function of the parameters a and b. Define a parameter x_∞ for this quantity and use it to eliminate a from the formula for $f(x)$. The result will be a model with parameters b and x_∞. The advantage of this change is that models are easier to understand when the parameters have a clear biological significance.

(b) Suppose an organism puts all of its resources into growth until it reaches a mature size of $x = X$ at time $t = T$. Then the organism's life history has two distinct phases. In the first phase, which lasts up to time T, all resources are used for growth, and none for reproduction. A model for this phase must use the function $f(x)$ to determine the relationship between X and T; hence, we must think of T as a function of X. This will be considered in Section 1.9. A model for the second phase, which lasts from time T until the death of the organism at age t_m, must use the function f evaluated at $x = X$ to determine the total amount of resources spent on reproduction. Construct a definite integral that represents the total amount of resources invested in reproduction in this second phase, assuming that the organism lives long enough to reach its mature size (so that $t_m > T$). Without loss of generality, you may set $b = 1$. Your integral will contain the quantities X, $T(X)$, t_m, and x_∞.

(c) Use the integral from part (b) to determine a fitness function $F(X)$. There are two things to keep in mind while doing this. First, the integrand is independent of the integration variable t and can be factored out of the integral, because the rate of resource accumulation is the same throughout the reproductive phase of the life history. Second, the final form of the fitness function includes the unknown function $T(X)$ as well as three parameters. We cannot use this new model to determine the mature size that maximizes fitness until we have obtained a formula for $T(X)$, and this must be deferred to Section 1.9.

(d) There is still a significant flaw in our fitness model. We have assumed t_m to be a fixed parameter, which is not realistic. Instead, t_m is necessarily probabilistic—individuals live to different ages. One idea is to average the reproduction over the individuals in a cohort. The reproduction computed in part (b) at any time t would be correct for the fraction of individuals still alive, while a reproduction of 0 is correct for those individuals who have already died. Compute a new fitness integral using this modification, with survival function ℓ as in Example 1.7.6. Rather than assuming a maximum age t_m, you may take the integral to ∞. The maximum age can be incorporated into the survival function by requiring $\ell(t_m) = 0$ if desired. Note that t in this model is time since birth, so there is no difference between t and a. Note also that the integrand is no longer independent of t, so we do not yet have the methods needed to evaluate the integral.

(This problem is continued in Problem 1.9.12.)

1.8 Properties of the Definite Integral

After studying this section, you should be able to:

- Use the elementary properties of definite integrals.
- Use the fundamental theorem to compute definite integrals of power functions, sine and cosine functions, and exponential functions.

Like the derivative, the definite integral is a *linear* operator. This means that we can break up sums and factor out constants.

Linearity Rule: Let a, b, A, and B be any constants. If the functions f and g have integrals on the interval $[a, b]$, then

$$\int_a^b [Af(x) + Bg(x)] \, dx = A \int_a^b f(x) \, dx + B \int_a^b g(x) \, dx. \qquad (1.8.1)$$

Example 1.8.1. Q) Suppose

$$\int_0^\pi f(x) \, dx = 2, \qquad \int_0^\pi g(x) \, dx = \frac{1}{4}\pi^4 + \pi^2.$$

Compute

$$\int_0^\pi [4g(t) - \pi^2 f(t)] \, dt.$$

A) It does not matter what symbol is used for the integration variable in a definite integral; hence, the integral formulas for f and g can be applied to the new computation, along with the linearity rules:

$$\int_0^\pi [4g(t) - \pi^2 f(t)] \, dt = 4 \int_0^\pi g(t) \, dt - \pi^2 \int_0^\pi f(t) \, dt$$

$$= 4\left[\frac{1}{4}\pi^4 + \pi^2\right] - \pi^2 \times 2 = \pi^4 + 4\pi^2 - 2\pi^2 = \pi^4 + 2\pi^2.$$

□

Because integrals represent area, they can be split into components over a partition of the interval.

Partition Rule: If the function f has integrals on the intervals $[a, b]$ and $[b, c]$, then

$$\int_a^c f(x) \, dx = \int_a^b f(x) \, dx + \int_b^c f(x) \, dx. \qquad (1.8.2)$$

Example 1.8.2. Given that

$$\int_0^\pi \sin x \, dx = 2,$$

find

$$\int_0^{\pi/2} \sin x \, dx.$$

Using the partition rule,

$$2 = \int_0^\pi \sin x\, dx = \int_0^{\pi/2} \sin x\, dx + \int_{\pi/2}^\pi \sin x\, dx.$$

From the graph of $\sin x$ (see Figure 1.8.1), the two integrals on the right side of the equation are equal; hence,

$$\int_0^{\pi/2} \sin x\, dx = 1.$$

□

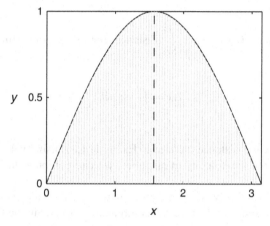

Fig. 1.8.1 The area represented by the integral $\int_0^\pi \sin x\, dx$, with *vertical line* at $x = \pi/2$

1.8.1 The Fundamental Theorem of Calculus

So far, we have treated the derivative and the definite integral as two important but unrelated concepts. In actuality, the connection between derivative and definite integral is so intimate that the formal statement of this connection is called the *fundamental theorem of calculus*.

The fundamental theorem is not difficult to derive from the definitions of the derivative and definite integral. However, in keeping with our emphasis on understanding and use of ideas, we choose instead to develop the fundamental theorem by solving the same problem twice, from two different points of view.

Suppose you begin to drive on a major highway. You travel in the direction of increasing mileposts for 1 h. How much change has there been in your position?

One method for determining total change in position is to examine the mileposts at the beginning and end of the hour.

Example 1.8.3. Let $x(t)$ be distance in miles at time t. If you start at time 0 and finish at time 1, then you have traveled from milepost $x(0)$ to milepost $x(1)$. The total change in your position[37]

---

[37] There is a subtle distinction between "total change in position" and "distance traveled" that matters if you travel both forward and backward. For simplicity, we assume that you only travel forward; however, we use the language that is still correct if backward travel is allowed. This is also why we use mileposts to measure distance rather than using the car's odometer.

is the difference between the starting and ending mileposts:

$$\Delta x = x(1) - x(0).$$

□

We can also determine the change in position from the car's velocity. Distance accumulates over time at the rate given by the velocity.

Example 1.8.4. Suppose your car has a device that records your velocity $v(t)$, with negative velocities corresponding to backward motion. Velocity is the rate of change of distance with respect to time, which is exactly what we need to integrate to determine change of position. In words,

$$\text{total distance (forward)} = \int_a^b (\text{distance (forward) per unit time}) \, dt.$$

In mathematical notation,

$$\Delta x = \int_0^1 v(t) \, dt.$$

□

We can approximate the distance traveled in Example 1.8.4 using velocity data at a discrete set of times. If we know the velocity at all times, we can compute the distance to arbitrary accuracy.

Examples 1.8.3 and 1.8.4 solve the same problem using different pieces of information. The solutions have to be the same, and so we can combine them to obtain the formula

$$\int_0^1 v(t) \, dt = x(1) - x(0).$$

We could have made our result more general by choosing arbitrary start and stop times a and b:

$$\int_a^b v(t) \, dt = x(b) - x(a). \tag{1.8.3}$$

The left side of this equation represents the area under the graph of the velocity function between the times a and b. The right side is the change in position during that time interval. Hence, (1.8.3) says that the total change in position is the area under the graph of the rate of change of position.

We derived (1.8.3) from a specific scenario involving a car driving on a highway, with variables indicating the mileposts and the velocity. But what details were really necessary for the mathematical result? The answer is that the narrative in which the mathematics is phrased is not important. All that matters are the relationships among the quantities. The primary requirement is that the quantity represented by v must be the rate of change of the quantity represented by x. Rewriting (1.8.3), with a little extra attention to mathematical issues, brings us to the first part of the fundamental theorem:

Theorem 1.8.1 (Fundamental Theorem of Calculus, Part 1). *Suppose f' is continuous on the interval $[a,b]$. Then*

$$\int_a^b f'(t) \, dt = f(b) - f(a). \tag{1.8.4}$$

Conceptually, this part of the fundamental theorem is really just a restatement of the general application of accumulation. In every accumulation problem, we express the total change in a quantity as the integral of the rate of change.

1.8.2 Computing Definite Integrals with the Fundamental Theorem

The fundamental theorem gives us a simple result for the integral of a *derivative*, but not for the integral of a general function. To calculate $\int_a^b f(x)\,dx$ with the fundamental theorem, we first need to be able to write the integrand $f(x)$ as the derivative of some known function.

Example 1.8.5. In Example 1.7.5, we conjectured the result

$$\int_0^1 x^2\,dx = \frac{1}{3}$$

based on approximation. We can calculate this integral exactly using the fundamental theorem, because we can express x^2 as the derivative of the function $f(x) = x^3/3$. We then have

$$\int_0^1 x^2\,dx = \int_0^1 f'(x)\,dx = f(1) - f(0) = \frac{1}{3} \times 1^3 - 0 = \frac{1}{3}.$$

$\square$

Check Your Understanding 1.8.1:

Compute $\displaystyle\int_0^1 2x^3\,dx$ by using the fundamental theorem, as in Example 1.8.5.

Example 1.8.6. In Example 1.8.2, we were given the result

$$\int_0^\pi \sin x\,dx = 2.$$

This result is easily calculated using the fundamental theorem and the derivative formula $[\cos x]' = -\sin x$. We have

$$\int_0^\pi \sin x\,dx = \int_0^\pi [-\cos x]'\,dx = -[\cos \pi - \cos 0] = -[-1 - 1] = 2.$$

$\square$

Each of the elementary derivative rules in Table 1.3.1 can easily be converted to a suitable rule for integration; these rules are summarized in Table 1.8.1.

Table 1.8.1 Elementary antiderivative formulas

$f'(x)$	$x^r \ [r \neq -1]$	e^{ax}	$\dfrac{1}{x}$	$\cos ax$	$\sin ax$	$\dfrac{1}{a^2 + x^2}$	$\dfrac{1}{\sqrt{a^2 - x^2}}$
$f(x)$	$\dfrac{x^{r+1}}{r+1}$	$\dfrac{1}{a}e^{ax}$	$\ln|x|$	$\dfrac{1}{a}\sin ax$	$-\dfrac{1}{a}\cos ax$	$\dfrac{1}{a}\arctan\dfrac{x}{a}$	$\arcsin\dfrac{x}{a}$

Example 1.8.7. In Example 1.7.7, we derived the integral

$$\int_0^{10} 1000e^{-0.1x} dx.$$

We can calculate this integral using the fundamental theorem because $-10000e^{-0.1x}$ is an antiderivative of $1000e^{-0.1x}$.

$$\int_0^{10} 1000e^{-0.1x} dx = \int_0^{10} \left[-10000e^{-0.1x}\right]' dx = \left[-10000e^{-1}\right] - \left[-10000e^0\right] = 10000(1 - e^{-1}).$$

$\square$

Check Your Understanding Answers

1. $\dfrac{1}{2}$.

Problems

Compute the definite integrals in Problems 1.8.1–1.8.8.

1.8.1.* $\displaystyle\int_1^2 (x^4 - 3x^2)\, dx.$

1.8.2. $\displaystyle\int_0^x (y^2 + 3y - 1)\, dy.$

1.8.3. $\displaystyle\int_1^4 \frac{1}{\sqrt{x}}\, dx.$

1.8.4.* $\displaystyle\int_1^e (x + x^{-1})\, dx.$

1.8.5. $\displaystyle\int_0^1 (e^{2t} + e^{-2t})\, dt.$

1.8.6. $\displaystyle\int_0^b e^{-kx}\, dx, \qquad k > 0.$

1.8.7.* $\displaystyle\int_0^{\pi/4} (\cos x + \sin 2x)\, dx.$

1.8.8. $\displaystyle\int_0^{\pi/6} (\cos 2x + \sin x)\, dx.$

1.8.9. (Continued from Problem 1.4.1.)
If $f(t)$ is the number of infected cells on day t of a disease epidemic, then the total extent of infection up to day T can be defined as the integral $F(T) = \int_0^T f(t)\, dt$.

(a) Compute the total extent of infection over the course of a bout of measles, using the function $f(t) = t(21 - t)(t + 1)$ and using the end of the infection for T.

(b) Repeat part (a), but without specifying T.

(c) Plot the graph of the function $F(T)$.

(d) Use the graph from part (c) to estimate the effective halfway point of the infection.

1.8.10. (a) Plot $f(x) = 1/\sqrt{1-x^2}$ on the interval $0 \leq x < 1$. Can you tell from the graph whether the area under this curve is finite?

(b) Use Table 1.8.1 to compute $g(b) = \int_0^b \dfrac{1}{\sqrt{1-x^2}}\, dx$ for $0 \leq b \leq 1$.

(c) Plot the function $g(b)$ from part (b).

(d) What is $\int_0^\infty \dfrac{1}{\sqrt{1-x^2}}\, dx$?

1.8.11. (a) Find the derivative of $F(x) = \dfrac{x}{a} e^{ax}$.

(b) Find the derivative of $F(x) + c e^{ax}$, where F is as in part (a).

(c) Find an antiderivative $G(x)$ for $g(x) = xe^{ax}$. [Hint: Choose the correct value of c for the function in part (b).]

(d) Use your antiderivative formula to compute $\int_0^1 xe^x\, dx$.

(e) Use your antiderivative formula to compute $\int_0^\infty xe^{-kx}\, dx$, where $k > 0$.

1.8.12. Consider the Euler–Lotka equation with age-independent death risk μ[38]:

$$\int_{a_1}^{a_2} e^{-(r+\mu)a} m(a)\, da = 1,$$

where r is the overall population growth rate, a_1 and a_2 are the ages at which reproduction begins and ends, and $m(a)$ is the reproduction rate per individual.

(a) Suppose an organism matures at age 1 and reproduces at constant per capita birth rate n until age 2. Show that n represents the total lifetime reproduction for individuals that live to age 2 or older.

(b) Our goal is to determine the reproduction rate needed to maintain the population. To that end, set $r = 0$, use the function m described in part (a), and evaluate the integral. Solve the resulting equation for n. This gives the necessary replacement rate needed to sustain the population with a given death rate.

(c) It is more meaningful to compare the replacement rate with the probability of survival to maturity at age 1. Given that $e^{-\mu a}$ is the probability of survival to age a, determine the relationship between s and μ. Use this relationship to replace μ in the equation of part (b), obtaining a formula that gives the necessary replacement rate as a function of survival probability.

(d) Suppose the probability of survival to maturity is 1/2. Common sense suggests that each surviving individual would need to have two offspring in order for the population to be maintained. Use your result from part (c) to obtain the correct replacement rate and offer a biological explanation of why the answer is not 2.

(e) Plot the resulting function $n(s)$. Choose a biologically reasonable range of s values for the plot. Discuss the biological significance of the result by considering the life histories of various animal species. For example, where would elephants be on the plot? How about fish?

---

[38] Equation (1.7.4).

1.9 Computing Antiderivatives and Definite Integrals

After studying this section, you should be able to:

- Use the method of substitution to compute definite integrals.
- Use the fundamental theorem of calculus to differentiate definite integrals.
- Use the fundamental theorem of calculus to construct antiderivatives.

In Section 1.3, we were able to use the definition of the derivative to create some basic derivative formulas and general rules, which we then used to develop a scheme for computing derivatives. We would now like to do the same thing for the definite integral, but there is no general method for doing so. We can use the fundamental theorem

$$\int_a^b f'(t)\,dt = f(b) - f(a), \tag{1.9.1}$$

but only if we know a function f whose derivative is the integrand in the definite integral. In other words, we can only compute $\int_a^b F(x)\,dx$ by the fundamental theorem if we can find a function $f(x)$ for which $f' = F$. Such a function f is called an **antiderivative** of F.

Example 1.9.1. Integrals of the form

$$\int_a^b e^{-x^2/2}\,dx$$

are important in probability. These integrals cannot be computed by the fundamental theorem because there is no elementary function $y(x)$ whose derivative is

$$y'(x) = e^{-x^2/2}.$$

$\square$

Even when we have an antiderivative, we have to be careful integrating over regions that are unbounded.

Example 1.9.2. In Example 1.7.4, we derived the integral

$$\int_0^\infty \frac{1}{1+x^2}\,dx\,;$$

however, it was unclear whether the area represented by the integral is finite. Table 1.8.1 shows that $\arctan x$ is a suitable antiderivative, so we have

$$\int_0^\infty \frac{1}{1+x^2}\,dx = \int_0^\infty [\arctan x]'\,dx = \lim_{x\to\infty} [\arctan x] - [\arctan 0] = \frac{\pi}{2}.$$

The limit exists, so the area is finite.

$\square$

Example 1.9.3. The integral

$$\int_0^1 \frac{1}{x}\,dx$$

represents the area between the graph of $1/x$ and the x axis from $x = 0$ to $x = 1$, as shown on the right half of the graph in Figure 1.9.1, provided that area is finite. If we try the same method as in Example 1.9.2, we have

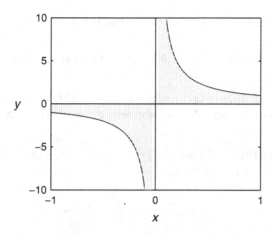

Fig. 1.9.1 The (infinite) area between $1/x$ and the x axis on the interval $[-1, 1]$

$$\int_0^1 \frac{1}{x}\,dx = \int_0^1 (\ln x)'\,dx = \ln 1 - \lim_{x \to 0} \ln x.$$

The limit is infinite, so the integral does not exist.

□

Example 1.9.4. The integral

$$\int_{-1}^1 \frac{1}{x}\,dx$$

represents the area between the graph of $1/x$ and the x axis from $x = -1$ to $x = 1$, as shown in Figure 1.9.1, provided that area is finite. By symmetry, we might think that the integral is 0, since the area below the x axis counts as negative. However, definite integrals must satisfy the partition rule, (1.8.2). Hence, we must have

$$\int_{-1}^1 \frac{1}{x}\,dx = \int_{-1}^0 \frac{1}{x}\,dx + \int_0^1 \frac{1}{x}\,dx$$

if the integral exists. We saw in Example 1.9.3 that the second of these integrals does not exist, so $\int_{-1}^1 \frac{1}{x}\,dx$ does not exist.

□

1.9.1 Substitution

Suppose we have a pair of functions $w(x)$ and $F(w)$, and $F'(w) = f(w)$. We can compute the derivative of F with respect to x using the chain rule:

$$\frac{d}{dx}[F(w)] = \frac{dF}{dw}\frac{dw}{dx} = f(w)\frac{dw}{dx}. \qquad (1.9.2)$$

By integrating (1.9.2) and applying the fundamental theorem, we obtain the result

$$\int_a^b f(w(x))\frac{dw}{dx}\,dx = \int_a^b \frac{d}{dx}[F(w)]\,dx = F(w(b)) - F(w(a)). \qquad (1.9.3)$$

We can also integrate the function $f(w)$ on the interval bounded by $w(a)$ and $w(b)$, obtaining

$$\int_{w(a)}^{w(b)} f(w)\,dw = \int_{w(a)}^{w(b)} F'(w)dw = F(w(b)) - F(w(a)). \qquad (1.9.4)$$

Combining (1.9.3) and (1.9.4) yields the substitution rule, which makes it possible to change variables in a definite integral, regardless of whether we actually know an antiderivative for it.

Theorem 1.9.1 (Substitution Rule). *If $w'(x)$ is continuous on $[a,b]$ and the area under $f(w(x))$ between $w(a)$ and $w(b)$ is finite, then*

$$\int_a^b f(w(x))\frac{dw}{dx}\,dx = \int_{w(a)}^{w(b)} f(w)\,dw.$$

It is helpful to think of the substitution rule as a matter of replacing $\dfrac{dw}{dx}\,dx$ by dw, provided that you are careful to change the limits of integration from the limits appropriate to the graph of $f(w(x))$ versus x to the limits appropriate to the graph of $f(w)$ versus w.

Example 1.9.5. To compute $\int_0^1 x\sqrt{2x^2+1}\,dx$, we let $w = 2x^2 + 1$. Then we have

$$\int_0^1 x\sqrt{2x^2+1}\,dx = \int_0^1 \frac{\sqrt{2x^2+1}}{4}\times 4x\;dx = \int_0^1 \frac{\sqrt{w(x)}}{4}\frac{dw}{dx}dx = \int_1^3 \frac{\sqrt{w}}{4}\,dw$$

$$= \int_1^3 \left(\frac{w^{3/2}}{6}\right)'\,dw = \frac{3^{3/2}-1}{6} \approx 0.699.$$

Note that the integration limits changed in the third equality from $0 \le x \le 1$ to $1 \le w \le 3$, which follows from the equation $w = 2x^2 + 1$. Also, the function $f(w)$ (in this case $\sqrt{w}/4$) is whatever remains of the integrand after $dw/dx = 4x$ is attached to dx. Figure 1.9.2 shows the integrals in both x and w coordinates. The axes have been scaled so that the entire area of each box is 2; hence, the shaded areas are each 35% of the area of the box. $\square$

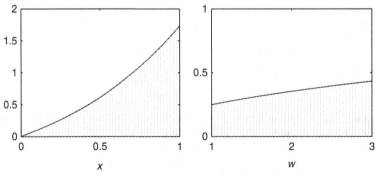

Fig. 1.9.2 The areas represented by the integrals $\int_0^1 x\sqrt{2x^2+1}\,dx$ and $\int_1^3 \frac{\sqrt{w}}{4}\,dw$

1.9.2 Constructing Antiderivatives with the Fundamental Theorem

Suppose we apply the first part of the fundamental theorem to an integral with a variable upper limit. We have

$$\int_a^x f'(t)\,dt = f(x) - f(a).$$

The quantities on both sides of this equation are the same function of x; hence, they have the same derivative:

$$\frac{d}{dx}\int_a^x f'(t)\,dt = f'(x).$$

This result indicates how to take the derivative of a definite integral. Its restatement, using F in place of f', constitutes the second part of the fundamental theorem of calculus.

> **Theorem 1.9.2 (Fundamental Theorem of Calculus, Part 2).** *Suppose a function $f(x)$ is defined by*
>
> $$f(x) = \int_a^x F(t)\,dt$$
>
> *where a is a constant and F is continuous. Then the derivative of $f(x)$ is $F(x)$.*

Example 1.9.6. In probability, we often want to compute quantities of the form

$$P_{a,b} = \frac{1}{\sqrt{2\pi}}\int_a^b e^{-z^2/2}\,dz.$$

These integrals *cannot* be computed *directly* using part 1 of the fundamental theorem, because the function $e^{-z^2/2}$ does not have an antiderivative that can be written in terms of elementary functions. The lack of an antiderivative with a simple formula doesn't mean there is no antiderivative at all. In fact, the second part of the fundamental theorem says that every continuous function has an antiderivative, even though that antiderivative might need to be written as a definite integral. Specifically, we can define a function N by

$$N(z) = \frac{1}{\sqrt{2\pi}}\int_{-\infty}^z e^{-t^2/2}\,dt.$$

We can then connect P to N using the fundamental theorem:

$$P_{a,b} = \frac{1}{\sqrt{2\pi}}\int_a^b e^{-z^2/2}\,dz = \int_a^b N'(z)\,dz = N(b) - N(a).$$

Values of N can be computed numerically with computers, and these values can also be tabulated. The function N is always part of the standard package of mathematical computer software, including scientific calculators. Its easy availability allows us to calculate a numerical approximation for any integral $P_{a,b}$.

Note that we could have chosen any constant for a. For reasons that will be discussed in Section 3.6, $a = -\infty$ is the standard choice. While the region $-\infty < t \le z$ is unbounded, the integral is always finite; indeed, the strange factor $1/\sqrt{2\pi}$ is used so that $N(\infty) = 1$, which is necessary if N is to represent a probability.[39]   □

---

[39] The function N represents the *standard normal distribution*. See Section 3.6.

1.9.3 Obtaining a Graph of f from a Graph of f′

We sometimes know more information about the derivative of a function than we know about the function itself. It is very helpful to be able to use such information to sketch the graph of the original function. The key to doing this is to keep in mind that f' indicates the rate of change of f, and hence the slope of the graph of f. The graph of f' therefore provides qualitative information that can be used to sketch the graph of f. Numerical values of f can be calculated from f' by means of the fundamental theorem, provided that one starting value of f is known.

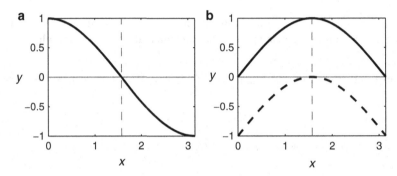

Fig. 1.9.3 The functions (**a**) $f'(x) = \cos x$ and (**b**) $f(x) = \sin x$ (*solid*) and $f(x) = \sin x - 1$ (*dashed*), with vertical line at $x = \pi/2$

Example 1.9.7. Suppose

$$f'(x) = \cos x.$$

Figure 1.9.3a shows the graph of $f'(x) = \cos x$ and Figure 1.9.3b shows the graphs of $\sin x$ and $\sin x - 1$ on the interval $[0, \pi]$. The value $f'(0) = 1$ means that f has a slope of 1 at $x = 0$. From $x = 0$ to $x = \pi/2$, f' remains positive, so f continues to have a positive slope. However, f' is becoming less positive in this interval, so the graph of f becomes flatter. Both the graphs of $\sin x$ and $\sin x - 1$ have the right general shape on the interval $[0, \pi/2]$ to be f. Quantitatively, we know that both $\sin x$ and $\sin x - 1$ increase by a total of 1 on the interval $[0, \pi/2]$. We can use the fundamental theorem to see just how much f increases over this interval:

$$f\left(\frac{\pi}{2}\right) - f(0) = \int_0^{\pi/2} f'(x)\,dx = \int_0^{\pi/2} \cos(x)\,dx.$$

By the symmetry of the cosine and sine functions, the areas under the graphs on the interval $[0, \pi/2]$ are the same. This area was determined in Example 1.9.2 to be 1. Thus,

$$f\left(\frac{\pi}{2}\right) - f(0) = 1.$$

This result is consistent with the guesses $f(x) = \sin x$ and $f(x) = \sin x - 1$.   □

Example 1.9.7 is somewhat contrived because we could have seen from the start that $f'(x) = \cos x$ implies $f(x) = \sin x + C$. However, we can often use this graphical method for cases where we *cannot* find a function f that has the desired derivative.

Example 1.9.8. In Example 1.9.6, we defined a function

$$N(z) = \frac{1}{\sqrt{2\pi}} \int_{-\infty}^z e^{-t^2/2}\,dt,$$

for use as an antiderivative of the function

$$N'(z) = \frac{1}{\sqrt{2\pi}} e^{-z^2/2}.$$

Figure 1.9.4 shows the functions N' and N. Note that

$$N''(z) = -\frac{z}{\sqrt{2\pi}} e^{-z^2/2}.$$

Thus, N' is always positive and N'' is positive when $z < 0$ and negative when $z > 0$. These calculations correspond to the qualitative features of the graph of N—this graph is increasing everywhere, with a slope that is greatest at $z = 0$. $\square$

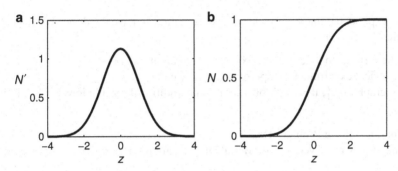

Fig. 1.9.4 The functions (**a**) $N'(z) = (1/\sqrt{2\pi})e^{-z^2/2}$ and (**b**) $N(z) = (1/\sqrt{2\pi}) \int_{-\infty}^{z} e^{-t^2/2} dt$

Problems

In Problems 1.9.1–1.9.8, compute the indicated integral by substitution.

1.9.1.* $\displaystyle\int_0^1 x^2 \sqrt{x^3 + 1}\, dx.$

1.9.2.* $\displaystyle\int_0^1 \frac{u}{(3u^2 + 1)^2}\, du.$

1.9.3. $\displaystyle\int_0^1 \frac{1}{x+h}\, dx.$

1.9.4. $\displaystyle\int_0^1 \sqrt{ax+b}\, dx.$

1.9.5. $\displaystyle\int_0^4 \frac{x}{x+4}\, dx.$ [Hint: You can do a substitution or you can do an algebraic operation that leaves you with two integrals—one of a constant and one of a quotient with a constant in the numerator.]

1.9.6. $\displaystyle\int_0^1 xe^{kx^2}\, dx.$

1.9.7.* $\displaystyle\int_0^{\sqrt{\pi}} x\sin\left(x^2 + \frac{\pi}{2}\right)dx.$

1.9.8. $\displaystyle\int_1^{\sqrt{e}} \frac{\ln x}{x}\,dx.$

1.9.9. The sum $\displaystyle\int_{-1}^0 (1+x)e^{-kx}\,dx + \int_0^1 (1-x)e^{-kx}\,dx$, where $k \neq 0$, can conveniently be written as a single integral by changing variables in the first integral.

(a) Use the substitution $y = 1 - x$ in the first definite integral.
(b) Reverse the limits in the new definite integral by using the fact that $\int_b^a f(x)\,dx = -\int_a^b f(x)\,dx$ for any function f for which the integrals are defined.
(c) Add the integrals together to obtain a single integral. Note that you can freely change the symbol used for the integration variable in an integral.

1.9.10. The exponential integral function $E_1(x)$ is defined to have the derivative $E_1'(x) = e^{-x}/x$ and the value $E_1(1) = 0$.

(a) Write the formula that defines the exponential integral function.
(b) Use numerical experiments to approximate the value $E_1(\infty)$.
(c) Use the graph of E_1', the result of part (b), and additional calculations to sketch the graph of E_1.

1.9.11. (Continued from Problem 1.6.5.)
In Problem 1.6.5, we derived the von Bertalanffy equation for growth of an organism:

$$\frac{dx}{dt} = r(x_\infty - x),$$

where $x(t)$ is a measurement of the linear size of the organism (length or diameter, for example), x_∞ is a parameter that represents the theoretical maximum size, and r is a rate constant. Our goal in this problem is to determine the amount of time t needed for the organism to reach size x. In effect, we will think of t as a function of x rather than the other way around. This is reasonable, given that our model organism is continuously growing. Each size is achieved only once at a specific time. The implicit function theorem says that if the variables x and t are related so that each could be thought of as a function of the other, then the derivatives dx/dt and dt/dx are reciprocals. Thus, we can rewrite the von Bertalanffy equation as

$$\frac{dt}{dx} = \frac{1}{r}\frac{1}{x_\infty - x}.$$

Use substitution to find the function $t(x)$ that has the correct derivative and has the value $t = 0$ when $x = 0$.

1.9.12. (Continued from Problem 1.7.10.)
In Problem 1.7.10, we derived a model for fitness of an organism that puts all resources into growth until it reaches size $x = X$ at time $t = T(X)$ and then uses all resources for reproduction until it dies at age $t_m > T$. A better model is that the fitness of an organism is given by

$$F(X) = \int_{T(X)}^{\infty} e^{-\mu t} \times bX^2(x_\infty - X)\,dt,$$

where X is the size at maturity, $T(X)$ is the time at which maturity is reached, x_∞ is the theoretical maximum size, $b > 0$ is a parameter that represents the collection rate of resources, and $e^{-\mu t}$ is the probability that the organism survives to age t, with $\mu > 0$ the so-called hazard rate.

(a) Compute the integral in the definition of F.

(b) Obtain a complete model for $F(X)$ by substituting the result from Problem 1.9.11 into the result of part (a). After algebraic simplification, your result should have the form

$$F(X) = CX^2(x_\infty - X)^{1+p},$$

where C and p are combinations of model parameters.

(c) Find the mature size X that maximizes the function $F(X)$ from part (b). Note that the optimal size does not depend on the parameter b.

(d) Check your answer to part (b) by examining the limit as $\mu \to 0$. This should agree with Example 1.5.1, where we used a simpler model. Why do we expect these answers to agree?

(e) Suppose $r = 1$. Plot the optimal size at maturity as a function of the hazard rate μ. What general prediction does the model make about the way mature size should depend on the daily risk of life?

(f) Suppose $\mu = 1$. Plot the optimal size at maturity as a function of the growth rate r. What general prediction does the model make about the way mature size should depend on the growth rate of immature organisms? Discuss the biological plausibility of the result.

References

1. Alexander, RM. *Optima for Animals*. Princeton University Press, Princeton, NJ (1996)
2. Atlantic magazine. http://www.theatlantic.com/health/archive/2012/06/science-confirms-bieber-fever-is-more-contagious-than-the-measles/258460/. CitedSep2012
3. Belovsky GE. Diet optimization in a generalist herbivore: The moose. *Theo. Pop. Bio.*, **14**, 105–134 (1978)
4. CDC and the World Health Organization. History and epidemiology of global smallpox eradication. In *Smallpox: Disease, Prevention, and Intervention*. http://www.bt.cdc.gov/agent/smallpox/training/overview/. CitedSep2012
5. Heffernan JM and MJ Keeling. An in-host model of acute infection: Measles as a case study. *Theo. Pop. Bio.*, **73**, 134–147 (1978)
6. Mahaffy J. http://www-rohan.sdsu.edu/~jmahaffy/courses/f00/math122/lectures/optimization/opt.html#Energy. CitedJul2012
7. National Incident Command, Interagency Solutions Group, Flow Rate Technical Group. *Assessment of Flow Rate Estimates for the Deepwater Horizon / Macondo Well Oil Spill*. U.S. Department of the Interior (2011)
8. National Oceanic and Atmospheric Administration. Mauna Loa CO_2 data. http://www.esrl.noaa.gov/gmd/ccgg/trends/#mlo_data. CitedSep2012
9. Pennings, TJ. Do dogs know calculus? *College Math. J.*, **34**, 178–182 (2003)
10. Perruchet P and J Gallego. Do dogs know related rates rather than optimization? *College Math. J.*, **37**, 16–18. (2006)
11. Stacey DA and MDE Fellowes. Temperature and the development rates of thrips: Evidence for a constraint on local adaptation? *Eur. J. Entomol.*, **99**, 399–404 (2002)
12. Zach R. Shell dropping: Decision-making and optimal foraging in Northwestern crows. *Behaviour*, **68**, 106–117 (1979)

Chapter 2
Mathematical Modeling

All mathematics texts include story problems. These are almost always of the sort that I call "applications":

- Applications are narrow in scope because they use fixed numbers rather than parameters and because the questions call for answers that are simply numbers. One example is "If a bacteria colony doubles every hour, how long does it take a single bacterium to become a population of one million?" Sometimes the parameter values must be calculated indirectly, as in "A jar initially contains 1 g of a radioactive substance X. After 1 h, the jar contains only 0.9 g of X. How much more time is required before the jar contains only 0.01 g of X?"
- The mathematical setting in an application is implicitly assumed to be exactly equivalent to the scientific setting. Hence, the mathematical answers are unquestioningly accepted as the answers to the scientific questions.

I use the term "applications" for problems with these characteristics because the emphasis is on the mathematics rather than the setting. Some effort may be required for modeling tasks such as determining parameter values and/or interpreting results in context, but all or most of the effort in application problems is in obtaining mathematical solutions. Even when word problems ask more sophisticated questions, they still often suffer from the common interpretation of mathematical models as "mathematical constructions that describe real phenomena in the physical world."

I view mathematical models as "mathematical constructions based on real world settings and created in the hope that the model behavior will resemble the real world behavior." With this interpretation, mathematical modeling is as much theoretical science as it is mathematics. In contrast to applications, model analysis requires us to determine the extent to which models are able to replicate real-world phenomena before we accept mathematical answers as meaningful. Mathematical modeling can be used for narrow questions that call for numerical answers, but it can also be used for broad questions about general behavior (e.g., for what ranges of parameter values will a population go extinct?).

Mathematical modeling requires interdisciplinary skills that are seldom taught in any courses, and this is one important reason why many students who have always had good grades in mathematics find themselves at a loss when they need to use mathematics to do work in science. The purpose of this chapter is to introduce modeling ideas and skills in a small number of simple settings. These ideas and skills are then utilized in the remainder of the text.

Sections 2.1 and 2.2 introduce the ideas of mathematical modeling. A brief example of scientific data is introduced in Section 2.1 to serve as a concrete focus for the development of the chapter. The section also discusses the interplay between data analysis and modeling

G. Ledder, *Mathematics for the Life Sciences: Calculus, Modeling, Probability, and Dynamical Systems*, Springer Undergraduate Texts in Mathematics and Technology, DOI 10.1007/978-1-4614-7276-6_2, © Springer Science+Business Media, LLC 2013

and the limitations of using deterministic models in biology. Section 2.2 presents a qualitative discussion of mathematical modeling, focusing on the differences and connections between mathematics, mathematical modeling, and theoretical science. These introductory ideas provide a framework with which the reader can construct his/her understanding of mathematical modeling.

In standard application problems, the reader is given a mathematical model. In actual practice, mathematical models are not given by some higher authority; hence, mathematical modeling requires skills in obtaining mathematical models, fitting models to data, and selecting among competing models. The remainder of the chapter is focused on these skills. The different topics are unified by the use of the data set of Section 2.1 and the framework of Section 2.2.

Models can be classified as *empirical* and *mechanistic*. Empirical models define a mathematical relationship between quantities in a data set. These are obtained from the general appearance of a data set without regard for underlying biological ideas. Mechanistic models attempt to show how certain quantities in a data set are causally linked to other quantities, independent of any links suggested by data.[1] The distinction between mechanistic modeling and empirical modeling is a central theme of this chapter.

While empirical models are "identified" rather than "derived," it does not follow that empirical modeling lacks mathematical validity. Empirical modeling requires determination of parameter values from data and selection among several candidate models. Both of these tasks can be done with statistical methods, which are developed in Sections 2.3, 2.4, and 2.7. Unlike in story problems, the data needed to determine parameters in mathematical modeling are not exact, and the experimenter collects a surplus of data to compensate for the uncertainty in each measurement. To assign values to parameters, we must identify a quantitative measure of fitting error and then solve the mathematics problem of minimizing that error. This topic is introduced in Section 2.3 on linear least squares and extended in Section 2.4 to models that are linear in one parameter and nonlinear in another. Selection among competing models is discussed in Section 2.7 in terms of the Akaike information criterion (AIC), a method for determining the statistical support provided for a given model by a specific data set. The method is simple to apply, although the mathematical justification is far beyond the scope of any undergraduate course—perhaps this accounts for the curious absence of AIC from elementary texts in statistics. The value that biologists gain from the use of AIC argues for its inclusion in any statistics text written for general use, as well as those written specifically for biologists.

Section 2.5 focuses on the primary task of mechanistic modeling: developing models from biological assumptions based on biological theory and observational data. In Section 2.6, we examine the use of different algebraic forms for the same mathematical model. Differences in form can be as simple as using different symbols for the same quantity, but they can be more complex as well. The modeler often has a variety of ways to define the parameters in a mathematical model, with the choice of parameters affecting the algebraic form. In particular, the work of modeling is almost always simplified by deriving a dimensionless version of the model. Readers of theoretical science papers will often encounter dimensionless models and need to understand what they are and how they relate to the original model. The point of these two sections is not to make the reader an expert on mechanistic modeling, but to give the reader a feel for how mechanistic modeling is done. In particular, Section 2.5 is long and contains some sophisticated modeling. Some readers will want to focus on the ideas, while readers who want to learn how to construct mechanistic models should devote extra time to understanding the details. Additional discussion and examples of mechanistic modeling appear in Chapters 5–7.

---

[1] Ideally, a model should be both empirically *and* mechanistically based, but the methods for the two types of modeling are distinct.

There are several sets of related problems:

Section	2.3	2.5			2.6	2.7
Fluorine at South Pole	2.3.4					2.7.7
Grape harvests	2.3.8					2.7.9
Lake ice duration	2.3.9					2.7.10
Chemostat		2.5.3	2.5.5	2.5.8	2.6.9	
SIR disease		2.5.4	2.5.6	2.5.7	2.6.10	

2.1 Mathematics in Biology

After studying this section, you should be able to:

- Identify the role of mathematical modeling in science.
- Discuss the concept of demographic stochasticity and apply this concept to biological experiments.
- Generate questions about an experiment that could possibly be addressed with a mathematical model.

The tremendous success of science stems from the interaction of two enterprises: theory and observation.[2] Theory without observation is nothing more than myth, and observation without theory is nothing more than a collection of disjointed facts. Progress in science is therefore possible only by combining them. Theory is used to explain and unify observations and to predict results of future experiments. Observations are used to motivate and validate theory.

The connection between theory and observation is the realm of mathematical modeling. Theory provides statements of scientific principles, while observation provides numerical data. Mathematics is a language that can bridge the gulf between the two. Metaphorically speaking, mathematical modeling is the tendon that connects the muscle of mathematics to the skeleton of science.

2.1.1 Biological Data

Pick up a calculus or precalculus book and find a story problem with a scientific setting. Most likely, the problem you find has exact data. Real scientific data is not exact, and this difference must be understood before we can do mathematical modeling. We can explain the difference here, but you will understand it much better if you discover it yourself. This section will be unnecessary for practicing biologists, but it will be helpful for those readers who have not collected research data themselves.

The real world is not an easy setting for the collection of biological data. Even ignoring the difficulties in getting a good data set, there is the problem that data collection takes a lot of time and effort. This effort is necessary if we are going to practice real science, but it is a distraction if our purpose is to learn mathematical modeling. An alternative to collecting data from an experiment in the real world is to collect data from an experiment in a virtual world. Virtual worlds can be studied in a comfortable chair in front of a computer, without having to wait for events to occur in natural time. If the virtual world is carefully designed, what we learn from it might even be helpful in understanding the real world.

---

[2] I am using the word "observation" to encompass both observation of the natural world and observation directed by experiments.

A famous virtual-world experiment in ecology was conducted by C. S. Holling in the late 1950s, before the capability of creating virtual worlds with computers. Holling was interested in understanding the relationship between the availability of prey and the amount of prey eaten by predators in a given amount of time. He set up a virtual world consisting of sandpaper discs tacked onto a plywood board. The discs represented insects and a blindfolded student represented a predatory bird. In each experimental run, a student tapped the board with a finger at a steady pace, moving randomly around the board. Each time the student touched a sandpaper disk, he removed it, placed it in a cup, and then returned to tapping. After 1 min, the student recorded the number of disks "eaten" in this manner. The data set consisted of pairs of numbers: disks available and disks "eaten." Holling used the data, along with his observations, to create the predation models that now bear his name [6].

The first time I taught mathematical modeling in biology, I recreated Holling's experiment with my class. It was only partly successful, both because the students focused as much on the activity itself as on careful collection of data and also because the virtual world of Holling's human experiment is not well regulated.[3] For the following year, I wrote a software application that creates a virtual world based on Holling's experiment [7]. The application, called BUGBOX-predator, consists of a Windows executable file with supplementary data files, all of which can be downloaded from my web page [8] or the web page for this book: http://www. springer.com/978-1-4614-7275-9.

The BUGBOX-predator world consists of a grid populated by x virtual aphids (the number x is chosen by the experimenter). In each experiment, a virtual coccinellid (ladybird beetle) moves randomly through the virtual-world environment, stopping to consume any virtual aphids in its path. The experiment outcome y is the number of prey animals eaten in 1 min.[4] The animation is unsophisticated, which has the advantage of helping the experimenter appreciate the extreme simplicity of the BUGBOX-predator world.

The problem set for this section consists largely of experiments using BUGBOX-predator. If you do not have a lot of experience collecting biological data, do these problems before reading the rest of the section.

2.1.2 Overall Patterns in a Random World

Table 2.1.1 BUGBOX-predator consumption (y) for various prey densities (x)

Prey density	x	0	10	20	30	40	50	60	70	80	90	100	110	120	130	140
P. steadius	y	0	2	7	10	9	14	21	20	25	20	30	25	29	35	38
P. speedius	y	0	7	11	19	19	22	25	21	25	26	23	27	29	30	29

Table 2.1.1 contains some data from trials with two of the predator "species" in the BUG-BOX world. The same data is plotted in Figure 2.1.1. Like real data, the BUGBOX predation data does not appear to fall on a perfectly smooth curve. The BUGBOX-predator world is extremely simple, but is based on rules for biological behavior rather than a mathematical model of the relationship between the variables. The distribution of prey animals is random, and there is also significant randomness built into the predator's movement algorithm. These elements create uncertainty in the outcome, just as in real experiments. It is important to understand

---

[3] Few students can resist the impulse to tap faster whenever they are having only minimal success.

[4] Given unit time for the experiment and unit area for the environment, we can interpret y as the consumption rate per predator, in prey animals per unit time, and x as the prey density, in prey animals per unit area.

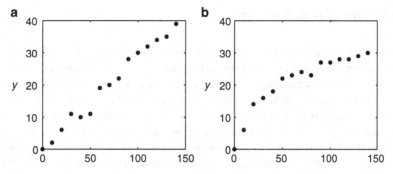

Fig. 2.1.1 Predation data for (**a**) *P. steadius* and (**b**) *P. speedius* from Table 2.1.1

the inevitability of this uncertainty. Real-world experiments can have uncertainty caused by difficulties with experiment design or measurement and by our inability to completely control an experimental environment; our virtual-world experiment is free from these sources of uncertainty. However, there is still variation among individuals. This random effect, called *demographic stochasticity*, is present whenever the number of individuals in an experiment is small and differences in individual behavior are significant. In contrast, experiments in chemistry are free of demographic stochasticity because of the extremely large number of particles in even a small amount of material.[5] This beneficial averaging does not occur in experiments with single individuals. Even if we designed an experiment with a population of 1,000 independently acting predators,[6] the number is insufficient to obtain a fully predictable average.[7]

At first thought, the inevitable randomness in biological data seems to suggest that mathematical modeling is pointless in biology. Mathematics is the most deterministic of disciplines, with many problems defined in such a way that there is a unique solution. If all biological events are affected by random factors, how can mathematics have any value in biology?

Obviously, the highly stochastic nature of biological events limits the possibilities for using mathematical methods in biology. It certainly *is* pointless to attempt to use mathematics to predict how many virtual aphids will be consumed by a virtual coccinellid in a single experimental run. However, mathematics can be used to study the patterns that arise when experiments are repeated many times.

If you run the same BUGBOX-predator experiment (the same predator species and the same initial prey population) hundreds of times, you will see that some outcomes are more likely than others. If your friend also runs the same experiment hundreds of times, the pattern of outcomes obtained by you and your friend will be very similar. In principle, if we repeat a BUGBOX-predator experiment millions of times, the average y for that given x for a particular species might be fully predictable. Averages, and also the expected distribution of results, can be estimated using *descriptive statistics*.[8]

---

[5] Demographic stochasticity of virus particles is not an issue in a model that tries to predict quantities of these particles in a person suffering from a communicable disease; however, demographic stochasticity in a population of people could be quite significant.

[6] An additional complication occurs if, as would usually be the case, the behavior of each individual is influenced by the behavior of the other individuals in the population.

[7] The connection between number of individuals and predictability is explored in Chapter 4.

[8] See Section 3.1.

2.1.3 Determining Relationships

So far we have only considered the patterns obtained in repetition of a specific experiment. We are often more interested in identifying patterns in relationships between quantities in an experiment. At its most elementary level, this is the goal of ANOVA (analysis of variance), a statistical extension of descriptive statistics. However, ANOVA only seeks to determine the significance of the relationship between quantities. A more ambitious goal is to search for a quantitative description of a relationship. Here again, we have to understand how the highly determined subject of mathematics can possibly be useful in the highly stochastic world of biology. The actual number we would obtain as the average of millions of repeated BUGBOX-predator experiments would be different for different initial numbers of prey. While the data from individual trials show clear evidence of stochasticity (as seen in Figure 2.1.1), it seems reasonable to expect that data from the averages of millions of trials would appear to lie on a smooth curve. That smooth curve is conceivably something that we could identify using mathematics. Think of an actual result of an individual trial as consisting of random variation superimposed on a predictable average y, which depends on the particular x for the experiments. We could then use our limited data set to model that predictable average. **This is the aim of this chapter—to develop methods for obtaining *deterministic* models for the *average* behavior of fundamentally *stochastic* biological processes.**

Examine the data in Figure 2.1.1. Allowing for the overall randomness of the data, there is an obvious difference between the two species. The data for *P. steadius* look approximately linear, but the data for *P. speedius* are definitely not. These kinds of qualitative differences occur in the real world as well as the virtual world. We can attempt to model each species individually; however, a more interesting question is whether we can develop a mathematical model that can explain the data for both predator species. A model that works for a collection of species would have much more value than a model that works for just one.

As noted above, the value of a model depends on the questions it is used to address. The modeler has to develop these questions along with the model. The possibilities are limited primarily by the creativity of the modeler, and the most worthwhile questions are generally not the obvious ones. Here is a brief list, by no means exclusive, of possible questions motivated by the BUGBOX-predator experiments. These are ordered from the most limited to the broadest.

1. If we want to represent the *P. steadius* data by a linear function (a straight line on Figure 2.1.1a), what parameters should we choose?
2. Is there a (clearly nonlinear) function we can use to represent the *P. speedius* data?
3. To what extent can we use the models to predict average predation for x values not given in the data?
4. If we find more than one model for the *P. speedius* data, is there some way to rank the models?
5. Can we identify biological characteristics that could account for the fact that different species have graphs of different shapes?
6. Can we use our predation models as part of a larger model of interacting populations?

The first five of these questions are addressed in the course of this chapter; the last one is addressed in Chapters 5 and 7.

Problems

These problems require BUGBOX-predator, which can be downloaded from http://www.math.unl.edu/~gledder1/BUGBOX/. Save the data sets from these problems, as they will be needed for problems in other sections.

2.1.1. Collect a predation data set for *P. steadius*, using the default choice of no replacement. Use prey values of *approximately* 10, 20, and so on up to 140. It is *not* necessary that the values be exact multiples of 10 (indeed, it is somewhat difficult to accomplish), but they should be roughly evenly spaced. Plot these data on a graph similar to Figure 2.1.1.

(This problem is continued in Problems 2.3.1 and 2.3.7.)

2.1.2. Collect a predation data set for *P. speedius*, using the default choice of no replacement. Use prey values of *approximately* 10, 20, and so on up to 140. It is *not* necessary that the values be exact multiples of 10 (indeed, it is somewhat difficult to accomplish), but they should be roughly evenly spaced. Plot these data on a graph similar to Figure 2.1.1.

(This problem is continued in Problem 2.4.6.)

2.1.3. Repeat Problem 2.1.1, but with replacement.

(This problem is continued in Problems 2.3.1 and 2.3.7.)

2.1.4. From your experience in Problems 2.1.1 and 2.1.3, discuss the significance of the replacement option. Which option would be easier to implement in an experiment with real organisms? Which option allows for unambiguous reporting of data (think about possible differences between what x is supposed to mean and the way we measure it)? This example illustrates the difficulty of designing biological experiments, given the need for practical implementation and the importance of avoiding ambiguity.

2.2 Basic Concepts of Modeling

After studying this section, you should be able to:

- Discuss the relationships between the real world and mathematical models.
- Discuss the distinctions between mechanistic and empirical modeling.
- Discuss the concepts of parameterization, simulation, and characterization with mathematical models.
- Discuss the concepts of the narrow and broad view of mathematical models and the function of parameters in each view.

A *mathematical model* is a self-contained set of formulas and/or equations based on an approximate quantitative description of real phenomena. This definition is useful in the semantic sense, but it fails to distinguish between models that are extremely useful and models that are totally worthless. Instead, I prefer to adopt a working definition that is necessarily vague:

> **Mathematical model**: *a self-contained set of formulas and/or equations based on an approximate quantitative description of real phenomena and created in the hope that the behavior it predicts will be consistent with the real behavior on which it is based.*

Note the tentative language of the added phrase. The emphasis is on the uncertainty in the connection between the mathematical model and the real-world setting to which it is applied. This emphasis means that modeling requires the theoretical science skills of approximation and validation, and it changes the focus of the mathematical skills from proof and solution to characterization (understanding the broad range of possible behaviors) and simulation (visualizing the behavior in specific examples). The thinking you need for mathematical modeling is therefore somewhat different from the thinking associated with mathematics per se and more like the thinking associated with theoretical science, as illustrated in Example 2.2.1.

Example 2.2.1. The Lotka–Volterra model tries to use a linear predation model[9] to explain the quantitative relationship between populations of predators and populations of prey. It was originally developed to explain changes in Mediterranean fish populations that occurred during and after World War I, which it succeeded in doing. Subsequently, it has been used (improperly, as explained in Example 2.2.6 below) in some differential equations textbooks to "prove" that hunting coyotes (to keep them from eating farm animals) increases the population of the coyotes' natural prey *without decreasing the coyote population.* This claim is unsupported by any biological data and is obviously incorrect. □

The coyote–rabbit setting does involve a predator and prey, but it does not follow that just any predator–prey model will be useful. The correct approach is to think of the Lotka–Volterra model as only one *possible* model.[10] Instead of accepting a ridiculous result, such as the impossibility of eliminating predators, we should conclude that the model is wrong.

The lesson of Example 2.2.1 bears frequent repetition:

> *The value of a model depends on the setting to which it is applied and the questions it is used to address.*

Mathematics has the benefit of certainty, as exemplified by proofs of theorems. This is of great value to mathematicians, because it minimizes the time spent arguing about facts. Once a mathematical claim has been proven, everyone is obligated to accept it. However, this certainty only applies to mathematical claims; it does not extend to mathematical modeling. Mathematical results about a model can be confirmed with mathematical proof, and proven results are correct—but only for the model. Conclusions drawn from models are only correct for the real-world setting to the extent that the behavior of the model reflects that of the real world. This question is not mathematical and must be addressed by other means. It is therefore misleading to think of models as "correct" or "true" for a real-world setting. At best, a mathematical model can be *valid*, in the sense of "giving meaningful results under a given set of real-world circumstances." There are almost certainly quantitative differences between model results and real-world empirical results, and there may be important qualitative differences as well. If the differences are small enough in the given setting, we judge the model to be valid and use it with confidence. The model may work for somewhat different settings as well, but we must worry about its validity in the new setting. Where the validation is not satisfactory, we must revise the model and try again.

Example 2.2.2. The exponential decay model

$$y = y_0 e^{-kt}, \qquad k, y_0 > 0 \tag{2.2.1}$$

is valid for a macroscopic amount of a single radioactive substance. The model can also be applied to other settings where a quantity is decreasing or increasing to a fixed value, such as the clearance of medication from the bloodstream of an animal. The ultimate value could be nonzero, in which case we can interpret y as the difference between the current value and the ultimate value. Whatever the context, we have to be careful that the model is appropriate. In lead poisoning, a significant portion of the lead is deposited in the bones, so a more sophisticated model[11] is needed to incorporate this physiological mechanism. Time-release medications are slow to absorb from the digestive system and require a more sophisticated model as well. □

---

[9] We might use this model for *P. steadius*, but we ought not use it for *P. speedius*.

[10] We return to this scenario later in this section. More appropriate models are presented in Chapter 7.

[11] See Section 7.1.

2.2.1 Mechanistic and Empirical Modeling

Mathematical models can be classified according to the method used to obtain them.

Mechanistic model: *a mathematical model based on assumptions about the scientific principles that underlie the phenomena being modeled.*

Empirical model: *a mathematical model based on examination of numerical data.*

The distinction between the two types of models is sharpened by separating the "approximate quantitative description" in the definition of a mathematical model into two distinct processes: that of approximation and that of quantitative description. To clarify this point, it is helpful to introduce the concept of the *conceptual model*.

Conceptual model: *an approximation of the real world that serves as a verbal description of a mathematical model.*

Conceptual models are seldom explicit in the presentation of mathematical models. Identifying the underlying conceptual model is necessary to understand biological literature that uses mathematics. Identification of conceptual models from examination of mathematical formulas is a recurring theme in the problems sets of this chapter and those of Part III.

Figure 2.2.1 illustrates the processes of mechanistic and empirical modeling. The flow of modeling is not unidirectional. Each of the components feeds into the others, but it is important to note the lack of a direct connection from the mathematical model to the real world. It is this point that distinguishes mathematical modeling from the "applications" of mathematics that appear in most textbooks.

Because of the lack of a clear direction in the flow, we describe these processes in alphabetical order before discussing some key issues in mathematical modeling.

Approximation: An intentional process of choosing features to include in models, analogous to drawing a political cartoon. Caricatures of President Obama always have ears that are much larger than any real person while omitting some of his more subtle features. Yet anyone who has seen the president can easily identify him in a political cartoon. Similarly, a conceptual model in mechanistic modeling focuses on features of the real world that the modeler believes are critical while omitting anything thought to be an unnecessary complication. The hope is that the resemblance of the simple model to the complicated real world will be unmistakable. Approximation is treated in more detail in Section 2.5 and parts of Chapters 5–7.

Characterization: Obtaining general results about a model. Sometimes we can reduce a model to an explicit solution formula. In other cases, we can use graphical or approximation methods to determine how the values chosen for the parameters influence the model behavior. Whatever we learn from characterization applies to the conceptual model, not the real world. Characterization uses techniques of calculus as well as advanced techniques discussed in Parts II and III.

Derivation: Constructing a mathematical model from a verbal description of assumptions and simplifying the model prior to analysis. Sections 2.5 and 2.6 contain examples of model derivation and simplification, as do Chapters 5 and 7. The associated problem sets focus on the reverse skill of identifying the underlying conceptual model from a given mathematical model. Model

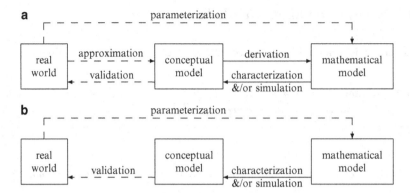

Fig. 2.2.1 Relationships between the real world, a conceptual model, and the corresponding mathematical model. *Solid arrows* indicate processes amenable to mathematical certainty, while *dashed arrows* indicate processes that must be viewed with scientific skepticism. (**a**) Mechanistic modeling. (**b**) Empirical modeling

derivation is a skill needed primarily by mathematical modelers, but anyone who wants to read quantitative biological literature needs to be able to understand the biological assumptions of a model, including assumptions that are only implied.

Model selection: Choosing a mathematical model from multiple options. In mechanistic modeling, construction of a conceptual model constitutes model selection. In empirical modeling, selection of a model should be done with the aid of the Akaike information criterion, a method for quantifying the statistical support a data set gives to a model. This topic is addressed in Section 2.7.

Parameterization: Using data to obtain values for the parameters in a model. Parameterization is necessary if the model analysis is to include simulation. If the analysis is to be general, then it is helpful to have ranges for parameter values; however, precise values fitted to data are unnecessary. Parameterization is addressed in Sections 2.3 and 2.4.

Simulation: Using mathematics and computation to visualize model behavior for a given set of parameters. If an explicit solution formula can be found, then simulation is just a matter of graphing the solution. However, in most cases, it is necessary to implement a numerical method with a computer. Parts II and III contain simulation techniques for specific problem types.

Validation: Determining whether a model reproduces real-world results well enough to be useful. The criteria for validation depend on the purpose of the model. This topic is addressed in more detail later in Section 2.2.2.

Few research projects in biology are simple enough to incorporate all of the processes in Figure 2.2.1a in a single study. One such project is summarized in Example 2.2.3.

Example 2.2.3. The University of Nebraska-Lincoln has a course called *Research Skills for Theoretical Ecology* that focuses on combining experiment and theory to understand the growth of a population of pea aphids.[12]

- Approximation: We assume that the population at day $t + 1$ depends only on the population at day t and make specific quantitative assumptions about this dependence.
- Derivation: The assumptions made in the approximation step lead us to a model consisting of formulas for determining day $t + 1$ populations from day t populations.

---

[12] Aphids are particularly convenient organisms for the study of population dynamics because many aphid species exist for long periods as asexual females that reproduce by cloning. Simple models are more likely to yield accurate results when applied to simple systems.

- Parameterization: We use laboratory experiments and statistical analysis to determine values for the various model parameters, such as the average number of offspring an adult produces in 1 day. These parameters are needed so that we can obtain quantitative results.
- Characterization: Independent of parameter values, the model predicts that the population will tend toward stable proportions of nymphs and adults with unrestricted population growth at some fixed rate. This rate is about 32 % per day for the parameter values determined from the laboratory data.[13]
- Simulation: We write a computer program to simulate an experiment in which a population consisting initially of a single adult is allowed to grow for several weeks. Using the parameters determined from laboratory data, the simulation predicts a variable growth rate that approaches 32 % per day over the first 2 weeks.
- Validation: The students measure the quantitative growth rate by experiment and obtain a result of approximately 32 %, as long as the food supply in the experiment is maintained at a high level. However, the unrestricted population growth predicted by the model is qualitatively wrong under conditions where food and space are limited. The model could be used to forecast future aphid populations subject to the condition of virtually unlimited resources and no predators, but not in more general circumstances.   □

2.2.2 Aims of Mathematical Modeling

Mathematical models can be used for different purposes, and the aim of the model plays a large role in determining the type of analysis and the criteria for validation. For example, sometimes the goal of modeling is to predict the results of hypothetical experiments.

Example 2.2.4. Mathematicians and biologists at the University of Tennessee have created a sophisticated computer simulation called Across Trophic Level System Simulation (ATLSS) that models populations of animal and plant species in the Florida Everglades [13]. This model has been used by the Army Corps of Engineers and other agencies to predict the effects of environmental policy on the Everglades ecosystem. For example, the model can be used to address the question "What effect would a new housing development in a particular area of the Florida Everglades have on the endangered Florida panther population?"   □

A model such as ATLSS needs specific geographic and climate data and specific values for many parameters, such as the average size of a litter of Florida panther cubs and the survival rate for newly hatched whooping cranes. The parameters are estimated for the model because the goal is to predict populations in a hypothetical experiment for a real scenario. Given this goal, the criteria for model validity are quantitative. The model is valid if the results it predicts for experiments are within an acceptable tolerance of the actual experiment results. Demonstrating validity of a model used for quantitative prediction can be difficult. In a laboratory setting, such as that of Example 2.2.3, the model simulation can be designed to match a specific experiment and the results can be directly compared. But we cannot conduct designed experiments for the Everglades. Instead, we look for historical events that can be thought of as experiments, in which case we can match the simulation to the historical event. If we know the effect of a historical housing development on the panther population, we can check to see that our model is quantitatively accurate for that specific case. If so, then we have evidence that our model will correctly predict the effect of a similar hypothetical event.

---

[13] One could not design a better organism for rapid population growth than the aphid. When reproduction is by cloning, individuals do not need to mature before they begin the reproduction process. Indeed, pea aphids are born pregnant and begin to give birth to live young within hours after becoming adults.

Quantitative prediction for hypothetical experiments is not all that can be done with mathematical models. In Section 1.5, we used a mathematical model to obtain a prediction for foraging behavior in a patchy environment. This required mathematical characterization rather than numerical simulation, and it resulted in the marginal value theorem, which we can think of as a qualitative prediction for a general setting. Broad questions, such as that of optimal foraging behavior, require mathematical characterization rather than numerical simulation. Other questions are less broad but require characterization of a general model because parameter values can vary.

Example 2.2.5. The side effects of chemotherapy and its efficacy against tumors depend on the dosage schedule. We could administer the medication at a constant rate over a fixed time interval, we could make the rate large initially but then decrease it to zero over some interval, or we could choose any other time-dependent dosage schedule. A reasonable goal for modeling is to identify a protocol that minimizes side effects while reducing the tumor at a desired rate. One way to attempt this is to run simulations with a chemotherapy model, but we could only test a few of the infinitely many dosing protocols. Instead, it is possible to use mathematical methods to obtain an approximate solution to the problem of optimizing the dosage schedule.

□

The validation of a model intended to address broad theoretical questions is different from that for a model intended for quantitative prediction. For a general model, the task of validation involves trying to confirm that the model behavior is qualitatively consistent with the behavior of the real biological system we are trying to model. Of course, we must specify what we mean by "consistent with the real behavior." For example, a model whose purpose is to study extinction risk for endangered species would need to be checked to ensure that it is actually capable of predicting extinction under some set of circumstances.

Example 2.2.6. Suppose we want to know what effect coyote hunting will have on a coyote population (the question of Example 2.2.1). From our reading of the biology literature, if not from direct experience, it should be obvious that predator extinction is a possibility. However, characterization of the Lotka–Volterra model shows it to be incapable of predicting this possibility. The proper response to this mathematical result is to immediately reject the use of the Lotka–Volterra model for the given setting. Careful examination of the conceptual model can identify the flaw that accounts for its failure to make correct qualitative predictions, which in turn can suggest a better model.[14]

□

2.2.3 The Narrow and Broad Views of Mathematical Models

In any particular instance of a mathematical model, we have one or more dependent variables and one or more independent variables, and a set of given values are assigned to the parameters. The focus of a simulation is on determining how the dependent variables depend on the independent variables. This is the *narrow view* of mathematical models. In contrast, there is a

---

[14] See Problem 2.2.1.

broad view of mathematical models, in which the objective is to understand the effect of the parameter values on the model behavior. The relationship between these views is illustrated in Figure 2.2.2. In the broad view, the role of "independent variable" is played by the parameters and the role of "dependent variable" is played by whatever aspects of the model behavior are of interest.

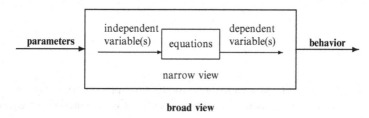

Fig. 2.2.2 Narrow and broad views of mathematical models

Example 2.2.7. Consider the family $y = \sin kt$, which is sometimes used as an empirical model for periodic data. When we choose a specific value for k and plot y as a function of t, we are working in the narrow view. Without choosing k, we can calculate the period to be the smallest time T such that $y(t + T) = y(t)$. Since the sine function repeats as the angle increases by 2π, the period is when $kT = 2\pi$, or $T = 2\pi/k$. If we plot T as a function of k, we are working in the broad view. ☐

As in Example 2.2.7, parameters function as constants in some aspects of model analysis and as variables in other aspects, corresponding to the narrow and broad views, respectively. This can be very confusing. Generally we are working with the narrow view for simulations and the broad view for characterization. Both are important. We can make use of the full power of computers for simulations, but we can address deeper questions when we retain the broad view.

2.2.4 Accuracy, Precision, and Interpretation of Results

Most people make little distinction in ordinary language between the terms "accuracy" and "precision," but these terms have very distinct meanings in science. *Accuracy* is the extent to which results are correct, while *precision* is the extent to which results are reproducible. Precision is easier than accuracy to measure, but of course it is accuracy that we really need. Nevertheless, one cannot be confident of accuracy in the absence of precision. A repeated theme of this book, starting with the discussion of biological data in Section 2.1 and continuing through the rest of the book, with special emphasis in Chapter 4 on probability distributions of samples, is that precision is limited in most areas of biology. Even where careful measurements are possible, results are not very reproducible. Have your blood pressure taken on five consecutive days, and you will see the point.

The lack of precision in most biological data has strong implications for how we interpret mathematical results. Computers give very precise results—divide 1.0 by 3 and you will not get 0.33, but 0.333333333333. This is alright if the numerator is certain to be very close to 1.0, but in biology it could be that you measured 1.0 when the "correct" value is 0.9. If the data is off by 10 %, then the additional digits beyond the second one are surely meaningless. Mathematics, of course, offers the possibility of infinite precision. In terms of biology, this does more harm than good. Apply infinitely precise methods to crude results and you get results that are infinitely precise in appearance without being reproducible. It is easy to take this apparent precision more seriously than it deserves. This is what I call the "measure it with your hand, mark it with a pencil, cut it with a laser" fallacy. The risk of this fallacy must be kept firmly in mind whenever we interpret results obtained from mathematical modeling applied to crude data.

Problems

2.2.1. Suppose we want to construct a realistic model for a predator–prey system. This model should allow for a variety of realistic results; in particular, it should predict three possible long-term results:

1. The predator and prey can coexist.
2. The prey can survive while the predator becomes locally extinct.
3. Both species can become locally extinct.

We look in a mathematical biology book and find the Lotka–Volterra model:

$$\frac{dx}{dt} = rx - qxy,$$

$$\frac{dy}{dt} = cqxy - my,$$

where x and y are the biomasses of the prey and predator, respectively,[15] r is the growth rate of the prey, m is the death rate of the predator, q measures the extent of predation, and c is a conversion factor for prey biomass into predator biomass. The model is presented as a predator–prey model, but we recognize the need to check that a model is appropriate for the setting we have in mind.

(a) Suppose the prey and predator populations stabilize to fixed biomasses $X \geq 0$ and $Y \geq 0$. If the biomasses at some time are $x = X$ and $y = Y$, then there should be no further change. This means that the fixed biomasses must make the right-hand sides of the differential equations be 0. Use this idea to find all possible pairs X and Y. Is this model capable of predicting all three possible long-term results? If not, which is missing?
(b) To help see what is wrong with the model,[16] write down the prey equation for the special case where the predator is absent. What does the model predict will happen?

2.2.2. Find an instance of a mathematical model in a biology book or research paper. Describe:

(a) The mathematical model itself,
(b) The conceptual model that corresponds to the mathematical model, and
(c) Features of the real-world setting that do not appear in the conceptual model.

2.2.3.* Some genetic traits are determined by a single gene having two variants (alleles), with one (**A**) dominating the other (**a**). This means that individuals who have two dominant alleles (**AA**) and individuals who have one of each type (**Aa**) both exhibit the physical

---

[15] Most descriptions of predator–prey models interpret the variables as the numbers of individuals, but the models are more realistic if the variables are viewed as being the total biomass of the individuals.

[16] The point here is that using the Lotka–Volterra model to demonstrate that something can't happen in the real world is a logical fallacy when the model itself contains the assumption that the thing can't happen.

characteristics (phenotype) of the dominant trait, while the recessive phenotype is only found among individuals who have two recessive alleles (**aa**). It is sometimes helpful in genetics to model inheritance as a two-step process: first, all of the parents' genes are assembled into a gene pool; then, pairs of genes are randomly withdrawn from the gene pool for individuals in the next generation.

(a) Suppose q is the fraction of recessive genes in the gene pool. Based on the two-step conceptual model, what will be the fraction of individuals in the next generation who exhibit the recessive trait? What will be the fraction of individuals who have one dominant allele and one recessive allele? What will be the fraction of individuals who do not have the recessive allele? (The combination of these results is called the *Hardy–Weinberg principle*.)

(b) About 13 % of the people of Scotland have red hair. Assuming that red hair is caused by a single recessive gene pair, what does the Hardy–Weinberg principle predict for the fraction of the recessive trait in the gene pool and the fraction of the population who do not have the recessive allele?

(c) Demographers estimate that 60 % of the people of Scotland do not have the recessive allele for red hair. What flaws in the conceptual model might account for the difference between this estimate and the estimate you obtained from the Hardy–Weinberg principle?

2.2.4. The model $y = y_0 e^{-kt}$, with $k > 0$, is often used to model radioactive decay, where y is the amount of radioactive material remaining and y_0 is the initial amount of the material. This model is derived in Section 2.5 using a conceptual model in which the decay rate is k times the quantity of material. Without calculus, we can get some sense of what this means. Consider the specific instance $y = e^{-2t}$. The average rate of decay over the interval $t_1 < t < t_1 + h$ is

$$r_h(t_1) = \frac{y(t_1 + h) - y(t_1)}{h}.$$

For the intervals $0 < t < 0.1, 0.1 < t < 0.2$, and so on up to $0.9 < t < 1.0$, calculate the average rate of decay and compare it to the average of the quantities of radioactive material at the beginning and end of the time interval. Explain why the results are consistent with the conceptual model as described here.

2.2.5. The model $y = y_0 e^{kt}$, with $k > 0$, is sometimes used to model bacterial growth.

(a) Describe the qualitative predictions made by the model. In particular, show that

$$G(t) = \frac{y(t+1)}{y(t)}$$

does not actually depend on t.

(b) Describe an experiment that tests the prediction of part (a).

(c) Describe a physical setting in which this model for population growth is clearly not appropriate.

(d) Describe a physical setting in which this model for population growth might be appropriate.

2.2.6. Of Problems 2.2.4 and 2.2.5, one works primarily with the narrow view of a model and the other primarily with the broad view. Match these descriptions with the problems, explaining why that view is the focus of the problem.

2.2.7. In this problem, we develop and study a model to predict the future effects of changes in the average number of children per adult and the average age for childbirth on human population growth rates in a highly developed country. The model is based on the Euler–Lotka equation, which was developed in Section 1.7:

$$\int_0^\infty e^{-rt}\ell(t)m(t)\,dt = 1\,,$$

where r is the unknown population growth rate resulting from a fecundity of $m(t)$ births per year per parent of age t, given a probability of $\ell(t)$ of survival to age t.

(a) For the sake of a simple thought experiment, we assume particularly simple forms for ℓ and m. We take $\ell = 1$ since survival to adulthood is high in highly developed countries. For m we assume a piecewise linear function with peak at age a, tapering to 0 at ages $a-5$ and $a+5$, and having a total (integrated over time without the factor e^{-rt}) of n. This will allow us to modify just the parameters a and n rather than the whole fecundity function m. Show that the function

$$m(t) = 0.04n \begin{cases} 0,\, t < a-5 \\ 5+t-a,\, a-5 < t < a \\ 5-t+a,\, a < t < a+5 \\ 0,\, t > a+5 \end{cases}$$

has all of the desired properties.

(b) Substitute $\ell = 1$ and the function m from part (a) into the Euler–Lotka equation to obtain the integral equation

$$0.2n \left[\int_{a-5}^a \left(1 + \frac{t-a}{5}\right) e^{-rt}\,dt + \int_a^{a+5} \left(1 - \frac{t-a}{5}\right) e^{-rt}\,dt \right] = 1.$$

(c) Make the substitution $x = (t-a)/5$ to simplify the integrals. You should now have one integral with $-1 < x < 0$ and one with $0 < x < 1$.

(d)* Make an additional substitution $y = -x$ in the integral on $-1 < x < 0$ and combine the two integrals into a single integral on the interval $(0, 1)$. See Problem 1.9.9.

(e) You should now have an equation of the form

$$ne^{-ra}F(r) = 1\,,$$

where $F(r)$ is a complicated definite integral. Show that $F(0) = 1$ and $F(0.02) < 1.001$. You can do the latter by calculating the integral (see Problem 1.8.12) or by numerical approximation (see Problem 1.7.9). The function F is strictly increasing, so the approximation $F(r) = 1$ has error less than 0.1 % if $r \le 0.02$.

(f) Indicate at least one biological assumption in this model that we can expect to introduce more than a 0.1 % error; conclude that the approximation $F = 1$ is fully justified.

(g) Plot a graph of r against n, with three curves using different values of a, and use this curve to discuss the effects of average number of children and average age of reproduction on population growth. To do this intelligently, you must choose reasonable low, medium, and high values for a and a reasonable range of n values for the horizontal axis.

(h) Repeat part (g), reversing the roles of n and a.

2.2.8. Suppose individuals of group X and individuals of group Y interact randomly. If x and y are the numbers of individuals in the respective groups, it is reasonable to expect each member of group Y to have kx interactions with members of group X, where $k > 0$ is a parameter. (This says that doubling the membership of group X should double the contact rate with group X for a member of group Y.)

(a) Use the information about contact rates for individuals in group Y to find a model for the overall rate at which members of the two groups interact.

(b) Suppose p is the (fixed) fraction of encounters between individuals of the two groups that results in some particular event occurring between the individuals. Use this assumption to create a model for the rate R at which the events occur in the population.

(c) If the model of part (b) is used for the rate of infection of human populations with some communicable disease, what do the groups X and Y represent?

(d) The model of part (b) was used successfully to model an influenza outbreak in a small boarding school in rural England. Why do you think the model worked well in this case?

(e) The Center for Disease Control in Atlanta did not use the model of part (b) to make predictions about the spread of the H1N1 virus in the United States in 2009. Explain why the model would not have been appropriate in this case.

(f) Describe some real-world settings other than epidemiology that could conceivably use this interaction model [Hint: This model finds common usage in chemistry and ecology as well as epidemiology.] How accurate do you expect the model to be in these different settings?

2.3 Empirical Modeling I: Fitting Linear Models to Data

After studying this section, you should be able to:

- Use the linear least squares method to obtain the best-fit parameter values for the linear models $y = mx$ and $y = b + mx$.
- Discuss the assumptions made in claiming that the results of the linear least squares method are the best parameter values for the data.

Simulations require values for the model parameters, which raises the question of how parameter values should be determined. Occasionally they can be measured directly, but more often they can only be inferred from their effects. This is done by collecting experimental data for the independent and dependent variables and then using a mathematical procedure to determine the parameter values that give the best fit for the data. In this section, we develop a parameterization method for two linear models:

$$y = mx, \qquad y = b + mx, \tag{2.3.1}$$

where m and b are parameters. These models, together with the exponential model,

$$y = Ae^{kx}, \tag{2.3.2}$$

and the power function model,

$$y = Ax^p, \tag{2.3.3}$$

comprise the principal empirical models commonly encountered.[17] Parameterization of exponential and power function models is considered in Section 2.4.

---

[17] The symbols in these models are generic; that is, they represent whatever actual variables are in a given model. For example, a model $H = CL^q$ is a power function model with independent variable L, dependent variable H, exponent parameter q, and coefficient parameter C. Most symbols in mathematics are not standard, so the reader must be able to identify models as equivalent when the only difference is the choice of symbols. This theme is extended much further in Section 2.6.

2.3.1 The Basic Linear Least Squares Method (y = mx)

Table 2.3.1 Predation rate y for prey density x

x	0	10	20	30	40	50	60	70	80	90	100	110	120	130	140
y	0	2	7	10	9	14	21	20	25	20	30	25	29	35	38

Table 2.3.1 reproduces the *P. steadius* data from Table 2.1.1. Because the data points appear to lie roughly on a straight line through the origin (Figure 2.1.1a), it makes sense to try a linear model without a parameter to represent the y intercept; that is, $y = mx$ rather than $y = b + mx$.[18] The parameterization process consists of finding and solving a mathematics problem to determine a value of m for this data set.

Obviously there is no single value of m for which the model $y = mx$ fits the data exactly. For any given value of m, some or all of the data points lie off the graph of the model. Figure 2.3.1 shows the data with several possible straight lines. Clearly, the slope m for the top line is too large and that for the bottom is too small.

2.3.1.1 Overview of the Method

Any optimization problem has a modeling step and an analysis step:

1. Determine a function that expresses the quantity to be maximized or minimized in terms of one or more variable quantities.
2. Determine the set of values for the variable quantities that yields the maximum or minimum function value among permissible values of the variables.

Optimization problems for mathematical models are conceptually more difficult than optimization problems in calculus because of the different roles played by variables and

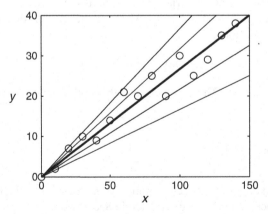

Fig. 2.3.1 Consumption rate y for prey density x from Table 2.3.1, showing several instances of the model $y = mx$; the *heavy line* is the instance that will emerge as the best fit

---

[18] There are two advantages to omitting the parameter b. Mathematically, it is much easier to find one parameter from data than two. More importantly, the model $y = mx$ may be more appropriate on biological and/or statistical grounds, as will be seen in Sections 2.5 and 2.7, respectively.

parameters. In our current example, the variables x and y represent specific data points, while the parameter m is unknown.

> When parameterizing a mathematical model from data, the *parameters* in the model are the *variables* in the optimization problem, while model variables appear in the optimization problem only as labels of values in the data set.

The imagery of the narrow and broad views of mathematical modeling (Figure 2.2.2) is helpful in thinking about the problem of determining a best value of m. In step 1, we assume a fixed value of the parameter m, generate a set of "theoretical" (x, y) data points using the model $y = mx$ with the x values from the actual data, and calculate some quantitative measure of the total discrepancy between the actual data and the data obtained using the model. This step occurs within the narrow view because m is fixed. Once we have a formula for calculating the total discrepancy, we change our perspective. Now we think of the *data* as fixed and the total discrepancy for that fixed data set as a function $F(m)$. We then obtain the optimal value of m using methods of ordinary calculus. Treating m as a variable locates step 2 within the broad view.

2.3.1.2 Quantifying Total Discrepancy

There are several reasonable ways to measure quantitative discrepancy. The standard choice is

$$F(m) = (\Delta y_1)^2 + (\Delta y_2)^2 + \cdots + (\Delta y_n)^2 \,, \tag{2.3.4}$$

where the *residuals* Δy_i are the vertical distances between the data points and the corresponding points from the model. For the model $y = mx$, we have

$$\Delta y_i = |mx_i - y_i| \,. \tag{2.3.5}$$

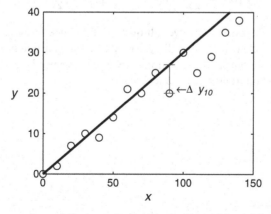

Fig. 2.3.2 Consumption rate y for prey density x, showing the model $y = 0.3x$ and the residual for $x = 90$

Example 2.3.1. Let $m = 0.3$. The data, model, and one of the residuals are shown in Figure 2.3.2. The total discrepancy is $F(0.3) = 218$, calculated as the sum of the bottom row of Table 2.3.2.

□

Table 2.3.2 Total discrepancy calculations for $y = 0.3x$ with the *P. steadius* data

x	0	10	20	30	40	50	60	70	80	90	100	110	120	130	140
$0.3x$	0	3	6	9	12	15	18	21	24	27	30	33	36	39	42
y	0	2	7	10	9	14	21	20	25	20	30	25	29	35	38
Δy	0	1	-1	-1	3	1	-3	1	-1	7	0	8	7	4	4
$(\Delta y)^2$	0	1	1	1	9	1	9	1	1	49	0	64	49	16	16

Check Your Understanding 2.3.1:
Repeat Example 2.3.1 for the model $y = 0.25x$. Is this model better or worse than $y = 0.3x$?

Total discrepancy calculations are easily automated with a spreadsheet; however, we cannot do the calculation for all possible values of m. We shall see that the optimal m can be determined mathematically, without actually computing any values of $F(m)$.

2.3.1.3 Minimizing Total Discrepancy

The problem of choosing m to minimize F is surprisingly easy to solve. Substituting (2.3.5) into (2.3.4) and expanding the squares, we have

$$F(m) = (m^2 x_1^2 - 2m x_1 y_1 + y_1^2) + \cdots + (m^2 x_n^2 - 2m x_n y_n + y_n^2),$$

which we can rearrange as

$$F(m) = (x_1^2 + x_2^2 + \cdots + x_n^2)m^2 - 2(x_1 y_1 + x_2 y_2 + \cdots + x_n y_n)m + (y_1^2 + y_2^2 + \cdots + y_n^2).$$

We can simplify this formula using summation notation[19]:

$$F(m) = \left(\sum x^2\right) m^2 - 2\left(\sum xy\right) m + \left(\sum y^2\right), \tag{2.3.6}$$

where the sums are understood to be over all of the data points. In the context for (2.3.6), the data points are known; hence, the total discrepancy formula is a function of a single variable m.[20] The function is a simple parabola pointing upward, so we need only find the vertex of that parabola to obtain the important mathematical result[21]:

---

[19] For ease of reading, I use a simplified form of summation notation. What I have as $\sum xy$, for example, is more properly given as $\sum_{i=1}^{n} x_i y_i$. In the given context, the extra notation decreases readability unnecessarily.

[20] Context is crucial. As noted earlier, the parameter m functions as a *constant* in the model $y = mx$ (narrow view) but as a *variable* in the total discrepancy function F (broad view). Meanwhile, x and y are variables in the model, but the data points (x_i, y_i) function as parameters in the total discrepancy calculation because we have a fixed set of data.

[21] The proof of Theorem 2.3.1 is given as Problem 2.3.10.

Theorem 2.3.1 (Linear Least Squares Fit for the Model $y = mx$). *Given a set of points (x_i, y_i) for $i = 1, 2, \cdots, n$, the value of m that minimizes the total discrepancy function for the model $y = mx$ is*

$$m = \frac{\sum xy}{\sum x^2}; \qquad (2.3.7)$$

the corresponding residual sum of squares is

$$RSS = \sum y^2 - \frac{(\sum xy)^2}{\sum x^2} = \sum y^2 - m \sum xy. \qquad (2.3.8)$$

The **residual sum of squares** is the total discrepancy for the model when the best value of m is used; that is, it is the minimum value of the function F. It will be needed for the semilinear data fitting scheme of Section 2.4 and the model selection scheme of Section 2.7.

We now have the mathematical tools needed to find the optimal value of m for the *P. steadius* data set.

Example 2.3.2. For the data of Table 2.3.1, we obtain the results

$$\sum x^2 = 101,500, \qquad \sum xy = 27,080, \qquad \sum y^2 = 7,331;$$

therefore, (2.3.7) and (2.3.8) yield the results $m \approx 0.267$ and RSS ≈ 106.1. The best-fit line is the heavy one in Figure 2.3.1. $\qquad \square$

Check Your Understanding 2.3.2:
Verify the values given in Example 2.3.2.

2.3.2 Adapting the Method to the General Linear Model

Most straight lines in a plane do not pass through the origin. While there are theoretical reasons for insisting that the predation model pass through the origin, this is obviously not valid for *all* linear models; hence, Theorem 2.3.1 would seem to be of limited use. However, the problem of fitting the model $y = b + mx$ to data can be reduced to the problem of fitting the model $y = mx$ to data. The derivation of the mathematical result for this two-parameter model is given as Problem 2.3.11.

Theorem 2.3.2 (Linear Least Squares Fit for the General Linear Model $y = b + mx$). *Let $\bar{x}$ be the mean of the values $x_1, x_2, \ldots, x_n$, let $\bar{y}$ be the mean of the values $y_1, y_2, \ldots, y_n$, and define shifted data points by*

$$X_i = x_i - \bar{x}, \quad Y_i = y_i - \bar{y}, \quad for\ i = 1, 2, \ldots, n. \qquad (2.3.9)$$

Then

1. *The best-fit slope and residual sum of squares for the model $y = b + mx$ can be found by fitting the XY data to the model $Y = mX$;*
2. *The best-fit intercept b is given by*

$$b = \bar{y} - m\bar{x}.$$ (2.3.10)

Example 2.3.3. To fit the model $y = b + mx$ to the data of Table 2.3.1, we first compute the means $\bar{x} = 70$ and $\bar{y} = 19$. Then we subtract the means from the original data set to obtain a shifted data set, as shown in Table 2.3.3. Applying Theorem 2.3.1 to the shifted data yields the results

$$m = 0.255, \qquad \text{RSS} \approx 100.4.$$

By Theorem 2.3.2, these results hold for the model $y = b + mx$ with the original data, and we calculate b from (2.3.10):

$$b = 1.175.$$

□

Table 2.3.3 Consumption rate y for prey density x, along with the shifted data $X = x - \bar{x}, Y = y - \bar{y}$

x	0	10	20	30	40	50	60	70	80	90	100	110	120	130	140
y	0	2	7	10	9	14	21	20	25	20	30	25	29	35	38
X	−70	−60	−50	−40	−30	−20	−10	0	10	20	30	40	50	60	70
Y	−19	−17	−12	−9	−10	−5	2	1	6	1	11	6	10	16	19

Check Your Understanding 2.3.3:
Verify the values given in Example 2.3.3.

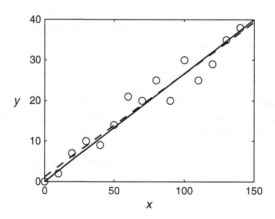

Fig. 2.3.3 Consumption rate y for prey density x, showing the linear least squares fits for the models $y = mx$ (*solid*) and $y = b + mx$ (*dashed*)

We now have two best-fit results for the Table 2.3.1 data: the line $y = 0.267x$ from Example 2.3.2, with residual sum of squares 106.1, and the line $y = 1.175 + 0.255x$ from Example 2.3.3, with residual sum of squares 100.4. Figure 2.3.3 shows both of these lines along with the data. Does the slightly lower residual sum of squares mean that the two-parameter model is better than the one-parameter model? Not necessarily. The calculation of the residual sum of squares treats all data equally. However, the data point $(0, 0)$ is free of experimental uncertainty, so perhaps we should be less tolerant of the discrepancy $\Delta y_1 = 1.175$ than the discrepancies at the other points. Perhaps we should insist that $y(0) = 0$ is a *requirement* for our model, even though doing so slightly increases the residual sum of squares. We take up this issue in Section 2.7 after we have laid more groundwork.

2.3.3 Implied Assumptions of Least Squares

The sum of squares of vertical residuals is not the only function that could be used to quantify the total discrepancy between an instance of a model and a set of data. We need some way to measure the distance between each point and the model line, but why not measure it horizontally or along a normal to the line? In choosing Δy as the measure of discrepancy, we are implicitly assuming that we have very accurate measurements of x and uncertainty primarily in y. This is frequently true, but not always. However, it is common to use $(\Delta y)^2$ even in cases where there is significant uncertainty in x values because the least squares formula is so easy to apply.[22]

Check Your Understanding Answers

Table 2.3.4 Total discrepancy calculations for $y = 0.25x$ with the *P. steadius* data

x	0	10	20	30	40	50	60	70	80	90	100	110	120	130	140
$0.25x$	0	2.5	5	7.5	10	12.5	15	17.5	20	22.5	25	27.5	30	32.5	35
y	0	2	7	10	9	14	21	20	25	20	30	25	29	35	38
Δy	0	0.5	2	2.5	1	1.5	6	2.5	5	2.5	5	2.5	1	2.5	3
$(\Delta y)^2$	0	0.25	4	6.25	1	2.25	36	6.25	25	6.25	25	6.25	1	6.25	9

1. The total discrepancy is $F(0.25) = 134.75$, which is less than $F(0.3) = 218$.

Problems

2.3.1. (Continued from Problems 2.1.1 and 2.1.3.)

(a) Fit the model $y = mx$ to your *P. steadius* data from Problem 2.1.1.
(b) Fit the model $y = mx$ to your *P. steadius* data from Problem 2.1.3.
(c) How much different (in percentage) are your results from the result in the text?
(d) Discuss whether or not replacement appears to be a significant source of differences in the results.

(This problem is continued in Problem 2.7.4.)

---

[22] See [10] for a much more complete discussion of this topic.

2.3.2. One of the data sets in Table 2.3.5 has the origin as its mean point. Find the equation of the straight line that best fits that data. Plot the data and the best-fit line together on a graph.

Table 2.3.5 Two xy data sets for Problem 2.3.2

x	-4	-1	0	2	3
y_1	-5	-2	0	2	4
y_2	-5	-2	1	2	4

2.3.3.* The data sets of Table 2.3.6 contain a parameter c that perturbs some of the data points away from the straight line $y = x$, while still maintaining an average y of 0. By examining the change in slope m as a function of c, we can measure the effect of measurement error on the result of the least squares procedure. To do this, plot the linear regression slope m as a function of the parameter c, where $0 \leq c \leq 1$, for each data set. How do measurement errors affect the least squares line? In particular, which errors are the least squares line more sensitive to?

Table 2.3.6 Two data sets for Problem 2.3.3

x_1	-2	-1	0	1	2
y_1	$-2+c$	-1	0	1	$2-c$
x_2	-2	-1	0	1	2
y_2	-2	$-1+c$	0	$1-c$	2

2.3.4.(a) Fit a linear model to the data of Table 2.3.7 (using a calculator to do the computations). The data gives the concentration C, in parts per trillion, of the trace gas F-12 at the South Pole from 1976 to 1980.
(b) Discuss the quality of the fit of the model to the data.
(This problem is continued in Problem 2.7.7.)

Table 2.3.7 Concentration of F-12 at the South Pole by year, with 1976 as year 0 [11]

t	0	1	2	3	4
C	195	216	244	260	284

2.3.5.*

(a) Fit a linear model to the data of Table 2.3.8 (using a calculator to do the computations).
(b) Discuss the quality of the fit of the model to the data.

(This problem is continued in Problem 2.7.8.)

Table 2.3.8 A data set for Problem 2.3.5

t	8.7	9	11	18	19	22	28
C	25	25	26	48	65	90	100

2.3.6. In a convenient programming environment, write a program that inputs a file with two columns of data and uses the least squares procedure to fit the model $y = b + mx$ using the first column for the x values and the second column for the y values. Test the program with the data from Problems 2.3.4 or 2.3.5.

2.3.7. (Continued from Problems 2.1.1 and 2.1.3.)
Repeat Problem 2.3.1 with the model $y = b + mx$ using the program of Problem 2.3.6.

2.3.8. The National Oceanographic and Atmospheric Administration (NOAA) has a data set on its web site that gives the dates of the beginning of the grape harvests in Burgundy from 1370 to 2003 [5]. This data offers a crude, but long-term, look at global climate change.

(a) Fit a linear model to the data using the program of Problem 2.3.6. Do three different calculations: (1) the years 1800–1950, (2) the years 1951–1977, and (3) the years 1979–2003. Of particular interest is the slope.
(b) On the average, grape harvest dates have been getting earlier since 1800. By how many days did the expected grape harvest date change in each of the three periods?
(c) Plot all of the data from 1800 to 2003 as points on a graph. Explain why it is a mistake to connect the points to make a dot-to-dot graph.
(d) Add the three linear regression lines to the plot, being careful to use only the appropriate time interval for each.
(e) What do the results appear to say about global climate change?
(f) Offer at least one possible explanation for the results that does not involve global climate change. [Hint: Think about possible biological explanations.]

(This problem is continued in Problem 2.7.9.)

2.3.9. The National Snow and Ice Data Center has a data set on its web site that gives the duration of ice cover for a number of Northern Hemisphere lakes dating back to the 1800s in some cases. These data sets offer a look at global climate change that is shorter term than the grape harvest data in Problem 2.3.8 but which has fewer confounding factors.

(a) Go to the search engine for the Global Lake and River Ice Phenology Database: http://nsidc. org/data/lake_river_ice/freezethaw.html
Type a name code in the appropriate box (in capitals). Suitable lakes for this study include JGL03 (Iowa); DMR1, JJM22, JJM27, and JJM33 (Wisconsin); KMS10, KMS11, and KMS14 (New York); GW240, GW341, GW369, GW512, and GW592 (Sweden); JK02, JK03, JK09, JK17, JK31, JK48 (Finland); and NG1 (Russia). In the output parameter options list, choose Ice Off Date, Ice Duration, Latitude, Longitude, Lake Name, and Country Name. In the output sort options list, choose Ice Season. Click the Submit Request button.
(b) Copy the data and paste it into a text file. Save the data with the file extension csv (comma-delimited).
Of course you will want to look at the lake name, latitude, longitude, and country name for context, but these columns are not part of the data to be analyzed. Open the file in a spreadsheet program and delete all but the first and fourth columns, which give the year and number of days of ice cover. Also delete any rows in which the ice duration is given as −999, which means that the data is unavailable. Save the file again.
(c) Fit a linear model to the data using the program of Problem 2.3.6. Do three different calculations: (1) the full data set, (2) the years 1930–1970, and (3) the years 1970 to the present. Of particular interest is the slope.
(d) Plot all of the data as points on a graph. Explain why it is a mistake to connect the points to make a dot-to-dot graph.

(e) Add the three linear regression lines to the plot, being careful to use only the appropriate time interval for each.

(f) What do the results appear to say about global climate change?

(This problem is continued in Problem 2.7.10.)

2.3.10. Derive the results of the linear least squares method for the model $y = mx$,

$$m = \frac{\sum xy}{\sum x^2}, \qquad \text{RSS} = \sum y^2 - \frac{(\sum xy)^2}{\sum x^2} = \sum y^2 - m \sum xy,$$

by applying optimization methods from calculus to the total discrepancy function

$$F(m) = \left(\sum x^2\right) m^2 - 2\left(\sum xy\right) m + \left(\sum y^2\right).$$

2.3.11. Derive the general linear least squares results (Theorem 2.3.2) by using the results for the $y = mx$ case.

2.4 Empirical Modeling II: Fitting Semilinear Models to Data

After studying this section, you should be able to:

- Use the least squares method to fit the linearized versions of the models $z = Ae^{\pm kt}$ and $y = Ax^p$ to data, where A, k, and p are parameters.
- Use a variant of the linear least squares method to fit models of the form $y = qf(x, p)$ to data, where q and p are parameters.

In Section 2.3, we learned the basic least squares method for linear models. The method can be adapted for some nonlinear models.

2.4.1 Fitting the Exponential Model by Linear Least Squares

Taking natural logarithms of both sides of the model

$$z = Ae^{-kt} \tag{2.4.1}$$

changes the model equation to

$$\ln z = \ln A - kt. \tag{2.4.2}$$

Now suppose we define a new set of variables and parameters by the equations

$$y = \ln z, \qquad x = t, \qquad m = -k, \qquad b = \ln A. \tag{2.4.3}$$

The result is the standard linear model

$$y = b + mx. \tag{2.4.4}$$

The algebraic equivalence of the original model (Equation (2.4.1)) and the linearized model (Equation (2.4.4)) allows us to fit the exponential model to data using linear least squares.[23]

---

[23] Equivalent models are the subject of Section 2.6.

Algorithm 2.4.1 *Linear least squares fit for the exponential model* $z = Ae^{-kt}$

1. *Convert the tz data to xy data using* $y = \ln z$ *and* $x = t$.
2. *Convert the xy data to XY data using* $X = x - \bar{x}$ *and* $Y = y - \bar{y}$, *where* $\bar{x}$ *and* $\bar{y}$ *are the means of the x and y values, respectively.*
3. *Find the parameters m and b for the xy model using the linear least squares formulas:*

$$m = \frac{\sum XY}{\sum X^2}, \qquad b = \bar{y} - m\bar{x}. \tag{2.4.5}$$

4. *Calculate the parameters for the exponential model:* $A = e^b$ *and* $k = -m$.

Example 2.4.1. Table 2.4.1 shows data from a radioactive decay simulation of 1,000 particles, each of which had an 8 % chance of decaying in any given time step. The original data consists of the time t and number of remaining particles z for each time. The xy and XY data sets were calculated in steps 1 and 2 of Algorithm 2.4.1, with $\bar{x} = 4.5$ and $\bar{y} = 6.542$. The linear least squares formulas (2.3.7) and (2.3.8) yield the results

$$m = -0.0835, \qquad b = 6.917,$$

from which we obtain

$$k = 0.0835, \qquad A = 1010.$$

Figure 2.4.1 shows the data and best-fit model in both the xy and tz planes. □

Table 2.4.1 Data sets for the exponential model in Example 2.4.1

t	0	1	2	3	4	5	6	7	8	9
z	1,000	929	855	785	731	664	616	568	515	471
x	0	1	2	3	4	5	6	7	8	9
y	6.908	6.834	6.751	6.666	6.594	6.498	6.423	6.342	6.244	6.155
X	−4.5	−3.5	−2.5	−1.5	−0.5	0.5	1.5	2.5	3.5	4.5
Y	0.366	0.292	0.209	0.124	0.053	−0.043	−0.118	−0.200	−0.297	−0.387

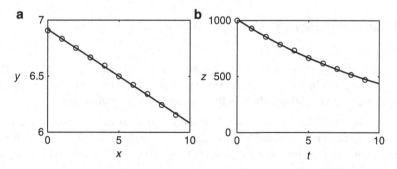

Fig. 2.4.1 The exponential model fit to the data of Table 2.4.1 using linear least squares (a) $y = b + mx$, (b) $z = Ae^{-kt}$

2.4.2 Linear Least Squares Fit for the Power Function Model $y = Ax^p$

The predation data for *P. speedius* (Table 2.1.1 and Figure 2.1.1) appears superficially to resemble a square root graph. This suggests a model of the form

$$y = Ax^p, \qquad A, p > 0. \tag{2.4.6}$$

This model can be fit using the same linearization technique used for the exponential model. However, we have to be careful about notation. Often a particular symbol has different meanings in two or more formulas needed to solve a particular problem. Here, the symbol x represents the number of prey animals in the biological setting (and hence in the model (2.4.6)), but it also represents the generic independent variable in the generic models $y = b + mx$ and $y = mx$. This kind of duplication is unavoidable because many formulas have their own standard notation. One way to avoid error in these cases is to rewrite the generic formulas using different symbols. In this case, let's use U, V, u, and v in place of X, Y, x, and y in the generic linear least squares formulation.

Taking a natural logarithm of (2.4.6) yields

$$\ln y = \ln A + p \ln x,$$

which is equivalent to the linear model $v = b + mu$ using the definitions

$$u = \ln x, \quad v = \ln y, \quad m = p, \quad b = \ln A.$$

We can then formulate an algorithm for fitting a power function model using linearized least squares.

Algorithm 2.4.2 *Linear least squares fit for the power function model $y = Ax^p$*

1. *Convert the xy data to uv data using $u = \ln x$ and $v = \ln y$.*
2. *Convert the uv data to UV data using $U = u - \bar{u}$ and $V = v - \bar{v}$, where $\bar{u}$ and $\bar{v}$ are the means of the u and v values, respectively.*
3. *Find the parameters m and b for the uv model using the linear least squares formulas:*

$$m = \frac{\sum UV}{\sum U^2}, \qquad b = \bar{v} - m\bar{u}. \tag{2.4.7}$$

4. *Calculate the parameters for the power function model: $A = e^b$ and $p = m$.*

Notice that all the symbols used in Algorithm 2.4.2 are defined within the algorithm statement. The meaning of a biological symbol is seldom clear from the context alone, so it is good modeling practice to define all symbols in the statement of a model or algorithm.

Example 2.4.2. In fitting the model $y = Ax^p$ to the *P. speedius* data from Table 2.1.1, we run into a problem. The change-of-variables formulas $u = \ln x$ and $v = \ln y$ do not work for the point $(0,0)$. This is not a serious problem, because $(0,0)$ satisfies $y = Ax^p$ exactly for any values of the parameters. Omitting $(0,0)$, we obtain the linearized least squares result

$$y = 2.71 x^{0.499}. \tag{2.4.8}$$

This model is plotted together with the data in Figure 2.4.2. The first plot shows the data and model as a plot of y versus x, while the second plot shows the linearized data and model as a plot of $\ln y$ versus $\ln x$. □

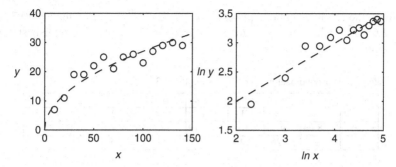

Fig. 2.4.2 The power function model fit to the *P. speedius* data [without $(0,0)$] using linear least squares

2.4.3 Semilinear Least Squares

In Example 2.4.2, the parameter values $A = 2.71$ and $p = 0.499$ minimize the fitting error on the graph of $\ln y$ versus $\ln x$. This is not the same thing as minimizing the fitting error on the graph of y versus x. It is tempting to accept the results as optimal, but this is not necessarily appropriate. If we want to minimize the fitting error in the original data, we must find a way to do so without linearizing the model. For this purpose, we need to adapt the linear least squares method to apply to what we can call *semilinear* models: models of the form $y = qf(x,p)$, where f is a nonlinear function of the independent variable x and one of the parameters, while the other parameter appears as a multiplicative factor.[24]

2.4.3.1 Finding the Best A for Known p

The semilinear regression method for the model $y = Ax^p$ involves two distinct mathematics problems: first we find the best A in terms of an arbitrary choice of p, and then we find the best p. The first of these problems can be solved for any particular p by using linear least squares on a data set that has been modified to account for the chosen value of p.

Example 2.4.3. Suppose we *assume* $p = 0.5$. Then, for each data point, we can calculate the quantity x^p exactly. Defining $z = x^{0.5}$, we can rewrite the model $y = Ax^{0.5}$ as $y = Az$. This allows us to convert the original xy data into the zy data of Table 2.4.2. The mathematical result for models of the form $y = mx$ (with z playing the role of x and A playing the role of m) then yields the slope and residual sum of squares:

$$A = \frac{\sum zy}{\sum z^2} \approx 2.67, \quad \text{RSS} = \sum y^2 - A\sum zy \approx 80.1.$$

---

[24] This could be done with the exponential model $z = Ae^{-kt}$ as well, if the goal is to minimize the fitting error in the original data. However, for reasons beyond the scope of this discussion, it is usually better to fit exponential models in the linearized form $\ln z = \ln A - kt$ rather than the original form.

In comparison, the residual sum of squares for $y = 2.71x^{0.499}$ is approximately 80.7. Thus, the model $y = 2.67x^{0.5}$ is a little more accurate on a graph in the xy plane than the best fit obtained by linearization.   □

Table 2.4.2 Data points for the model $y = Ax^{0.5}$ of Example 2.4.3

$z = x^{0.5}$	0	3.16	4.47	5.48	6.32	7.07	7.75	8.37
y	0	7	11	19	19	22	25	21
$z = x^{0.5}$	8.94	9.49	10.0	10.49	10.95	11.40	11.83	
y	25	26	23	27	29	30	29	

The very slight improvement we found in Example 2.4.3 is not enough to justify the more complicated procedure for finding the parameter values. However, we only *guessed* the value $p = 0.5$; what we really need is a way to find the *best p*.

2.4.3.2 Finding the Best *p*

When fitting the model $y = Ax^p$ to data, we define a residual sum of squares in terms of the parameters A and p. The goal is to choose the pair (p, A) that minimizes the residual sum of squares. Optimization problems for two-parameter nonlinear models are usually very difficult, but in this case we already know how to find the best value of A for any given choice of p. If we assume that we will always use the best A, then we can think of the residual sum of squares as a function of p only.

Formally, define a residual sum of squares function F by

$$F(p) = \min_A (\text{RSS}(p, A)) = \text{RSS}(p, \hat{A}(p)), \qquad (2.4.9)$$

where $\hat{A}(p)$ is found as in Example 2.4.3. From the calculation in the example, we have $\hat{A}(0.5) = 2.67$ and $\text{RSS}(0.5, 2.67) = 80.1$; therefore, $F(0.5) = 80.1$. For any given value of p, we have to create a modified data set and use the linear least squares formulas to get the corresponding F for that p. That is a lot of work, but it is a type of work for which computers are ideally suited. The simplest way to use a computer to identify the best parameters is to calculate a lot of points on the graph of F and identify (approximately) the smallest F from that set of points.

> **Check Your Understanding 2.4.2:**
> Repeat the calculation of Example 2.4.3 to obtain the result $F(0.4) = 63.1$. This result shows that $p = 0.4$ is closer to the optimal value than our original guess of $p = 0.5$.

Example 2.4.4. Table 2.4.3 shows some values of $F(p)$ for the model $y = Ax^p$ using the *P. speedius* data from Table 2.1.1. From this table, we can see that the optimal p value is somewhere in the interval $0.35 \le p \le 0.45$. We then compute the values of $F(p)$ for $p = 0.350, 0.351, 0.352, \ldots 0.450$. These are plotted in Figure 2.4.3. By examining the list of F values, we see that the optimal p value, to three significant figures, is 0.409, with a corresponding residual sum of squares of 62.9. We then obtain A by linear least squares, as in Example 2.4.3, leading to the result

$$y = 4.01x^{0.409}, \qquad \text{RSS} = 62.9. \qquad (2.4.10)$$

This new model is plotted along with that of Example 2.4.2 in Figure 2.4.4.   □

A visual examination of Figure 2.4.4 shows the clear superiority of the semilinear least squares method compared to the linear least squares method for power function models. The

Table 2.4.3 Some values of the residual sum of squares function $F(p)$ for the model $y = Ax^p$ using the *P. speedius* data from Table 2.1.1

p	0.30	0.35	0.40	0.45	0.50	0.55	0.60	0.65	0.70
$F(p)$	92.3	71.2	63.1	66.6	80.1	102.4	132.3	168.7	210.7

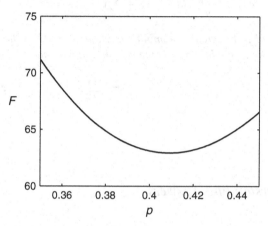

Fig. 2.4.3 The minimum residual sum of squares (F in (2.4.9)) for the *P. speedius* predation data with the power function model (Example 2.4.4)

semilinear method is also superior to the Lineweaver–Burk linearization that is commonly used to determine parameters for Michaelis–Menten reactions in biochemistry.[25]

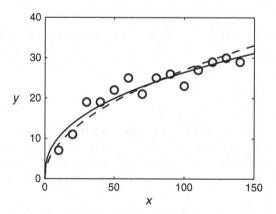

Fig. 2.4.4 The power function model fit to the *P. speedius* data (without $(0,0)$) using linear least squares (*dashed*) and semilinear least squares (*solid*)

---

[25] See Problem 2.4.8 for an illustration of how important this is. Other authors (see [10], for example) also state that one should use nonlinear regression rather than using linear regression on a linearized model, but they don't always explain the reason carefully or present an illustrative example.

The procedure described here works for any model of the general form $y = qf(x, p)$, where x is the independent variable, y is the dependent variable, and q and p are parameters.[26] We summarize the result here.

Theorem 2.4.1 (Semilinear Least Squares). *Given a model* $y = qf(x, p)$ *with data points* (x_1, y_1), (x_2, y_2), ..., (x_n, y_n), *define a function F by*

$$F(p) = \min_q (RSS(p, q)) = \sum y^2 - \frac{[\sum y f(x, p)]^2}{\sum f^2(x, p)}.$$

Let $\hat{p}$ *be the value of p that yields the minimum value of F. Then the minimum residual sum of squares on the graph in the xy plane is achieved with parameter values*

$$p = \hat{p}, \qquad q = \frac{\sum y f(x, \hat{p})}{\sum f^2(x, \hat{p})}.$$

Check Your Understanding Answers

Table 2.4.4 Total discrepancy calculations for $y = 0.25x$ with the *P. speedius* data

x	0	10	20	30	40	50	60	70	80	90	100	110	120	130	140
$0.25x$	0	2.5	5	7.5	10	12.5	15	17.5	20	22.5	25	27.5	30	32.5	35
y	0	2	7	10	9	14	21	20	25	20	30	25	29	35	38
Δy	0	0.5	2	2.5	1	1.5	6	2.5	5	2.5	5	2.5	1	2.5	3
$(\Delta y)^2$	0	0.25	4	6.25	1	2.25	36	6.25	25	6.25	25	6.25	1	6.25	9

1. The total discrepancy is $F(0.25) = 134.75$, which is less than $F(0.3)$ (Table 2.4.4).

Problems

2.4.1.* The data of Table 2.4.5 gives the population of a bacteria colony as a function of time. Find the exponential function that best fits the data set using the linearization method. Plot the linearized data with the best-fit line. Plot the original data with the exponential curve corresponding to the best-fit line.

Table 2.4.5 Population of bacteria after t hours

t	0	1	2	3	4
N	6.0	9.0	13.0	21.0	29.0

2.4.2.* Use semilinear least squares with the data in Problem 2.4.1 to fit the exponential model without linearization. Compare the parameter results from the two calculations. Plot the two resulting models as y versus t and again as $\ln y$ versus t. Discuss the results.

---

[26] In Example 2.7.3, we will use this method with a model derived in Section 2.5.

2.4.3. (Continued from Problem 2.3.4.)

(a) Use the semilinear method to fit the model $y = Ax^p$ to the data from Problem 2.3.4 and determine the residual sum of squares.

(b) What is clearly wrong with using the model $y = Ax^p$ for this data set?

2.4.4. (Continued from Problem 2.3.5.)

(a) Use the linearization method to fit the model $y = Ae^{-kt}$ to the data from Problem 2.3.5.

(b) Use the semilinear method to fit the model and data from part (a).

(c) Find the residual sums of squares for the results of parts (a) and (b) on graphs of y versus t and on graphs of $\ln y$ versus t.

(d) Which is better, the linearization result or the semilinear result?

2.4.5.* Use semilinear least squares with the data in Table 2.4.1 to fit the exponential model without linearization. Compare the parameter results with Example 2.4.1, which used the same data. Why are the results different? Plot the two resulting models as y versus t and again as $\ln y$ versus t. Draw a reasonable conclusion from your observations.

2.4.6. (Continued from Problem 2.1.2.)

(a) Fit the model $y = Ax^p$ to your *P. speedius* data from Problem 2.1.2 using linearized least squares.

(b) Fit the model $y = Ax^p$ to the same data using the semilinear method.

(c) Plot the model from part (a) along with the data on a graph of $\ln y$ versus $\ln x$. Repeat for the model from part (b). Compare the visual appearances of the two plots.

(d) Plot the model from part (a) along with the data on a graph of y versus x. Repeat for the model from part (b). Compare the visual appearances of the two plots.

(e) Describe and explain any conclusions you can draw from this set of graphs.

(This problem is continued in Problem 2.7.5.)

2.4.7. Table 2.4.6 shows data for average lengths in centimeters of Atlantic croakers (a species of fish) caught off the coasts of three states. Use this data to fit the von Bertalanffy growth equation,

$$x(t) = x_\infty(1 - e^{-rt}),$$

where $x(t)$ is the length of the fish, x_∞ is the asymptotic maximum length, and r is a positive parameter.[27] How well does the model fit the data?

Table 2.4.6 Average length in centimeters of Atlantic croakers from New Jersey, Virginia, and North Carolina [4]

Age	1	2	3	4	5	6	7	8	9	10
NJ	30.3	31.1	32.4	34.2	35.0	34.8	37.4	36.6	36.1	37.4
VA	25.8	28.9	31.8	34.0	35.2	36.1	37.4	40.2	40.2	40.3
NC	24.6	27.3	29.7	33.1	35.2	37.2	37.8	38.4	37.7	38.1

2.4.8. Michaelis–Menten reactions are enzyme-catalyzed reactions in biochemistry. The initial reaction rate v depends on the concentration S of the principal reactant (called the *substrate*), according to the Briggs–Haldane model

---

[27] It would be better to fit the data for individual lengths rather than averages; however, the raw data sets are quite large and not generally available.

$$v = \frac{v_m S}{K_M + S}, \qquad S, v_m, K_M > 0,$$

where v_m is the maximum rate (corresponding to a very large substrate concentration) and K_M is the semisaturation parameter (see Section 2.6). Lineweaver and Burk rewrite the reaction rate equation as

$$\frac{1}{v} = \frac{1}{v_m} + \frac{K_M}{v_m} \frac{1}{S}$$

and then fit the data to this form [9].[28]

(a) Derive the Lineweaver–Burk formula from the original model.
(b) Use the Lineweaver–Burk linearization to determine the parameter values using the data in Table 2.4.7; that is, use the data to create a data set for variables $y = 1/v$ and $x = 1/S$. After fitting the model $y = b + mx$ to this data using linear least squares, determine the corresponding parameters v_m and K_M from the original model.
(c) Find the parameter values for the model using the data in Table 2.4.7 with the semilinear least squares method.
(d) Prepare a plot that includes:

    a. The data points (S, v);
    b. The Briggs–Haldane curve, using the parameter values obtained in part (b); and
    c. The Briggs–Haldane curve, using the parameter values obtained in part (c).

(e) Calculate the residual sum of squares for each parameter set. Discuss the results. In particular, which of the curves seems to fit the data better, and why is this the case?

Table 2.4.7 Substrate concentration S and reaction velocity v for nicotinamide mononucleotide adenylyltransferase in pig livers [3]

S	0.138	0.220	0.291	0.560	0.766	1.46
v	0.148	0.171	0.234	0.324	0.390	0.493

2.4.9. (Continued from Problem 1.1.13.)
In Problem 1.1.13, we considered the Sinclair coefficient used to compare weightlifters of different sizes:

$$C(m) = 10^{A[\log_{10}(m/b)]^2},$$

---

[28] A 1975 paper presents compelling evidence that other methods in use in the mid-1970s were preferable to the Lineweaver–Burk method [2]. Unfortunately, Lineweaver–Burk was entrenched by then, and scientific progress has failed in this case to overcome the inertia of standard practice. The Lineweaver–Burk method remains in common usage today. The 1975 tests included an implementation due to Wilkinson of the nonlinear method [14], which of course produced the best results with simulated data under reasonable assumptions about the types of error in the data. There can be no question that the semilinear least squares method produces the best fit on a plot of the original data, nor is there any reason in the world of fast computing to settle for a method that is not as good simply because it is faster for hand computation. In general, a good understanding of the theories of various methods for solving problems helps us to identify cases, such as this one, where older methods should be replaced by newer computer-intensive methods.

where m is the mass, b is the mass of the heavyweight world record holder, and A is a positive parameter of unclear meaning. The formula is based on the assumption that the two-lift total should fit the model

$$T(m) = C\,10^{-A[\log_{10}(m/b)]^2} .$$

(a) Although this model could be fit as a three-parameter model, which would be outside the scope of our semilinear method, the parameter b in practice is fixed at $b = 173.961$ as of this writing. The value of the parameter A should be chosen to make the model fit the data optimally. To test this point, use the semilinear method to fit the model for $T(m)$ to the data in Table 2.4.8, which gives the official post-1998 world records for each weight class as of June 2012. Report A to the nearest four decimal digits and compare with the official value $A = 0.7848$.

(b) Redo Problem 1.1.13 using the new value of A. Does this change the top three spots in the overall ranking?

Table 2.4.8 Post-1998 world records for men's weightlifting as of June 2012, with T the total amount lifted in two events by lifters of mass m kg

m	56	62	69	77	85	94	105	174
T	305	326	357	378	394	412	436	472

2.5 Mechanistic Modeling I: Creating Models from Biological Principles

After studying this section, you should be able to:

- Discuss the relationship between biological observations, a conceptual model, and a mechanistic mathematical model.
- Discuss the conceptual model for the Holling type II predation function.
- Explain the mathematical derivation of the Holling type II predation function.

In Section 2.4, we modeled radioactive decay with an exponential model. We were able to fit the data quite well, but the empirical justification for the model limits its explanatory value. Would an exponential model be a good fit with a different data set for the same substance? What about a data set that extends the total time of the experiment or a data set for a different radioactive substance? Empirical modeling cannot answer these questions, because empirical reasoning must *begin* with the data.

The alternative approach to modeling is *mechanistic modeling*, in which we obtain a model from assumptions based on theoretical principles. Sometimes a mechanistic justification can be found for a model we have already identified empirically, as we will see with our exponential model for radioactive decay. In this case, the model gains explanatory value. In other cases, we may be able to discover a model not previously identified empirically.

2.5.1 Constructing Mechanistic Models

Textbooks in algebra and calculus include story problems, where you have to derive mathematical models from a verbal statement. These verbal statements are conceptual models, as discussed in Section 2.2. They can be translated into mathematical statements simply by following some basic rules. This process is routine, in the sense that mastery of the rules is all that is needed for it. The hard part of model construction is getting the conceptual model in the first place—this work is done by the author of the story problem rather than the student.

How do we write our own conceptual models? In practice, we typically use whatever we have learned from prior experience in modeling. More fundamentally, conceptual models come from qualitative observation and measurement of data. The observations and data have to be generalized, and the conceptual model is obtained by restating these generalizations as fact. As an example, we consider the development of a model for radioactive decay. Imagine that radioactivity has only just recently been discovered. We have observations and data from experimental scientists, and we must use this information to construct a conceptual model.

Example 2.5.1. From examining data on radium decay rates, we obtain a specific observation.

Specific Observation:
All samples of radium-226 lose approximately 0.043 % of their number in 1 year.

By itself, this observation is interesting enough to suggest more studies with other time intervals and other radioactive isotopes. All such studies yield similar observations, except that the percentage loss and time unit are different for each. Taken together, these experiments allow us to write a much stronger statement.

General Observation[29]:
The percentage of particles lost by a given radioactive substance in one time unit is approximately constant.

Idealizing the general observation yields a mechanistic assumption. In this simple example, only one mechanistic assumption is needed to complete the conceptual model.

Conceptual Model 1:
The fraction of particles lost by a given radioactive substance in one time unit is fixed.

This conceptual model differs only slightly from the corresponding general observation: the latter uses the qualifier "approximately," while the former reads as a statement of fact. Models are not necessarily correct, but they are always specific.

Now that we have a conceptual model, we need only rewrite it in mathematical notation.

Mathematical Model 1:
If $y(t)$ is the amount of radioactive material at time t, then

$$\frac{y(t) - y(t+1)}{y(t)} = K, \tag{2.5.1}$$

for some $K > 0$.

Equation (2.5.1) is a direct translation of the conceptual model. The quantity $y(t) - y(t+1)$ is the number of particles lost between time t and time $t+1$. Dividing this quantity by $y(t)$ gives the fraction of particles lost between time t and time $t+1$. It is easier to work with fractions than percentages, so we simply assign the parameter K to be the constant fraction lost in one time unit. In a specific example, we need to know the value of K; otherwise we keep it unspecified so that we can apply the mathematical model to any radioactive substance.

Equation (2.5.1) is a discrete mathematical model for radioactive decay. We can rearrange the model so that it makes a prediction about the number of particles at time $t+1$:

$$y(t+1) = (1-K)y(t).$$

Given y_0 particles at time 0, the model predicts

---

[29] The simulated experiment in Example 2.4.1 was based on this general observation, with 8 % used as the percentage for the hypothetical substance.

$$y(0) = y_0, \quad y(1) = y_0(1-K), \quad y(2) = y_1(1-K) = y_0(1-K)^2, \quad \cdots.$$

There is a clear pattern—the number of particles at time t is[30]

$$y(t) = y_0(1-K)^t. \tag{2.5.2}$$

$\square$

Note that the mathematical model in Example 2.5.1 did not come *directly* from the data of one or more experiments, but *indirectly* from a set of assumptions suggested by the results of many experiments. These assumptions are the conceptual model.

The result (2.5.2) of the discrete model in Example 2.5.1 is not the usual exponential function $y = y_0 e^{-kt}$; however, we can obtain the exponential function model by means of algebraic manipulation.

Example 2.5.2. Beginning with the model $y = y_0(1-K)^t$, we can replace the factor $1-K$ with the factor e^{-k}. There is no harm in doing this; K is an arbitrary parameter, and it is alright to use a different arbitrary parameter if we prefer the new one. The advantage of k over K is that we can then write the model as $y = y_0(e^{-k})^t = y_0 e^{-kt}$. If we already know the value of K for a specific substance, we can use the equation $1 - K = e^{-k}$ to calculate the corresponding value of k. Solving for k gives us the formula

$$k = -\ln(1-K) \tag{2.5.3}$$

which gives the exponential rate parameter in terms of the fraction of particles lost in one time unit. For the simulation used for Example 2.4.3, we have $K = 0.08$, so $k = 0.0834$. The empirical result $k = 0.0835$ obtained in the example agrees very nicely with this theoretical result. $\square$

The model $y = y_0 e^{-kt}$ of Example 2.5.2 can also be obtained using calculus.

Example 2.5.3. To derive the model $y = y_0 e^{-kt}$ directly, we need a more sophisticated conceptual model. Suppose we could measure the instantaneous rate of radioactive decay (particles per time) rather than the average rate of decay (particles in one unit of time). Because equal fractions of particles decay in equal units of time, we expect a sample of 100 atoms to have twice as many decays in a given amount of time as a sample of 50 atoms. Thus, the rate of decay for a sample should be proportional to the sample size.

Conceptual Model 2:
   In any radioactive decay process, the rate of decay is proportional to the amount of the substance.

Translating this into mathematical notation gives us the corresponding mathematical model.

Mathematical Model 2:
   If $y(t)$ is the amount of radioactive material at time t, then

$$\frac{dy}{dt} = -ky \tag{2.5.4}$$

for some $k > 0$.

Note that we can identify the family of functions $y = y_0 e^{-kt}$ as the functions that satisfy (2.5.3).

---

[30] Formally, this result can be proved by *mathematical induction*.

To understand the meaning of the parameter k, we can rewrite the model as

$$-\frac{1}{y}\frac{dy}{dt} = k .$$
(2.5.5)

Thus, k is the decay rate $-dy/dt$ divided by the amount y, which is called the *relative rate of decay*. An alternative statement of Conceptual Model 2:
The relative decay rate of a radioactive substance is constant.   □

The derivation of Example 2.5.3 is preferable to that of Examples 2.5.1 and 2.5.2 on general principles:

Conceptual models for continuous processes should be based on rates of change rather than discrete change.

2.5.2 Dimensional Analysis

Dimensional analysis is a powerful tool that simplifies the construction of mathematical models. The idea is that all models must be dimensionally consistent; colloquially, "You can't add apples to oranges." There are three requirements for dimensional consistency.[31]

Rules for Dimensional Consistency

1. Quantities can be added together or set equal to each other only if they have the same dimension.
2. The dimension of a product is the product of the dimensions.
3. The argument of a transcendental function, such as a trigonometric, exponential, or logarithmic function, must be dimensionless.

Example 2.5.4. The circumference of a circle of radius r has length $2\pi r$. The ratio of circumference to radius is 2π, which is, by definition, the radian measure of a full circle. The radian measure of any other angle is similarly defined as the ratio of the length of the corresponding circular arc to the radius of the circle. Because both circumference and radius are lengths, the radian measure of an angle is dimensionless. The rule that arguments of trigonometric functions must be dimensionless suggests that these arguments should be given in radians.   □

Example 2.5.5. In functions such as $\sin kt$, the argument does not necessarily correspond to a geometric angle, so the term "radians" does not really apply. Nevertheless, dimensional consistency requires the quantity k to have dimension 1/time.   □

Example 2.5.6. The mathematical model $\frac{dy}{dt} = -ky$ must be dimensionally consistent. The independent variable t is a time. The dependent variable y can be considered to be either a mass or a count of atoms; to be specific, let's take it to be a mass. Hence, the derivative dy/dt has dimension mass per time. The right side of the model must also be mass per time, and so the dimension of k must be 1/time.

---

[31] Note the distinction between dimensions, such as length, and the associated units of measurement, such as meters, feet, and light-years.

The related mathematical model, $y(t) = y_0 (1 - K)^t$, also has to be dimensionally consistent, so the parameter K and the variable t must both be dimensionless. In general, time is dimensionless in any discrete model. □

2.5.3 A Mechanistic Model for Resource Consumption

We are now ready to develop mathematical models for resource consumption, which we hope will explain the differences between the predation data for the two species of predator introduced in Section 2.1.

Consider a situation in which an organism is placed in an environment with a fixed concentration of resource x. Since the organism consumes the resource, it is therefore necessary for the resource to be replenished. We measure the rate y with which we have to replenish the resource; this is equivalent to measuring the rate at which the resource is being consumed. Our goal is to construct a mathematical model that relates the intake rate y to the resource concentration x. Note that it doesn't make any difference how the organism obtains the resource. The organism could be a predator that feeds by hunting, an herbivore that feeds by grazing, or a single cell that feeds by absorbing nutrients through its surface.

To develop a conceptual model for this experiment, it helps to create a narrative version that appeals to human experience. Imagine yourself as the organism in the experiment. You live alone in a building with numerous hiding places that could contain servings of food. A total of x servings of food are distributed randomly throughout the building. Each time you eat a serving, a restaurant worker hides another serving in some randomly chosen location within the building, thereby keeping the food concentration at a constant level.[32] In this setting, our question is "How much do you eat per unit time in an environment with a constant food supply x per unit area?"[33] We denote the amount eaten per unit time by y. Thus, we seek a model to predict y as a function of x.

We begin with the simplest conceptual model.

Example 2.5.7. You have to get to the food in order to eat it. For the simplest conceptual model, we imagine that you continually search the building, eating the food as you find it. It seems reasonable that you will search some fixed amount of space per unit of time. The amount of space you can search in one unit of time could vary a lot, depending on the size of the rooms and the possible number of hiding places. However, it does *not* depend on the amount of food available. Every experiment in a given building with a given person can use the same search rate, regardless of x. Thus, the search rate is a parameter, which we designate as s. The food density x could be measured in servings per square meter, and the intake rate y could be measured in servings consumed per hour. The rate at which you locate food should be the product of the food density and the search rate; thus, we have the model

$$y = sx . \tag{2.5.6}$$

Note that the model also makes sense if we replace the quantities by their dimensions:

$$\frac{\text{food}}{\text{time}} = \frac{\text{area}}{\text{time}} \times \frac{\text{food}}{\text{area}}.$$

□

---

[32] Of course, this does not happen in a real feeding scenario; however, this is an assumption in the conceptual model. In practice, this discrepancy between the real experiment and the conceptual model causes difficulties in the measurement of the parameters.

[33] Note that x need not be very large. In our human version, we could have a restaurant the size of a shopping mall with an average of one serving of food in each of the stores.

2.5.4 A More Sophisticated Model for Food Consumption

We now have a mechanistic formulation for the linear model that we used in Section 2.3 for *P. steadius*. The model is obviously not useful for *P. speedius* because the plot of the data (Figure 2.1.1) is not approximately linear. We must conclude that the conceptual model of Example 2.5.7 is missing some biologically significant features. The crucial step in finding a better model for *P. speedius* is the search for a more realistic conceptual model.

Example 2.5.8. The model (2.5.6) is based on the assumption that you spend all of your time searching for food. This might be the case if food is extremely scarce. However, if you live in a buffet restaurant, where food is plentiful and easy to find, you spend only a tiny fraction of your time searching for it. You don't keep eating just because you can find more food! Instead, you spend nearly all of your time digesting and hardly any of it searching for more. The time required for digestion is a feature of the actual experiment that our first conceptual model lacks. Run another BUGBOX-predator trial with *P. speedius* and a large prey density. Notice that the predator spends only a small portion of the experiment time searching, because it pauses whenever it locates prey.

We need a conceptual model in which time is partitioned into two types: search time and "handling" time. With this distinction between two types of time, our first model is no longer dimensionally consistent; the "time" in y is total time, while the "time" in s is search time. Our dimensional equation needs an additional factor to account for the distinction:

$$\frac{\text{resource}}{\text{total time}} = \frac{\text{search time}}{\text{total time}} \times \frac{\text{area}}{\text{search time}} \times \frac{\text{resource}}{\text{area}}.$$

Using f to denote the fraction of time spent searching, we have

$$y = fsx. \tag{2.5.7}$$

If f is a parameter, like s, then we are done. However, this is not the case. We've already observed that f is approximately 1 when the resource is scarce, but approximately 0 when it is plentiful. Thus, f is a second dependent variable. We therefore need a second equation. From our conceptual model,

$$\text{search time} + \text{handling time} = \text{total time}.$$

Dividing by total time, we have

$$\frac{\text{search time}}{\text{total time}} + \frac{\text{handling time}}{\text{total time}} = 1.$$

It seems reasonable that the amount of handling time should be proportional to the amount of resource consumed, so we define a new parameter h to be the time required to handle one unit of the resource. Now we can think of the dimensional equation as

$$\frac{\text{search time}}{\text{total time}} + \left(\frac{\text{handling time}}{\text{resource}} \times \frac{\text{resource}}{\text{total time}} \right) = 1.$$

In symbols, this is

$$f + hy = 1. \tag{2.5.8}$$

Equations (2.5.7) and (2.5.8) are a pair of equations for a pair of dependent variables. Thus, we have the right number of equations for a complete model. Substituting from Equation (2.5.8) into (2.5.7), we have

$$y = (1 - hy)sx = sx - shxy ,$$

or

$$y + shxy = sx .$$

Thus, we arrive at the model

$$y = \frac{sx}{1 + shx} . \tag{2.5.9}$$

One instance of this model is shown in Figure 2.5.1. The model shows the right general shape to fit the *P. speedius* data. □

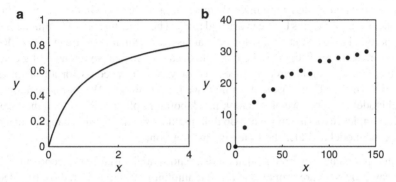

Fig. 2.5.1 Side-by-side comparison of (**a**) the Holling type II model $y = x/(1 + x)$ and (**b**) the predation data for *P. speedius*, showing the same general shape

Equation (2.5.9) is one form of the model known in ecology as the *Holling type II* predation function and in microbiology as the *Monod growth function*. In its current form, the model is not semilinear, so we cannot use the method of Section 2.4 to fit it to data. In Section 2.6, we obtain the more common semilinear form of the model and fit this form to the *P. speedius* data. The model is also mathematically equivalent to the Briggs–Haldane model used to determine reaction rates for Michaelis–Menten reactions.

2.5.5 A Compartment Model for Pollution in a Lake

Suppose we want to model the concentration of pollutants in a lake. There is a general modeling framework, called *compartment analysis*, that is used when the primary structure of the model consists of accounting for changes in one or more quantities. For simplicity, we consider a single body of water and a single pollutant. We can think of the amount of pollutant in two ways, as the total mass Q and as the concentration C, which is the mass per unit volume. It is easier to think of mass while doing the accounting, but it is more meaningful to think of concentration if we want to understand the biological significance of the pollution.

The idea of compartment analysis is that the net rate of increase (or decrease) of a quantity is the sum of the various processes that work to increase or decrease that quantity. In other words, we have

$$\begin{array}{c} \text{net rate of} \\ \text{increase} \end{array} = \begin{array}{c} \text{input} \\ \text{rate} \end{array} - \begin{array}{c} \text{output} \\ \text{rate} \end{array} + \begin{array}{c} \text{production} \\ \text{rate} \end{array} - \begin{array}{c} \text{decay} \\ \text{rate} \end{array} ,$$

$$\tag{2.5.10}$$

where each of the rates is expressed as mass of pollutant per unit time. Lakes are continually fed by some streams and drained by others, so we can expect to have inflow and outflow of water. If the inflow is polluted, then the rate equation will include an input rate of pollutant, which we will have to express in terms of basic quantities. The outflow will, of course, carry some of the pollutant along with the water. There is probably no mechanism for production of pollutants within the lake, but there may be decay because of chemical instability or a chemical reaction with something else in the water. The term "decay" should be broadly interpreted as any process that removes pollutant from the lake other than flow with the water, which could include processes whereby the pollutant is absorbed onto rocks or sinks into the sediment. The exact mechanism is unimportant as long as we can make a simple assumption about the rates of the processes collected together as pollutant decay.

Real lakes are complicated, with multiple input streams that vary in flow rates as the seasons change and the weather varies between dry and rainy. The concentration of pollutant is higher in areas near polluted inflow streams and lower in areas near clean inflow streams, so C depends on location as well as time. There may be multiple decay processes, some involving complicated chemical reactions. In some circumstances, we may want to account for all of these details, but our model will be of more general value if we keep it simple. We therefore adopt a simple conceptual model. Suppose we have a tank in a laboratory playing the role of the lake, and we imagine the simplest experiment we can do that might be similar to what happens in a real lake. Our conceptual model includes the following assumptions:

1. The water in the tank is stirred so that there are no spatial variations in concentration.
2. Any decay processes are proportional to the amount present, as in radioactive decay. This assumption requires a proportionality constant, which we will call R.
3. The input and output flow rates are constant in time and equal so that the volume in the tank is constant. We take V to be the volume and F to be the flow rate (volume/time).
4. The input stream has a fixed pollutant concentration C_{in} (mass/volume).

The net rate of increase is simply dQ/dT, where we are using T for time to facilitate some additional analysis that will appear in Section 2.6. Since Q is a mass, we know by the requirement of dimensional consistency that each of the other terms in the equation must have the dimension mass/time. This observation helps us quantify the rates of the three nonzero processes on the right side of the equation. The decay rate, by assumption 2, is RQ, with R having the dimension of 1/time. The input rate of pollutant should be larger for larger water flow rates as well as for larger concentrations. The product FC_{in} has a dimension of (volume/time)*(mass/volume) = (mass/time), so this quantity has the right dimension to be the input rate. Similarly, the output rate is $FC(T)$, where the assumption of adequate mixing justifies equating the concentration of the pollutant in the output stream with that in the lake. Combining all of these formulas, we have the equation

$$\frac{dQ}{dT} = FC_{in} - FC(T) - RQ(T) .$$

This equation is not quite what we need, because it contains both of the related quantities Q and C. Given that C is mass of pollutant per volume, these quantities are related by the equation $C(T) = Q(T)/V$. We can use this to eliminate either of the variables; given that the concentration is of greater biological importance, we replace Q by VC and divide by V to get

$$\frac{dC}{dT} = \frac{F}{V}C_{in} - \frac{F}{V}C(T) - RC(T) . \tag{2.5.11}$$

We will do some preliminary analysis with this model in Problem 2.6.8.

2.5.6 *"Let the Buyer Beware"*

As noted in Section 2.2, all mathematical models come with a disclaimer. Results obtained from their analysis are only guaranteed to be true for the corresponding conceptual models. Whether they are also true for the real-world setting depends on the quality of the approximation process, which is non-mathematical. Each of the models developed in this section requires validation to justify application to a real problem. The radioactive decay model has been validated innumerable times and is accurate for any macroscopic quantity to within any degree of accuracy that has been achieved by any experiment. Likewise, Figure 2.5.1 serves as a qualitative validation of the Holling type II predation model for the BUGBOX data, though similar data is needed to validate the model for a real biological setting. The pollutant model (2.5.11) also requires experimental validation. It would be easy to validate it for a laboratory experiment, but that would not guarantee its validity for use as a model of pollutants in a real lake. Most likely, historical data collected to monitor improvement in polluted lakes would serve for validation, but this would depend on the extent to which the assumptions in the conceptual model are actually true for that lake. A lake that is poorly mixed and has highly variable pollutant inflow and water flow rates will behave less like the model than one that has steady currents and nearly constant pollutant inflow and water flow rates.

Problems

2.5.1. Two possible models for the dynamics of a renewable resource (biotic or abiotic) are

$$\frac{dx}{dt} = 0.1 - \frac{xy}{1+x} \quad \text{and} \quad \frac{dx}{dt} = 0.1x - \frac{xy}{1+x},$$

where $x(t)$ is the amount of resource present at time t and y is the number of consumers.

(a) For each of these models, describe a mechanism that accounts for the growth of the resource in a way that is consistent with the model.
(b) Explain the assumption the models make about the consumers.

2.5.2. The populations $x(t)$ and $y(t)$ of two interacting species are modeled using the equations

$$\frac{dx}{dt} = ax + bxy, \qquad \frac{dy}{dt} = cx + dxy,$$

where a, b, c, and d are parameters, not necessarily positive.

(a) Suppose the species are herbivores that compete for a common plant resource. Which of the four parameters should be positive and which negative? Explain.
(b) Repeat (a) for the case where x is an herbivore and y is a predator that eats the herbivore.

Problems 2.5.3–2.5.8 are based on models for a chemostat and an SIR disease.

- A *chemostat* is a device that grows bacteria in a steady environment by continuously adding fresh nutrient solution at rate Q and removing the whole mixture at the same rate. One possible model for a chemostat is given by the equation

$$\frac{dN}{dT} = RN\left(1 - \frac{N}{K}\right) - QN, \qquad R,K,Q > 0,$$

where N is the population of bacteria at time T, R is the maximum bacterial growth rate, K is the maximum population that the chemostat can support, and Q is the flow rate of the mixture.

- The *SIR disease model* tries to predict the sizes of three subgroups—Susceptible, Infective, and Recovered—of a population of constant size N subjected to a disease pathogen. The model is

$$\frac{dS}{dT} = -pBSI, \qquad \frac{dI}{dT} = pBSI - KI, \frac{dR}{dT} = KI, \qquad B, K > 0, \quad 0 < p \leq 1,$$

where S, I, and R are the sizes of the three subgroup populations at time T, B is a parameter that quantifies the rate at which encounters among population members occurs, p is the probability of transmission in an encounter between an infected and a susceptible individual, and K is the recovery rate parameter.

2.5.3. Explain the chemostat model by comparing it with the exponential growth model

$$\frac{dN}{dT} = RN .$$

Specifically,

(a) What is the effect of the extra factor $1 - N/K$ in the growth term?
(b) What physical process in the chemostat is represented by the term $-QN$?
(c) What assumption does the algebraic form $-QN$ make about the process it describes?
(d) Fluid flows out of the chemostat at the rate Q, but the model does not include the volume as a dependent variable. Why not?

2.5.4. Explain the SIR model. Specifically,

(a) Why is the change in S proportional to both S and I?
(b) Why do the terms $pBSI$ and KI appear in two different equations, and why are they positive in one instance and negative in the other?
(c) The model

$$\frac{dR}{dT} = KR$$

represents exponential growth of R. Why does

$$\frac{dR}{dT} = KI$$

not represent exponential growth?
(d) Does the model clearly limit how large R can be? What do you expect will happen to I after a long time?

2.5.5. Consider a conceptual model similar to the chemostat model, but with two differences:

1. There is no fluid flow.
2. There is a predator that feeds on the bacteria. Assume that the population of predators is fixed at P and that the rate of predation per predator is given by the Holling type II model:

$$y(N) = \frac{sN}{1 + shN}, \qquad s, h > 0 .$$

Write down the mathematical model that corresponds to this conceptual model.

2.5.6.* Suppose scientists discover that immunity to a disease is lost at a rate proportional to the recovered population with rate constant Q. Rewrite the SIR model equations to include this additional mechanistic process.

2.5.7. Suppose the original SIR model is changed by including vaccination of susceptibles at a rate VS, where V is the vaccination rate constant. Rewrite the model equations for this revised conceptual model.

2.5.8. For the chemostat model:

(a) Explain how we know that the dimension of K is the same as the dimension of N.
(b) Explain how we know that the dimension of R is 1/time.
(c) Determine the dimension of Q.

2.6 Mechanistic Modeling II: Equivalent Forms

After studying this section, you should be able to:

- Identify equivalent forms of a mathematical model.
- Convert a mathematical model to dimensionless form, given appropriate choices for reference quantities.
- Explain why equivalent forms do not always produce equivalent parameter values when fit to a data set.

In physics, almost everyone chooses the same way to write Newton's Second Law of Motion: $F = ma$. The model could be written in other ways, but this form has become standard. This uniformity makes it easy for a reader to compare material written by different authors. In contrast, uniformity of model appearance is rare in biology. Different authors inevitably choose different ways to write the same model. Hence, the reader who wants to understand mathematical models in biology needs to develop the skill of discriminating between different models and different forms of the same model. This is a skill that mathematicians take for granted, but it is problematic for most students. In this section, we consider different ways of writing the same model, working our way up from forms that differ only in notation to forms with more substantive differences.

2.6.1 Notation

Any given model can be written with a variety of notations. Sometimes these differences are due to the different settings in which the model can arise; other times, they are due simply to lack of standardization.

Example 2.6.1. Many important biochemical reactions are of the Michaelis–Menten type. The initial rate of reaction depends on the concentration of the principal reactant, called the substrate. In Chapter 7, we derive the well-known Briggs–Haldane model that relates the rate of reaction and substrate concentration. This model is generally written as

$$v_0 = \frac{v_{\max}[S]}{K_M + [S]}, \qquad [S], v_{max}, K_M > 0,$$

where v_0 is the initial rate of the chemical reaction, $[S]$ is the concentration of one of the chemical reactants, v_{max} is the initial rate of chemical reaction for infinite $[S]$, and K_M is called the semisaturation parameter.

The Monod growth function in microbiology is used to model the rate of nutrient uptake by a microorganism as a function of the concentration of nutrients in the environment. The Monod model, in one of several common notations, is

$$r = \frac{qS}{A+S}, \qquad S, q, A > 0,$$

where r is the rate of nutrient uptake, S is the concentration of the nutrient, q is the maximum uptake rate, and A is the semisaturation parameter.

The Briggs–Haldane and Monod models were designed for different contexts. Although the interpretation of each model is based on its own context, the models are mathematically identical, differing only in notation.   □

The notation for the Briggs–Haldane equation is now standard, while the Monod growth function has almost as many systems of notation as the number of authors who have written about it. Although the two formulas use different symbols, they are identical in mathematical form. *Both formulas say that the dependent variable is a rational function of the independent variable, with the numerator of the form "parameter times independent variable" and the denominator of the form "parameter plus independent variable."* If we focus on the *symbols*, we are misled by the apparent differences. If we focus on the *roles* of the symbols, we see that the formulas are identical. Few symbols in mathematics have a fixed meaning. Even π, which is the universal symbol for the ratio of circle circumference to circle diameter, is occasionally used for other purposes. This lack of standardization makes it necessary to provide the meaning of each symbol for any given context.

2.6.2 Algebraic Equivalence

Occasionally, it is possible to write the same model in two different ways by performing algebraic operations. This is most common with formulas that include a logarithmic or exponential function.

Example 2.6.2. In fitting an exponential model by linearization in Section 2.4, we converted the model

$$z = Ae^{-kt}$$

to the form

$$\ln z = \ln A - kt .$$

These forms are equivalent in the sense that they both produce the same z values for any given t value.   □

Example 2.6.3. A certain problem in the life history of plants eventually reduces to the algebraic equation

$$Pe^{-P} = e^{-R} ,$$

where R is the independent variable and P is the dependent variable. Hence, the formula defines P implicitly in terms of R. Various algebraic operations can be performed to change this equation. For example, we can multiply by e^P and e^R to obtain the equation

$$Pe^R = e^P .$$

We can then take the natural logarithm on both sides, obtaining

$$\ln P + R = P .$$

Further rearrangement yields

$$P - \ln P = R .$$

All of these forms are algebraically equivalent, although they appear different at first glance. □

None of the formulas in Example 2.6.3 allows us to solve algebraically for P. The choice among them is a matter of taste. I prefer the last one, because that form allows us to graph the relationship by calculating values of R for given values of P.

2.6.3 Different Parameters

More subtle than differences in algebraic presentation are differences resulting from parameter choices. Sometimes it requires some algebraic manipulation to see that two similar models are actually equivalent.

Example 2.6.4. In Section 2.5, we developed the Holling type II model

$$y(x) = \frac{sx}{1 + hsx} , \qquad x, s, h > 0 , \tag{2.6.1}$$

for the relationship between the number of prey animals eaten by a predator in a unit amount of time (y) and the number of prey animals available in a specific region (x), where s is the amount of habitat that a predator can search per unit time and h is the time required for the predator to process one prey animal. The same model can also be written as

$$y(x) = \frac{qx}{a + x} , \qquad x, q, a > 0 , \tag{2.6.2}$$

where the parameters q and a are different from the original s and h. The *functions* in (2.6.1) and (2.6.2) are mathematically different, so the models are not *identical*. However, the *models* represented by the functions are *equivalent* if we define the parameters in (2.6.2) by $q = 1/h$ and $a = 1/(hs)$. Note that the Holling type II model is also equivalent to the Briggs–Haldane and Monod models, with yet another context. □

> **Check Your Understanding 2.6.1:**
> Substitute $q = 1/h$ and $a = 1/(hs)$ into (2.6.2) and simplify the result to obtain (2.6.1).

Each of the equivalent forms (2.6.1) and (2.6.2) is preferable from a particular point of view. Form (2.6.2) is preferable from a graphical point of view, because the parameters q and a indicate properties directly visible on the graph (see Problem 1.1.10). It has the additional advantage of being semilinear (as defined in Section 2.4), so we can fit it to empirical data by the semilinear least squares method.[34] Form (2.6.1) is better from a biological point of view, because it allows us to study the effects of search speed and maximum consumption rate separately.

---

[34] See Example 2.7.2.

2.6.4 *Visualizing Models with Graphs*

The appearance of a graph depends on the ranges chosen for the variables. In making a choice, it is important to have a purpose in mind. Consider three instances of the predation model (2.6.2):

$$y_1 = \frac{2x}{0.5+x}, \qquad y_2 = \frac{2x}{2+x}, \qquad y_3 = \frac{0.5x}{2+x}.$$

If our purpose is to see what effect the different choices of q and a have on the graph, we should plot the three models together, as in Figure 2.6.1a. But suppose instead that we want to plot each curve separately in a way that best shows the behavior of the function. The ranges chosen for Figure 2.6.1a look best for y_2, so we plot this curve with the same ranges (Figure 2.6.1d).

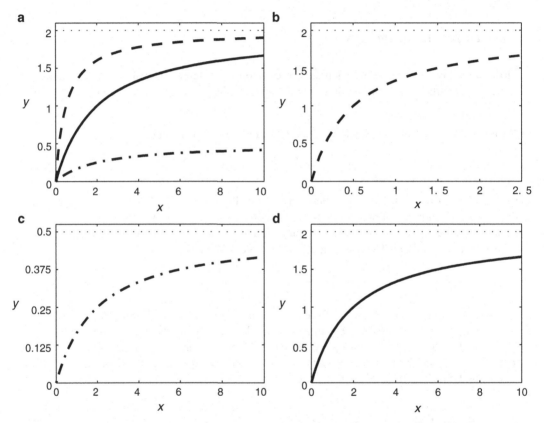

Fig. 2.6.1 The model $y(x) = qx/(a+x)$, with (q,a) values of $(2,0.5)$ (*dashed*), $(2,2)$ (*solid*), $(0.5,2)$ (*dash-dotted*)

The function y_3 has a much lower maximum value than y_2, so we might choose a narrower range for the vertical axis (Figure 2.6.1c). On the other hand, most of the variation in y_1 occurs on the left side of Figure 2.6.1a, so we might choose a narrower range for the horizontal axis (Figure 2.6.1b). Notice that the three individual graphs are now identical, except for the numbers on the axes.

2.6.5 Dimensionless Variables

Look again at Figure 2.6.1. Notice that the vertical axis range is $0 \le y \le q$ for each of the single-curve plots. Similarly, the horizontal axis range for each of the curves is $0 \le x \le 5a$. We can make use of this observation to produce one plot with axis values that are correct for these three cases, and all others as well. One way is to incorporate the parameters q and a into the axis values (Figure 2.6.2a). Alternatively, we could incorporate the parameter values into the variables themselves, as in Figure 2.6.2b.

The quantities y/q and x/a are dimensionless versions of the original quantities y and x. This means that they represent the same things (food consumption rate and food density, respectively), but are measured differently. Where y is the food consumption rate measured in

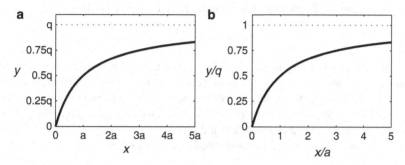

Fig. 2.6.2 The model $y(x) = qx/(a+x)$, using two different labeling schemes: (**a**) the factors a and q are in the axis values; (**b**) the factors a and q are in the axis labels

terms of some convenient but arbitrary unit of measurement (prey animals per week or grams per day, for example), y/q is the food consumption rate measured as a fraction of the maximum food consumption rate q. Where x is the nutrient concentration in a convenient but arbitrary unit, x/a is the nutrient concentration relative to the semisaturation concentration a.

2.6.6 Dimensionless Forms

The quantities y/q and x/a have an algebraic benefit as well as a graphical benefit. To see this benefit, we define dimensionless variables Y and X by[35]

$$Y = \frac{y}{q}, \qquad X = \frac{x}{a}. \tag{2.6.3}$$

Rearranging these definitions yields substitution formulas:

$$y = qY, \qquad x = aX. \tag{2.6.4}$$

Now we replace y and x in (2.6.2) using the formulas of (2.6.4). This yields the equation

---

[35] In this text, we will usually adopt the practice of using one case for all of the original variables in a model and the opposite case for the corresponding dimensionless variables. Other systems, such as those that add accent marks for either the dimensional or the dimensionless quantities, have greater flexibility but other disadvantages. Distinguishing by case has the advantage of easy identification of corresponding quantities without the clumsiness of accent marks.

$$qY = \frac{qaX}{a + aX} . \qquad (2.6.5)$$

Removing common factors yields the dimensionless form of the predation model:

$$Y = \frac{X}{1 + X} . \qquad (2.6.6)$$

The graph of the dimensionless model is the same as Figure 2.6.2b, except that the axis labels are X and Y instead of x/a and y/q. The dimensionless variables measure quantities in terms of units that are intrinsic to the physical setting. The statement $y = 0.5$ cannot be understood without a unit of measurement or a value of q for comparison. The statement $Y = 0.5$ can immediately be understood as a predation rate that is half of the maximum predation rate.

Example 2.6.5. Consider the energy budget model from Example 1.5.1:

$$q = ax^2 - bx^3 , \qquad (2.6.7)$$

where x is the length of an organism, q is the surplus energy available for growth or repro-duction, and a and b are parameters. Suppose we want to obtain a dimensionless form for this model. One way to do this is to identify quantities made from the parameters that have the same dimensions as x and q. We would first have to find the dimensions of a and b using the rules for dimensional consistency introduced in Section 2.5.

An alternative method is to assume a substitution formula of the form

dimensional variable = dimensional reference quantity × dimensionless variable

using an unknown dimensional reference quantity, which can be chosen later. With this in mind, we write the substitution formulas

$$q = q_r Q, \qquad x = x_r X ,$$

where Q and X are to be the dimensionless forms of q and x, and q_r and x_r are the yet-to-be-determined reference quantities. With these substitutions, the model becomes

$$q_r Q = ax_r^2 X^2 - bx_r^3 X^3 .$$

The three terms in this equation must all have the same dimension. Since Q and X are dimen-sionless, this means that q_r, ax_r^2, and bx_r^3 all have the same dimension as q. At this point, the choices of q_r and x_r are free, so we can either choose them directly or we can specify two equa-tions that they must satisfy. The simplest option is to make all three coefficients equal. Thus, we have

$$q_r = ax_r^2 = bx_r^3 .$$

From these equations, we get the results

$$x_r = \frac{a}{b} , \qquad q_r = \frac{a^3}{b^2} . \qquad (2.6.8)$$

Given these choices, the final dimensionless model is

$$Q = X^2 - X^3 . \qquad (2.6.9)$$

$\square$

Differential equation models can also be nondimensionalized. This requires substitution formulas for the independent variable of derivatives. It is easiest to write these substitution formulas in terms of the differentials.

Example 2.6.6. Consider the differential equation

$$\frac{dy}{dt} = -ky \, .$$

We can define new variables Y and T by

$$Y = \frac{y}{y_0} \, , \qquad T = kt \, .$$

These formulas corresponds to the differential substitution formulas $dy = y_0 \, dY$ and $dT = k \, dt$, so we have the substitution formulas

$$\frac{d}{dt} = k\frac{d}{dT} \, , \qquad \frac{dy}{dt} = ky_0\frac{dY}{dT} \, .$$

The final result is

$$\frac{dY}{dT} = -Y \, . \tag{2.6.10}$$

□

Dimensionless models are not always parameter-free like (2.6.6), (2.6.9), and (2.6.10), but they always have fewer parameters than the original dimensional model. That alone makes them preferable for model characterization. Sometimes making a model dimensionless provides valuable insight into the behavior of the model, even before any analysis is performed.[36]

Problems

2.6.1. Derive (2.6.2) from (2.6.1). In so doing, find the algebraic formulas that define h and s in terms of the parameters q and a.

2.6.2. Explain the graphical significance of the parameter a in the model

$$y = \frac{qx}{a+x} \, .$$

[Hint: What is y if you set $x = a$?]

2.6.3.* Rewrite the model $y = y_0 e^{-kt}$ in dimensionless form by choosing appropriate dimensionless quantities to replace y and t.

2.6.4. Create a set of four plots similar to Figure 2.6.1 using the functions

$$y_1 = \frac{4x}{2+x} \, , \qquad y_2 = \frac{2x}{2+x} \, , \qquad y_3 = \frac{2x}{4+x} \, .$$

Your plots of the individual functions should be identical to those of Figure 2.6.1, except for the numerical values on the axes.

---

[36] See Chapter 7.

2.6.5. (a) Redo the nondimensionalization of the energy budget model $q = ax^2 - x^3$ using as reference quantities the value x^* that maximizes q and the corresponding maximum value $q^* = q(x^*)$.

(b) Create a plot of the model $q = ax^2 - x^3$ similar to Figure 2.6.2, using the same reference quantities as part (a).

2.6.6. (Continued from Problem 1.5.3.)
Redo Problem 1.5.3, but nondimensionalize the function in part (c) before solving the optimization problem in part (d).

2.6.7. (Continued from Problem 1.6.7.)
Repeat Problem 1.6.7, but without assuming $r = 1$. Instead, replace the quantities $x(t)$, $z(t)$, and s using $x = YX$, $z = YZ$, $t = T/r$, and $s = rS$ and determine the optimal jumping point X^* in terms of the dimensionless swim speed S. Note that the key equations for the model are $z^2 = x^2 + Y^2$ and $dx/dt = r$, which are used to determine the rate of change of z with respect to t while running, and $dz/dt = s$, which defines the rate of change of z with respect to t while swimming.

2.6.8.* In Section 2.5, we derived the model

$$\frac{dC}{dT} = \frac{F}{V}C_{\text{in}} - \frac{F}{V}C(T) - RC(T)$$

for the concentration $C(T)$ of pollutant in a lake of volume V, where F is the flow rate of water in and out of the lake and R is the decay rate of the pollutant. Nondimensionalize the differential equation using C_{in} for the reference concentration and V/F for the reference time. Your model will contain one dimensionless parameter. Explain the biological meaning of that parameter.

Problems 2.6.9–2.6.11 use the chemostat and SIR models that were introduced in the Section 2.5 problem set.

2.6.9. (a) Dimensionless variables for the chemostat model can be defined by

$$n = \frac{N}{K}, \qquad t = RT .$$

Use these definitions to write appropriate substitution formulas to replace N and T in the model.

(b) Obtain a dimensionless model by making the substitutions for N and T. Note that you will have to divide through by RK so that the left side of the model will be dn/dt. The last term of the equation should include a dimensionless parameter q, which you will need to define.

2.6.10. (a) Determine the dimension of B and the dimension of K in the SIR model. Show that the quantities

$$s = \frac{S}{N}, \qquad i = \frac{I}{N}, \qquad t = KT$$

are dimensionless. (Note that p is dimensionless.)

(b) Use the definitions of s, i, and t to obtain substitution formulas for S, I, and d/dT.

(c) Use the substitution formulas of part (b) to derive a dimensionless model for the variables s and i. (You need not consider the R equation, as the first two form a self-contained model and $R = N - S - I$.) Your model should include a single dimensionless parameter, b, which represents the rate of spread of the infection.

2.6.11. An SI epidemic model is similar to an SIR epidemic model, except that recovered individuals are assumed to be susceptible again. The model can be written as

$$\frac{dS}{dT} = -pBSI + KI, \qquad \frac{dI}{dT} = pBSI - KI, B, K > 0, \quad 0 < p \le 1 ,$$

where S and I are the sizes of the subgroups at time T, B is the encounter rate parameter, p is the transmission probability, and K is the recovery rate parameter.

(a) Let $N = S + I$. Show that $dN/dT = 0$.
(b) Use the fact that N is constant to eliminate the variable S from the I equation. The result is a model consisting of a single equation for I with four parameters (p, B, K, and N).
(c) Determine the dimension of B and the dimension of K. Show that the quantities

$$i = \frac{I}{N} , \qquad t = KT$$

are dimensionless. (Note that p is dimensionless.)
(d) Use the definitions of i and t to obtain substitution formulas for I and d/dT.
(e) Use the substitution formulas of part (c) to derive a dimensionless model for the variable i. Your model should include a single dimensionless parameter, b, which represents the rate of spread of the infection.

2.7 Empirical Modeling III: Choosing Among Models

After studying this section, you should be able to:

- Use the Akaike information criterion to compare the statistical validity of models as applied to a given set of data.
- Choose among models based on a variety of criteria.

In this section, we consider the issue of how to choose from among several different models for a given setting. There are three key criteria: quantitative accuracy, complexity, and the availability of a mechanistic model.

2.7.1 Quantitative Accuracy

Quantitative accuracy is clearly an important criterion for choosing among possible models. It can easily be measured by the residual sum of squares, but comparisons can only be made when the residual sums of squares are measured on the same graph.

Example 2.7.1. In Examples 2.4.2 and 2.4.4, we fit a power function model using linear least squares on the linearized data and semilinear least squares on the original data. Each set of parameter values wins on its "home field"; the semilinear result is clearly better when viewed on a plot of y versus x, while the linearized result is clearly better when viewed on a plot of $\ln y$ versus $\ln x$. The question of better fit in this case is not settled mathematically, but by the choice of which graph to use to measure errors. Unless we have a scientific reason for considering a graph of logarithms to be more meaningful than a graph of original values, we should use the semilinear method.[37]   ☐

Example 2.7.2. Using the semilinear least squares method to fit the Holling type II model

$$y = \frac{qx}{a+x}$$

to the *P. speedius* data from Table 2.1.1, we obtain

$$a = 36.4 , \qquad q = 36.3 , \qquad \text{RSS} = 43.2 . \tag{2.7.1}$$

Previously we fit the power function model $y = Ax^p$ to the same data, with the result RSS = 62.9.[38] The Holling type II model has a significantly smaller residual sum of squares than the power function model. Quantitative accuracy favors the Holling model; other criteria will be considered below.   ☐

2.7.2 Complexity

Before we conclude that model selection is usually a simple matter, we turn to the question "Shouldn't we *always* use the model that has the smallest residual sum of squares?" The answer to this question is an emphatic *NO*, as the following example makes clear.

Table 2.7.1 A small partial data set for consumption rate y as a function of prey density x

x	0	20	40	60	80
y	0	7	9	21	25

Table 2.7.2 A larger partial data set for consumption rate y as a function of prey density x

x	0	10	20	30	40	50	60	70	80
y	0	2	7	10	9	14	21	20	25

Example 2.7.3. Suppose we collected the *P. steadius* data from Section 2.1 in stages. The initial data set, in Table 2.7.1, consists of every other point from the first part of the full data set. We could fit the models $y = mx$ and $y = b + mx$ to this data set, and we would find that the second model yields a smaller residual sum of squares than the first model. In fact, it is possible to show that adding a parameter to a model *always* decreases the residual sum of squares until it falls to 0. So if our goal is minimum residual sum of squares, we can just add more parameters. A fourth degree polynomial has five parameters, which means that we can find one whose graph

---

[37] For models that have more than one nonlinear parameter, one can use a fully nonlinear method. These can be found in any mathematical software package, such as R or Maple. Other packages, such as spreadsheets, that fit exponential or power function models to data use the linearization method.

[38] See Example 2.4.4.

passes through all five data points, giving us a residual sum of squares of 0. The result is the model

$$y \approx 1.1375x - 0.0628x^2 + 0.00134x^3 - 0.00000859x^4.$$

On the basis of quantitative accuracy for the data points used to obtain the parameters, this model is perfect. However, the graph of this model shows a problem. Figure 2.7.1a shows the five points used for the fit, the fourth degree polynomial that passes through the points, and the least squares line $y = 0.313x$. The sharp curves in the polynomial are suspicious.

Table 2.7.2 includes the full set of data. A useful model should remain reasonably good when more data is added; however, Figure 2.7.1b shows that the fourth degree polynomial has very little predictive value for the extra data in this case. If we fit a fourth degree polynomial to the larger set of data, the graph of the best fit will be significantly changed. Meanwhile, the model $y = 0.313x$ looks reasonably good with the larger set of data. □

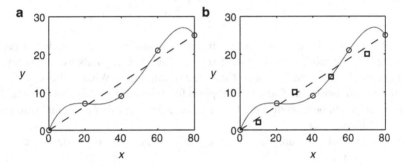

Fig. 2.7.1 The linear and fourth degree polynomial models fit to the data of Table 2.7.1, with the additional data of Table 2.7.2 included in (**b**)

Example 2.7.3 shows that there is such a thing as too much accuracy in fitting a data set. Adding parameters to a model increases the accuracy by allowing the model graph to conform more closely to the individual data points. However, real data has measurement uncertainty as well as the demographic stochasticity discussed in Section 2.1. This means that bending the graph to make it conform to the data can be overdone, resulting in a loss of predictive power. We can use this discovery to formulate a general principle of model selection[39]:

> *Additional parameters should not be added to models unless the increased accuracy justifies the increase in complexity.*

Note that we are not arguing that simpler models are better than complex models for philosophical or aesthetic reasons. We are saying that greater complexity can sometimes lead to an undesirable loss of predictive power.

---

[39] This statement is a mathematical version of *Occam's razor*, a well-known scientific principle attributed to the fourteenth-century philosopher William of Ockham, although not found in his extant writings and actually appearing in some form before Ockham. The most common form was written by John Punch in 1639 and translates literally from the Latin original as "Entities must not be multiplied beyond necessity." My interpretation is much more in keeping with the actual statement than its more common renderings in English.

2.7.3 The Akaike Information Criterion

We have now established the idea that quantitative accuracy needs to be balanced against complexity. We can quantify accuracy using the residual sum of squares and complexity using the number of parameters. The challenge is to combine these measures in a meaningful way. As an example, consider the *P. steadius* data set, which we fit in Section 2.1 using the linear models $y = mx$ and $y = b + mx$, with residual sums of squares 106.1 and 100.4, respectively. Is the improvement in accuracy for the model $y = b + mx$ worth the added complexity?

The issue of balancing accuracy and complexity was addressed by Hirotugu Akaike using information theory and statistics in a 1974 paper, which defined what has come to be known as the Akaike information criterion, or AIC [1]. An alternative measure, the corrected Akaike information criterion (AICc), is also in common use. Based on recent work of Shane Richards [12], I recommend using the original version, which we can write as

$$\text{AIC} = n \ln\left(\frac{\text{RSS}}{n}\right) + 2(k+1), \tag{2.7.2}$$

where RSS is the residual sum of squares for the specific model, k is the number of parameters that are in the specific model, and n is the fixed number of data points used to determine the model parameters.[40] The actual values of AIC are unimportant. What matters is that smaller values represent a higher level of statistical support for the corresponding model. A comparison can only be made when both models are fit to the same data set using a residual sum of squares defined in the same manner.

Armed with the AIC, we are now ready to address the issue of model selection for the *P. steadius* predation data.

Example 2.7.4. In Section 2.3, we obtained two results from linear least squares for the full *P. steadius* data set[41]:

$$y = 0.267x,$$

$$y = 1.175 + 0.255x.$$

As an additional empirical option, we can also determine (using methods beyond the scope of this section) the best-fit parabola to be

$$y = 0.197 + 0.300x - 0.00032x^2.$$

Finally, the best fit for the Holling Type II model is

$$y = \frac{196x}{628 + x}.$$

Table 2.7.3 summarizes the results and Figure 2.7.2 displays the data and the various models. The model $y = mx$ has the smallest AIC value. The very slight improvements in accuracy for the other models do not quite compensate for the additional complexity.   □

---

[40] The usual formula for AIC ends with $2K$ rather than $2(k+1)$, where K is the number of parameters that have to be fit using statistics. This includes the statistical variance along with the k model parameters. For those readers not well versed in statistics, it is easier to count the number of model parameters than the number of statistical parameters, so our version builds the extra parameter into the formula.

[41] See Table 2.1.1.

Table 2.7.3 Comparison of three models for the *P. steadius* data of Table 2.1.1 using AIC

Model	RSS	n	k	$n \ln \left(\dfrac{\text{RSS}}{n} \right)$	$2(k+1)$	AIC
$y = mx$	106.1	15	1	29.3	4	33.3
$y = b + mx$	100.4	15	2	28.5	6	34.5
$y = b + mx + ax^2$	96.1	15	3	27.9	8	35.9
$y = qx/(a+x)$	95.5	15	2	27.8	6	33.8

2.7.4 Choosing Among Models

Choosing a model is a matter of informed judgment. The best choice using empirical criteria is determined by the AIC; however, AIC results are only as good as the data set being fit. When two models have very close AIC scores with a given data set, there is a possibility that the rankings would be reversed with a different data set. It is reasonable to ask how much of a difference in AIC should be considered definitive. We must consider two separate cases.

A pair of models is said to be *nested* if one can be obtained from the other merely by setting one of its parameters to zero. For example, the pair $y = mx$ and $y = b + mx$ is nested, as is the pair $y = mx$ and $y = qx/(a+x)$; however, the pair $y = b + mx$ and $y = qx/(a+x)$ is not nested. For a nested pair, the best the simpler model can do is to achieve the same residual sum of squares. In this case, the complexity penalty gives the simpler model an AIC advantage of 2. Thus, a difference of 2 must be considered significant for a nested pair. As a rule of thumb, an AIC difference of 1 in a nested pair should probably be taken as definitive.

In contrast, it is harder to be confident in the significance of AIC differences for models that are not a nested pair. A difference of 6 is generally taken to be definitive, but this seems far too conservative, as a difference of 6 is achieved when two models have the same residual

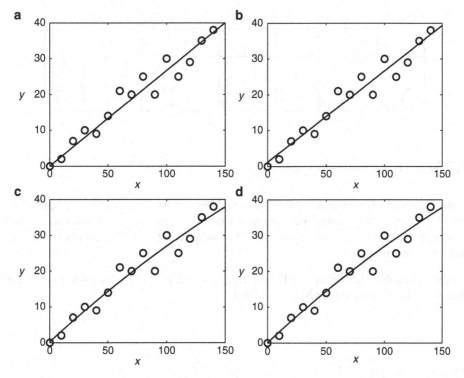

Fig. 2.7.2 The data and models of Example 2.7.4: (**a**) $y = mx$, (**b**) linear, (**c**) quadratic, (**d**) Holling type II

sum of squares but one has three more parameters than the other. It seems intuitively clear that the model with three fewer parameters is the better choice, pending new information such as additional data. In practice, a difference of 2 is probably enough in most cases. It is hard to argue on empirical grounds that a model with an extra parameter that fails to achieve a lower residual sum of squares than a simpler model can be the best choice, even when the models are not nested.

AIC has the advantage of being quantitative, but we should not overestimate its value. Non-quantitative criteria must often be considered. We should usually prefer models that have a mechanistic justification, even if such models have AIC values that are not significantly better than the best empirical model.

Example 2.7.5. Q) Which model should we use for *P. steadius*?

A) In Example 2.7.4, we found the lowest AIC score for the model $y = mx$, but with only a slightly higher score for the Holling type II model. A different data set for the same experiment could yield a lower AIC score for the Holling model, or the difference could be greater than in the example. Both models have a mechanistic justification, which the other two models lack. Either is arguably the best choice. In practice, it is generally wise not to use a more complicated model than necessary, so the simple $y = mx$ is probably the better choice.   □

Example 2.7.6. Q) Which model should we use for *P. speedius*?

A) We have already obtained[42] best fits for the power function and Holling type II models. We can also try polynomials of various degrees. Table 2.7.4 summarizes the results and Figure 2.7.3 illustrates the four models with the original data. According to the AIC values, the Holling type II model is the best. The cubic curve scores close to the Holling type II; however, the shape of the cubic curve is clearly not quite right, nor is it supported by a mechanistic justification. Note how the curve turns sharply upward at the right end of the interval.   □

Table 2.7.4 Comparison of four models for the *P. speedius* data of Table 2.1.1 using AIC

Model	RSS	n	k	$n \ln\left(\dfrac{RSS}{n}\right)$	$2(k+1)$	AIC
$y = ax^p$	62.9	15	2	21.5	6	27.5
$y = qx/(a+x)$	43.2	15	2	15.9	6	21.9
$y = b + mx + ax^2$	88.1	15	3	26.6	8	34.6
$y = b + mx + ax^2 + cx^3$	38.5	15	4	14.1	10	24.1

2.7.5 Some Recommendations

The Akaike information criterion should be thought of as a theoretical tool for empirical modeling. When we want to determine calculated values to represent theoretical data for an experiment, we should usually choose the model with the smallest AIC. Nevertheless, there are pitfalls that we must avoid:

• Empirical models that fit data well over a given range may not work over a larger range. Empirical models should not be extrapolated.

---

[42] See Examples 2.4.4 and 2.7.3.

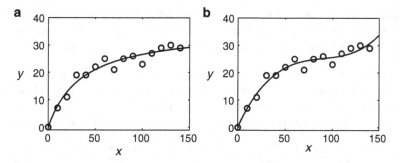

Fig. 2.7.3 The data and the two best models of Example 2.7.6: (**a**) Holling type II, (**b**) cubic

- The ranking of models for a set of data depends on the location of the points on a graph, which can be expected to vary because of random factors. If we collect a new set of data from the same experiment, the model selected by AIC for the first data set may not be the best fit for the second set.
- Part of the amazing success of models in the physical sciences owes to their mechanistic derivation from physical principles. Empirical models, by definition, do not attempt to do this. We can have more confidence in a model that can be obtained mechanistically than in one that is strictly empirical.

For theoretical work, we should always prefer a mechanistic model, even if its AIC value for a given data set is a little larger than that of a strictly empirical model. Of course, we must also be careful with mechanistic models. If the data from experiments fits an empirical model significantly better than it fits a mechanistic model, then either the experimental data is severely flawed or the mechanistic model is not valid for that experimental setting.

Problems

2.7.1.(a) Compute the residual sum of squares for the model

$$y \approx 1.1375x - 0.0628x^2 + 0.00134x^3 - 0.00000859x^4$$

with the full data set of Table 2.7.2.
(b) Compute the residual sum of squares for the model $y = 0.313x$ with the full data set of Table 2.7.2.
(c) Both of the models used in parts (a) and (b) were obtained by fitting to the partial data set of Table 2.7.1. Which has the better quantitative accuracy for the full data set?

2.7.2.* The points corresponding to x values of 0, 10, 20, and 30 in Table 2.7.2 seem to lie close to the least squares line $y = 0.313x$. We can find a polynomial of the form $y = ax + bx^2 + cx^3$ that passes through the points $(0,0)$, (h,y_1), $(2h,y_2)$, and $(3h,y_3)$ by the *method of successive differences*. The method yields simple formulas for the three coefficients:

$$c = \frac{y_3 - 3y_2 + 3y_1}{6h^3}, \qquad b = \frac{y_2 - 2y_1}{2h^2} - 3hc, \qquad a = \frac{y_1}{h} - hb - h^2c.$$

(a) Use these coefficient formulas to fit the data from Table 2.7.2 for the x values 10, 20, and 30.
(b) Plot the polynomial from part (a), along with the three data points, on a common graph.
(c) Does the cubic polynomial model work well for these data? Why or why not?

2.7.3. Using the *method of successive differences*, relatively simple formulas can be found to determine the coefficients of a polynomial of degree n that passes through $n+1$ points having equally spaced x values. Suppose the points are (x_1, y_1), (x_2, y_2), (x_3, y_3), and (x_4, y_4), and $x_4 - x_3 = x_3 - x_2 = x_2 - x_1 = h$. Then the polynomial $y = a + bx + cx^2 + dx^3$ has coefficients given by

$$d = \frac{y_4 - 3y_3 + 3y_2 - y_1}{6h^3}, \qquad c = \frac{y_3 - 2y_2 + y_1}{2h^2} - 3x_2 d,$$

$$b = \frac{y_2 - y_1}{h} - (x_1 + x_2)c - (3x_1x_2 + h^2)d, \qquad a = y_1 - x_1 b - x_1^2 c - x_1^3 d.$$

(a) Use these coefficient formulas to fit the data from Table 2.7.2 for the x values 10, 30, 50, and 70.
(b) Repeat part (a) using the data for the x values 20, 40, 60, and 80.
(c) Plot the polynomials from parts (a) and (b), along with all of the data, on a common graph.
(d) How good are these two models for the full data set?

2.7.4. (Continued from Problems 2.3.1 and 2.3.7.)

(a) Use your *P. steadius* data set from Problem 2.1.1 to fit the Holling type II model.
(b) Compute the AIC for the Holling type II model, the model $y = mx$ (Problem 2.3.1a), and the model $y = b + mx$ (Problem 2.3.7a).
(c) Which model gives the lowest AIC with your data? Combining these results with Example 2.7.4, which model do you recommend should be used for *P. steadius*? [Note: Comparison of AIC values is only meaningful when those values were obtained from one data set.]

2.7.5. (Continued from Problem 2.4.6.)

(a) Use your *P. speedius* data set from Problem 2.1.2 to fit the Holling type II model.
(b) Compute the AIC for the Holling type II model and the model $y = Ax^p$ (Problem 2.4.6b).
(c) Which model gives the lowest AIC with your data? Combining these results with Example 2.7.5, which model do you recommend should be used for *P. speedius*? [Note: Comparison of AIC values is only meaningful when those values were obtained from one data set.]

2.7.6.* Suppose we have a set of data points (x_i, y_i), which we would like to fit to a model $y = b + mx + ax^2$.

(a) Suppose further that we construct a new data set (X_i, Y_i) with $X_i = x_i - \bar{x}$ and $Y_i = y_i - \bar{y}$. By substituting $x = X + \bar{x}$ and $y = Y + \bar{y}$ into the model $y = b + mx + ax^2$, find the parameters B and M, in terms of $b, m, c, \bar{x}$, and $\bar{y}$, such that the model is equivalent to $Y = B + MX + aX^2$.
(b) Use the result from part (a) to obtain formulas for b and m in terms of $B, M, a, \bar{x}$, and $\bar{y}$.
(c) Set up a spreadsheet to fit the model $y = b + mx + ax^2$ to a set of data by converting the data to X and Y as in part (a), calculating B, M, and a, and then finding b and m as in part (b). The formulas for B, M, and a can be shown (using calculus) to be

$$a = \frac{\sum X^2 Y \sum X^2 - \sum XY \sum X^3}{\sum X^2 \sum X^4 - (\sum X^3)^2 - (\sum X^2)^3 / n}, \qquad M = \frac{\sum XY - a \sum X^3}{\sum X^2}, \qquad B = -\frac{a \sum X^2}{n},$$

where n is the number of data points. Test your spreadsheet by confirming the results of Example 2.7.4.

2.7.7. (Continued from Problems 2.3.4 and 2.4.3.)

(a) Use the spreadsheet from Problem 2.7.6 to fit the model $y = b + mx + ax^2$ to the data from Problem 2.3.4.
(b) Use the AIC to compare the results of part (a) with the linear model $y = b + mx$ and the results from Problem 2.4.3 for the model $y = Ax^p$. Are the AIC differences significant? Is the best model very good?

2.7.8. (Continued from Problems 2.3.5 and 2.4.4.)

(a) Use the spreadsheet from Problem 2.7.6 to fit the model $y = b + mx + ax^2$ to the data from Problem 2.3.5.
(b) Use the AIC to compare the results of parts (a) with the linear model $y = b + mx$ and the results from Problem 2.4.4 for the models $y = Ax^p$ and $y = Ae^{kt}$. Are the AIC differences significant? Is the best model very good?

2.7.9. (Continued from Problem 2.3.8.)

(a) Use the spreadsheet from Problem 2.7.6 to fit the model $y = b + mx + ax^2$ to the data from Problem 2.3.8, from 1979 to 2003.
(b) Use the AIC to compare the results with the linear model.

2.7.10. (Continued from Problem 2.3.9.)

(a) Use the spreadsheet from Problem 2.7.6 to fit the model $y = b + mx + ax^2$ to the data from Problem 2.3.9, from 1970 to the present.
(b) Use the AIC to compare the results with the linear model.

References

1. Akaike H. A new look at the statistical model identification. *IEEE Transactions on Automatic Control*, **19**: 716–723 (1974)
2. Atkins GL and IA Nimmo. A comparison of seven methods for fitting the Michaelis–Menten equation. *Biochem J.*, **149**, 775–777 (1975)
3. Atkinson, MR, JF Jackson, and RK Morton. Nicotinamide mononucleotide adenylyltransferase of pig-liver nuclei: The effects of nicotinamide mononucleotide concentration and pH on dinucleotide synthesis. *Biochem J.*, **80**, 318–323 (1980)
4. Atlantic States Marine Fisheries Commission. Atlantic Croaker 2010 Stock Assessment Report. *Southeast Fisheries Science Center, National Oceanic and Atmospheric Administration* (2010). http://www.sefsc.noaa.gov/sedar/Sedar_Workshops.jsp?WorkshopNum=20 Cited in Nov 2012
5. Chuine I, P Yiou, N Viovy, B Seguin, V Daux, and EL Ladurie. Grape ripening as a past climate indicator. *Nature*, **432**, 18 (2004)
6. Holling CS. Some characteristics of simple types of predation and parasitism. *Canadian Entomologist*, **91**: 385–398 (1959)
7. Ledder G. An experimental approach to mathematical modeling in biology. *PRIMUS*, **18**, 119–138 (2007)
8. Ledder G. BUGBOX-predator (2007). http://www.math.unl.edu/~gledder1/BUGBOX/ Cited Sep 2012
9. Lineweaver H and D Burk. The determination of enzyme dissociation constants. *Journal of the American Chemical Society*, **56**, 658–666 (1934)
10. Motulsky H and A Christopoulos. *Fitting Models to Biological Data Using Linear and Nonlinear Regression*. Oxford University Press, Oxford, UK (2004)
11. Rasmussen RA. Atmospheric trace gases in Antarctica. *Science*, **211**, 285–287 (1981)
12. Richards S. Testing ecological theory using the information-theoretic approach: Examples and cautionary results. *Ecology*, **86**, 2805–2814 (2005)
13. University of Tennessee. Across Trophic Level System Simulation (1996). http://atlss.org Cited in Nov 2012
14. Wilkinson GN. Statistical estimations in enzyme kinetics. *Biochem J.*, **80**, 324–332 (1961)

Part II
Probability

Chapters 3 and 4 provide a foundation in probability and a small amount of inferential statistics. In most books, probability appears in the context of statistics. I prefer to think of probability as a subject in its own right. Descriptive statistics is essential background for probability and appears in Section 3.1. I can then present probability as the study of mathematical models of data sets. This viewpoint places the focus on probability distributions, which play a prominent role in the characterization of data and are needed to run simulations for stochastic dynamical system models.

Much of statistics is concerned with the assessment of inferences, and one can see Chapter 3 and Sections 4.1–4.3 as providing the necessary background for the brief treatment of inferential statistics that occupies Section 4.4. A full treatment of this subject lies beyond the scope of this book. I do not present analysis of variation, which asks questions such as, "Does this characteristic explain variation in that characteristic?" My treatment of inferential statistics is limited to questions such as, "Is the population from which this data was collected different from the general population?," which can be addressed using the standard distributions that characterize simple data sets. Section 4.4 focuses on using probability models, with minimal attention to the statistical rules of thumb that convert probability results into answers of "yes" or "no."

Conditional probability (Sections 4.6 and 4.7) is normally presented before probability distributions and could be studied immediately after Section 3.2. I have placed this material at the end of Part II so as not to interrupt the flow toward Section 4.4. I also believe that it is better pedagogy to focus on independent random variables in detail before examining dependence.

The problem sets include several case studies that are split over multiple sections:

Section	3.1, 3.4	3.5		3.6	3.6, 3.7	4.2	4.4	4.5
Dopamine and psychosis	3.1.3					4.2.1	4.4.2	4.5.1
Fruit fly egg production	3.1.4					4.2.2	4.4.3	4.5.2
Weight gain in rats	3.1.5					4.2.3	4.4.4	4.5.3
Blood pressure medication	3.1.6						4.4.6	
Resting pulse rates	3.1.7			3.6.9		4.2.5		
Left-handed presidents	3.4.5						4.4.8	
Chest sizes of soldiers		3.5.2		3.6.6				
Cuckoo egg measurements		3.5.3		3.6.7				
Butterfat in cow milk		3.5.4		3.6.12	3.6.13	4.2.4	4.4.5	4.5.4
Iris sepal lengths		3.5.5	3.5.6	3.6.10	3.6.11	4.2.7		4.5.5
Malaria parasite counts					3.7.6			4.5.9
Spiders under boards					3.7.7			4.5.10
Nerve pulse times					3.7.13	4.2.8		

The accompanying sketch shows the interdependencies of the sections in Part II and connections to prerequisite topics.

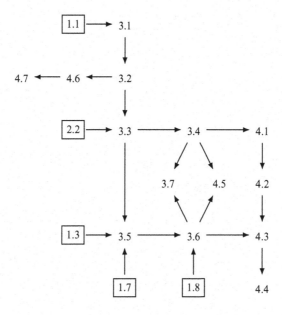

Chapter 3
Probability Distributions

I have yet to see a problem, however complicated, which, when you looked at it in the right way, did not become still more complicated.

Poul Anderson, 1969.

Poul Anderson might have been thinking about probability when he wrote this quip. The concepts are more abstract than in many other areas of mathematics, and the notation can be quite confusing. In an effort to minimize this confusion, this chapter focuses on the central concept of the probability distribution, with other probability topics deferred to Chapter 4. We can approach the probability distribution concept by beginning with descriptive statistics and then conceptualizing it as a model for a large population of data.

Understanding a few distributions thoroughly makes it much easier to understand other distributions. Hence, this chapter focuses on the basic ideas of discrete and continuous probability distributions rather than attempting to present a large catalog of useful distributions.

In Section 3.1, we examine a simple data set and show how to visualize it with histograms and characterize it with the mean and standard deviation, which are measures of central tendency and variation, respectively. In Section 3.2, we introduce the notion of a probability distribution as an infinite population in which the fractions falling in any particular interval are given. We can then think of probability as the fraction of times in which a desired result would occur over infinitely many experiments, thereby generalizing what we know about properties of finite data sets.

Section 3.3 introduces the general concepts for distributions of discrete numerical outcomes, using examples in which the number of outcomes is finite. These distributions can be characterized by a mean and standard deviation, similar to the characterization of populations of real data. Section 3.4 deals with binomial distributions, using elementary genetics as a source of examples.

Section 3.5 extends the probability distribution concept to distributions of continuous numerical outcomes. Such distributions can be defined using a cumulative distribution function or a probability density function. We emphasize the relationship between these functions and use probability density functions to define the mean and standard deviation of a continuous distribution. Section 3.6 deals with normal distributions, the most ubiquitous of all probability distributions.

The chapter concludes with two other distributions of common use in biology: Poisson and exponential distributions. These distributions provide an opportunity to review the similarities and differences between discrete and continuous distributions; placing them in the same section allows for a focus on the relationship between them.

G. Ledder, *Mathematics for the Life Sciences: Calculus, Modeling, Probability, and Dynamical Systems*, 149
Springer Undergraduate Texts in Mathematics and Technology, DOI 10.1007/978-1-4614-7276-6_3,
© Springer Science+Business Media, LLC 2013

There is a wealth of named probability distributions besides the four presented in this chapter. Some of these are included in the problem sets. Also, a table showing connections among problems in this chapter and Chapter 4 can be found in the introduction to Part II.

3.1 Characterizing Data

After studying this section, you should be able to:

- Classify data by type.
- Display data in a histogram.
- Compute the mean of a set of data.
- Compute the variance and standard deviation of a set of data.

3.1.1 Types of Data

Different classes of data must be treated differently, so a first step in data analysis is data classification. We begin by dividing all data into two large categories, *arithmetic data* and *non-arithmetic data*.[1]

Arithmetic data: *data consisting of numerical values that can be averaged in a meaningful way.*

Non-arithmetic data can be further subdivided into several classes:

Categorical or nominal data: *data consisting of descriptive terms, such as "male" and "female," or "married," "single," "divorced," and "widowed," etc.*

Ordinal data: *numerical data in which the ordering of categories is significant, but the distance between the category descriptions is ambiguous. An example of ordinal data is the data from a survey in which respondents are asked to rate a statement with an integer from 1 through 5, with 1 for "strongly agree" and 5 for "strongly disagree."*[2]

Arithmetic data can be further categorized according to relationships with other variables.

Univariate data: *data consisting of values of a single arithmetic quantity, such as a listing of the ages of all members of a population.*

---

[1] The word "arithmetic" used here is the adjective, pronounced "eh-rith-MEH-tic," not the noun of the same spelling that is pronounced "a-RITH-muh-tic."

[2] If one person gives a rating of 3 and another gives a rating of 1, we can *say* that the "average" rating is 2, but this average has no clear meaning in terms of the actual categories. There is no objective reason to think that one "agree" and one "disagree" are interchangeable with two "neither agree nor disagree."

> **Multivariate data**: *data consisting of sets of values of more than one arithmetic quantity; for example, multivariate data used for public health might include the height, weight, and age of each subject.*

Multivariate data is **independent** if there is no mathematical relationship between the values of the different quantities. For example, the heights, weights, and ages of people are clearly not independent,[3] but the number of letters in the hometown of each person is probably independent of that person's height, weight, and age. Independent data collected as part of a multivariate set can be treated as univariate.

Both univariate and multivariate data may also be **stratified**, meaning that the data are split into several different subsets according to a categorical variable. For example, a data set on the height of adult Americans might be subdivided into male and female subsets.

In the remainder of this section, we consider only univariate arithmetic data.

3.1.2 Displaying Data

In a study published in 1904, A.R. Cushny and A.R. Peebles studied the effects of the drug hyoscine hydrobromide on the sleep of ten subjects. The researchers measured the average increase in hours of sleep per night on each of two forms of the drug, as compared with a placebo. This data set appears in Table 3.1.1.[4]

Table 3.1.1 Hours of sleep gained on D- and L-hyoscine hydrobromide

Number	1	2	3	4	5	6	7	8	9	10
D-hyoscine hydrobromide, X	0.7	−1.6	−0.2	−1.2	−1.0	3.4	3.7	0.8	0.0	2.0
L-hyoscine hydrobromide, X	1.9	0.8	1.1	0.1	−0.1	4.4	5.5	1.6	4.6	3.4

How should we make sense of this data? The first important point is that, like all univariate data, there is no significance to the order in which the numbers are listed. We could just as easily have given the data as a list: 0.7, −1.6, −0.2, ... We could also have sorted the data to present it in ascending or descending order.

The primary difficulty in understanding a mass of data is that the crucial information is overwhelmed by the level of detail. To see this crucial information, we need a way to filter out the excess detail. One way to do this is to partition the range of the data into subintervals of equal width, called **classes** or **bins**. We can then think of each item of data as a member of a class rather than a specific value.

---

[3] On the whole, taller people weigh more than shorter people, and children weigh less than adults.

[4] The drug tested in the study is more commonly known as scopolamine; today it is used in a patch applied to the skin to prevent motion sickness. The subjects were not insomnia patients; they were inmates at the Michigan Asylum for the Insane and obviously not volunteers. (Medical ethics has improved a lot since the early 1900s!) The clinicians who administered the test first checked the drugs' safety by trying them on themselves. (Experimental protocols have also improved a lot since the early 1900s!) The data was subsequently used by William Sealy Gosset, better known by the pseudonym "Student," in a classic 1908 paper that introduced the test now known as "Student"'s t test [19]. Gosset incorrectly copied the column headings, so he did not reach the correct conclusion from the data. (Publication standards in science have improved a lot since the early 1900s as well!) For a more complete history of this interesting story, see [15].

Example 3.1.1. Suppose we partition the data of Table 3.1.1 into the bins

$$(-0.5, 0.5), (0.5, 1.5), (1.5, 2.5), \ldots, (5.5, 6.5).$$

We can name these bins "0," "1," …, "6." In effect, we are rounding off the original data to the nearest hour.[5] With the data grouped by class, we can display it as a table that gives the number of occurrences, or **frequency**, of each class. The resulting **frequency table** for the L-hyoscine data appears in Table 3.1.2. Note that the value 5.5 could be counted as either "6" or "5." Typically, these borderline values are rounded up, but there is no valid reason for choosing to do so. I prefer to count 5.5 as 0.5 occurrence of each of the values "5" and "6."   □

Table 3.1.2 Frequency table for the number of hours of sleep gained on L-hyoscine hydrobromide

Hours	0	1	2	3	4	5	6
Frequency	2	2	2	1	1	1.5	0.5

Check Your Understanding 3.1.1:
Prepare a frequency table for the D-hyoscine hydrobromide data, using bins
$(-2.5, -1.5), (-1.5, -0.5), \ldots.$

The obvious advantage of a frequency table over a list of data is that it looks simpler. A more important advantage is that frequency tables can be meaningfully plotted, using the classes on the horizontal axis and the frequencies on the vertical axis. Frequency data is visualized as a **histogram**—a bar graph in which the heights of the bars give the frequencies of the data classes. Figure 3.1.1a is a histogram for the L-hyoscine hydrobromide data using the frequencies in Table 3.1.2.

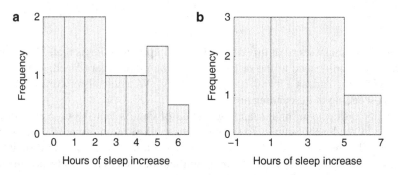

Fig. 3.1.1 Hours of sleep gained on L-hyoscine hydrobromide: (**a**) bin width of 1 hr; (**b**) bin width of 2 hr

Check Your Understanding 3.1.2:
Prepare a histogram for the frequency table in Check Your Understanding 3.1.1.

There are a number of different ways to collate data for any given data set; hence, different histograms can be used to represent the same data.

---

[5] If the data set contained 100 values instead of 10, we might prefer rounding off to the nearest half-hour instead.

Example 3.1.2. Table 3.1.3 and Figure 3.1.1b show the L-hyoscine hydrobromide data using a different set of bins. □

Table 3.1.3 Frequency table for the number of hours of sleep gained on L-hyoscine hydrobromide

Hours	$(-1,1)$	$(1,3)$	$(3,5)$	$(5,7)$
Frequency	3	3	3	1

3.1.3 Measures of Central Tendency

Suppose we want to summarize the L-hyoscine hydrobromide sleep data with just a single number that represents the "average" sleep gain. There are two reasonable ideas for how to do this.

Example 3.1.3. One idea of average is "the average of the individual values." We simply add up the values and divide by the total:

$$\frac{1.9 + 0.8 + 1.1 + 0.1 - 0.1 + 4.4 + 5.5 + 1.6 + 4.6 + 3.4}{10} = 2.33.$$

The other idea of average is "the value of the average individual." Here, we discard the values in pairs, always discarding the largest and smallest of the remaining values, until only one or two values are left. If one value remains, we have found "the value of the average individual." If two values remain, we find their average. With ten values, we discard the largest four and the smallest four, leaving the fifth and sixth, which are 1.6 and 1.9. The "value of the average individual" is

$$\frac{1.6 + 1.9}{2} = 1.75.$$ □

"The average of the individual values" is more properly called the *mean*, while "the value of the average individual" is more properly called the *median*:

Mean: *the arithmetic average of the data values.*

Median: *the numeric value separating the upper half of the data values from the lower half, taken to be the middle value when the number of values is odd and the average of the two middle values when the number of values is even.*

As in linear least squares,[6] it is customary to use $\bar{x}$ for the mean of x data.

Check Your Understanding 3.1.3:
Find the mean and median of the D-hyoscine hydrobromide data.

---

[6] See Section 2.3.

The choice of whether to use the median or the mean depends on the purpose of the data. To cite a non-biological example, consider a set of data that gives the yearly salaries of a Major League Baseball team. The owner has to pay all of the salaries, regardless of how the money is distributed among the players, so he is most concerned about the mean of the salaries. The players' union represents the views of the players as individuals, so it is most likely concerned with the median salary. This difference of viewpoint helps explain some of the perceptual differences in labor negotiations—salaries in any organization usually include a small number of larger salaries and a large number of smaller salaries, a case for which the mean is larger than the median. If management thinks the "average" is too high and labor thinks it is too low, we could justifiably argue that both are right.

Another issue to consider in measuring central tendency is the importance of **outliers**, which are values that seem rather far from the others.

Example 3.1.4. Suppose we decide that the data for the subject whose sleep increased by 5.5 h per night cannot be correct. Omitting this point changes the sum of the hours from 23.3 to 17.8. This causes the average to drop from 2.3 to 2.0. The median becomes the sixth largest from the original set of 10, which is 1.6 rather than 1.75. The change in the mean for this example is twice the change in the median. This means that more extreme outliers have a proportionately greater effect on the mean, sometimes without changing the median at all. For example, if we replace 5.5 with 6.5, this raises the mean by 0.1 but leaves the median unchanged.   □

While the mean is almost always the correct measure to use for scientific work, its sensitivity to outliers raises the question of whether outliers should be included or omitted from a set of scientific data. There is no definitive answer to this question, but it is possible to make an informed choice based on probability. This issue is addressed in Section 4.2.

3.1.4 Measuring Variability

To characterize the variability of a set of data, we need to look at the differences between the data values and the mean. Table 3.1.4 extends the original L-hyoscine hydrobromide data table to include these deviations.

Table 3.1.4 Hours of sleep gained on L-hyoscine hydrobromide, along with the deviations from the mean

Number	1	2	3	4	5	6	7	8	9	10
Hours gained, x	1.9	0.8	1.1	0.1	−0.1	4.4	5.5	1.6	4.6	3.4
Deviations, $x - \bar{x}$	−0.43	−1.53	−1.23	−2.23	−2.43	2.07	3.17	−0.73	2.27	1.07

We could suggest several different ways to combine the deviations of the data values from the mean. A good starting point is the sum of the squares of the deviations, which is consistent with the use of the residual sum of squares to measure the total discrepancy between data values and values predicted by a mathematical model.[7] This logical choice requires some modifications. In the first place, the *sum* of a number of data values depends on how many data values are being added. For example, if every data value is counted twice, then the sum of the square deviations from the mean will be twice as large. Clearly, we do not want to say that a data set in which all of the values appear twice has more variability than the corresponding set without doubled values. A good measure of variability should depend only on the way the values are

---

[7] See Section 2.3.

distributed, not on the size of the data set. This suggests that we should use the mean of the square deviations. The *standard deviation*, which is the accepted measure of variability in data, is only slightly different from the mean of the square deviations.

Standard deviation of a set of values $x_1, x_2, \ldots, x_n$: *the quantity given by*

$$s = \sqrt{\frac{(x_1 - \bar{x})^2 + (x_2 - \bar{x})^2 + \cdots + (x_n - \bar{x})^2}{n - 1}},$$

where $\bar{x}$ is the mean.

Example 3.1.5. The standard deviation of the L-hyoscine hydrobromide data is

$$s = \sqrt{\frac{(-0.43)^2 + (-1.53)^2 + (-1.23)^2 + \cdots + (1.07)^2}{9}} \approx 2.00.$$

$\square$

Check Your Understanding 3.1.4:
Find the standard deviation of the D-hyoscine hydrobromide data.

The definition of the standard deviation indicates two differences from the more natural choice of the mean of the square deviations. The first is that the definition includes a square root so that the standard deviation has the same units of measurement as the data values themselves. Each data value is a time measured in hours. The mean of the square deviations calculated from the data set has a unit of h^2, so the standard deviation ends up being measured in hours, just like the data values. The other difference is that the definition of the standard deviation has denominator $n - 1$ rather than n. The standard deviation is thus a little larger than the square root of the mean of the square deviations.[8]

Obviously a data set with a larger standard deviation has more variability than a data set with a smaller standard deviation. Other than this simple comparison of data set variability, it is difficult to ascribe a specific meaning to a standard deviation value. The exact interpretation is only clear under certain circumstances. We'll return to this point in Section 3.6 when we study the normal distribution.

Check Your Understanding Answers

1. See Table 3.1.5.
2. See Figure 3.1.2.
3. $\bar{x} = 0.66$ and the median is 0.35.
4. $s = 1.86$

---

[8] The correct denominator actually is n when the data represents the full population of interest. The choice $n - 1$ is appropriate for the much more common case where data for a sample is being used to try to characterize a larger population. The reason for this change is beyond the scope of an elementary treatment. As the sample size increases, the distinction makes less of a difference.

Table 3.1.5 Frequency table for the number of hours of sleep gained on D-hyoscine hydrobromide

Height	−2	−1	0	1	2	3	4
Frequency	1	2	3	1	1	1	1

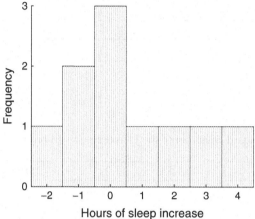

Fig. 3.1.2 Hours of sleep gained on D-hyoscine hydrobromide

Problems

The data sets for Problems 3.1.2–3.1.8 are from [5] and can be found in both csv and xlsx form at http://www.springer.com/978-1-4614-7275-9.

3.1.1. The actual purpose of the experiment that yielded the data of Table 3.1.1 was to measure differences between the D and L versions of pharmaceuticals. Determine the mean and standard deviation of the differences between the D and L data. Also prepare a histogram to display the data. Without doing any statistical tests, do you think the data support the claim that the biological activity of the D and L versions is different?

3.1.2.* The file JACKAL.* lists lengths for golden jackal (*Canis aureus*) mandibles in the British Museum.

(a) Determine separate means and standard deviations for the male mandibles and the female mandibles
(b) Prepare a histogram of the full data set.
(c) Without doing any statistical tests, do you think the data support the claim that mandible lengths differ between the sexes?

3.1.3. The file DOPAMINE.* lists the activity levels of an enzyme that indicates dopamine activity in the brains of patients being treated for schizophrenia. The data are stratified into two groups, depending on whether the hospital staff judged the patients to be psychotic.

(a) Determine the mean and standard deviation for each of the two groups.
(b) Prepare a histogram of the full data set.
(c) Without doing any statistical tests, do you think the data support the claim that dopamine activity can be used to diagnose schizophrenia?

(This problem is continued in Problem 4.2.1.)

3.1.4. The file FRUITFLY.* lists the fecundity (average daily egg yield) for the first 14 days of maturity for female fruit flies. The data are segregated by genetic markers, one group having been bred for resistance to the pesticide DDT, one group for DDT susceptibility, and one group not selectively bred.

(a) Determine the mean and standard deviation for each of the three groups.
(b) Prepare a histogram of the full data set.
(c) Without doing any statistical tests, do you think the data support the claim that selective breeding for either resistance or susceptibility reduces fecundity?
(d) Suggest a reason why it is plausible for selective breeding for DDT resistance or suscepti-bility to reduce fecundity.
(e) Suppose the DDT-resistant group is used in a follow-up experiment in which they are se-lectively bred for fecundity. What do you think would happen to the DDT resistance of the resulting population?

(This problem is continued in Problem 4.2.2.)

3.1.5. The file WEIGHT.* lists weight gains for rats fed four different diets.

(a) Determine the mean and standard deviation for the weight gains for each of the four diets.
(b) Prepare a histogram of the full data set.
(c) Without doing any statistical tests, do you think the data support a claim that some diets are better than others?

(This problem is continued in Problem 4.2.3.)

3.1.6. The file BLOOD.* lists the systolic and diastolic blood pressures for 15 patients, both before and after treatment with the drug captopril.

(a) Determine the decrease in systolic blood pressure for each patient. Find the mean and stan-dard deviation. Prepare a histogram of relative frequency.
(b) Repeat part (a) for diastolic blood pressure.
(c) Prepare a histogram that includes the data from both pairs.
(d) Without doing any statistical tests, do you think the data support the claim that captopril is effective in reducing blood pressure?

(This problem is continued in Problem 4.4.6.)

3.1.7. The file PULSE.* lists resting pulse rates for a group of native Americans from Peru.

(a) Find the mean and standard deviation for the pulse rate data. Prepare a histogram of relative frequency.
(b) Based on part (a) without any statistical tests, what do you think is the "normal" range of resting pulse rates in this population?

(This problem is continued in Problem 3.6.9.)

3.1.8. The file HEIGHT.* lists the heights and weights of a group of 11-year-old girls attending Heaton Middle School in Bradford, England.

(a) Find the mean and standard deviation for the height data. Prepare a histogram of relative frequency.
(b) Find the mean and standard deviation for the weight data. Prepare a histogram of relative frequency.
(c) Use linear least squares (See Section 2.3) to fit a linear function to the data.

3.2 Concepts of Probability

After studying this section, you should be able to:

- Identify experiments, outcomes, and events in a probability setting.
- Compute probabilities for sequences of experiments.
- Compute probabilities using the elementary probability rules.

DNA is a polymer, which means that it is a chain of indeterminate length composed of individual units.[9] Each of these individual units of DNA incorporates one of four amino acids, collectively called **nucleotides**: adenine (A), guanine (G), cytosine (C), and thymine (T). For the moment, assume that all of the nucleotides are stripped out of the DNA of a cell and collected together in a test tube. This collection of nucleotides can be represented as a population of the corresponding letters of the alphabet. In theory, we could take a census of this population. The result would be a set of categorical data,[10] which we could describe in two ways. We could report the actual count, but these numbers would be so large that it would be hard to understand them. Alternatively, we could report the *frequency* of each letter, which gives each count as a fraction of the total rather than as a number. For example, if there are 1,500 occurrences of the letter A out of a total of 6,000 letters, then the frequency of A is 25 %. If we randomly select one nucleotide from the population, we expect to choose A approximately 25 % of the time. We could also say that 25 % is the *probability* that the chosen letter is A. For a known population, the probability of a characteristic is simply the fraction of the population that exhibits that characteristic. More generally, we can define the probability of some statement about the population.

Probability: *the (theoretical) frequency at which some statement about the outcome of a random process is true.*

Example 3.2.1.

Q) Suppose we have a population consisting of equal numbers of each nucleotide. What is the probability that a randomly selected nucleotide from this population is either C or G?

A) If we randomly select a nucleotide from this population, we expect to select each of the nucleotides C or G 25 % of the time. Thus, we will satisfy the requirement "either C or G" 50 % of the time. The probability is 0.5.   □

The analogy of DNA nucleotides with letters of the alphabet can be extended to the genetic code that is "written" in DNA. Sequences of three nucleotides combine together to make a *codon*, and it is the codons that give the genetic code its meaning. The codon "dictionary" appears in Table 3.2.1. Each of the 64 codons functions as either a word or a period in the DNA language. The "words" are the 20 amino acids, represented by 61 of the 64 codons. The "periods" are the three stop codons. A gene is a long DNA "sentence" of amino acid words, ending in a stop codon. With 64 different code words and only 21 possible meanings, there is some redundancy in the DNA dictionary. Surprisingly, there are six different ways to "spell" leucine, arginine, and serine, but only one way to "spell" methionine and tryptophan. Another

---

[9] A chromosome is a large DNA molecule.

[10] See Section 3.1.

Table 3.2.1 The 64 DNA codons

TTT	Phenylalanine	TCT	Serine	TAT	Tyrosine	TGT	Cysteine
TTC	Phenylalanine	TCC	Serine	TAC	Tyrosine	TGC	Cysteine
TTA	Leucine	TCA	Serine	TAA	Stop	TGA	Stop
TTG	Leucine	TCG	Serine	TAG	Stop	TGG	Tryptophan
CTT	Leucine	CCT	Proline	CAT	Histodine	CGT	Arginine
CTC	Leucine	CCC	Proline	CAC	Histodine	CGC	Arginine
CTA	Leucine	CCA	Proline	CAA	Glutamine	CGA	Arginine
CTG	Leucine	CCG	Proline	CAG	Glutamine	CGG	Arginine
ATT	Isoleucine	ACT	Threonine	AAT	Asparagine	AGT	Serine
ATC	Isoleucine	ACC	Threonine	AAC	Asparagine	AGC	Serine
ATA	Isoleucine	ACA	Threonine	AAA	Lysine	AGA	Arginine
ATG	Methionine	ACG	Threonine	AAG	Lysine	AGG	Arginine
GTT	Valine	GCT	Alanine	GAT	Aspartate	GGT	Glycine
GTC	Valine	GCC	Alanine	GAC	Aspartate	GGC	Glycine
GTA	Valine	GCA	Alanine	GAA	Glutamate	GGA	Glycine
GTG	Valine	GCG	Alanine	GAG	Glutamate	GGG	Glycine

curiosity is that the third nucleotide in a codon doesn't always matter. Out of the 16 possible combinations of the first two nucleotides, there are 8 cases (TC, CT, CC, CG, AC, GT, GC, GG) in which the same amino acid is obtained no matter what the third nucleotide is.

Imagine an experiment in which we randomly choose a sequence of three nucleotides from a hypothetical population consisting of equal numbers of each nucleotide. We can determine the probabilities for obtaining various amino acids simply by counting the number of associated codons.

Example 3.2.2.

Q) If we randomly choose a sequence of three nucleotides from a population of equal numbers, what is the probability that we choose (a) a glycine codon? (b) the tryptophan codon?

A) By counting the numbers of codons in Table 3.2.1, the probability of glycine is 4/64 or 1/16, while the probability of tryptophan is 1/64.   □

Table 3.2.2 Set theory notation

Statement	Meaning
$x \in \Omega$	x is an element of the set Ω.
$A \subseteq \Omega$	A is a subset of Ω (a set consisting of some or all of the elements of Ω).
$A \subset \Omega$	A is a proper subset of Ω (a subset that does not include all of the elements of Ω).
A^c	The *complement* of A, consisting of all elements of Ω not in A.

3.2.1 Experiments, Outcomes, and Random Variables

Probability theory makes use of some notation taken from set theory, summarized in Table 3.2.2, as well as terms whose meanings differ somewhat from their meanings in standard English usage. The random selection of one nucleotide from a population of DNA in Example 3.2.1 is an *experiment* with four possible *outcomes*: A, C, G, and T. The set $\{A, C, G, T\}$ of possible outcomes is the *sample space* for the experiment. Any set of possible outcomes (a subset of the sample space) is called an *event*; for example, the set of codons representing leucine is an event for the experiment of drawing three successive nucleotide bases.

Experiment: *any process by which one or more individuals are chosen from some population and used to determine the value of some quantity.*

Outcome: *any of the various values that can be obtained from a given experiment.*

Sample space: *the set of all outcomes for a given experiment.*

Event: *a set consisting of some of the outcomes in the sample space of a given experiment.*

Many of the statements whose probability we want to calculate are of the form "An experiment is conducted and some specific outcome is the result." The statement in Example 3.2.2b is "Three nucleotides are chosen randomly from a population of equal numbers of nucleotides, and the result is the sequence TGG." In Example 3.2.2a, the third nucleotide is irrelevant, we can write the statement as "Two nucleotides are chosen randomly from a population of equal numbers of nucleotides, and the result is the sequence GG." Structurally, we can represent these statements by the form

experiment result = specific outcome.

This general statement is more concise than the full verbal statements, but it is still clumsy. A mathematical notation is necessary. Because the experimental result cannot be known without doing the experiment, it must be denoted by a variable

Random variable: *a variable whose value is obtained from an experiment.*

The statement "A nucleotide is chosen randomly from the uniform DNA distribution, and the result is C" corresponds to the mathematical notation

$$X = \text{C},$$

where X is understood to be the experiment result. The four statements $X = \text{A}$, $X = \text{C}$, $X = \text{G}$, and $X = \text{T}$ are collectively referenced as

$$X = x,$$

where x refers to a specific, but unspecified, outcome. Note the subtle difference between X and x. We are using X to denote the randomly determined experimental outcome, while we are using x to represent a specific outcome that we prefer not to identify. We extend the notation to probability by using

$$P[X = x]$$

to represent the probability that the statement $X = x$ is true.

Example 3.2.3. Let X be a randomly selected sequence of three nucleotides from a population of equal numbers of the four nucleotides. The result of Example 3.2.2b is written in mathematical notation as

$$P[X = \text{TGG}] = 1/64. \qquad \Box$$

The notation $X = x$ suffices for all statements that refer to only one of the possible outcomes. This is not adequate for Example 3.2.1, where the statement refers to both outcomes C and G. For this example, the statement is of the form

$$\text{experiment result} \in \text{event}.$$

The notation $X \in A$ is used to indicate that the result of an experiment is an outcome defined to be in the desired event.

Example 3.2.4. Let CG be the event consisting of the outcomes C and G. The statement $X \in CG$ means "A randomly selected nucleotide from the uniform DNA distribution is either C or G." Under the assumption that each nucleotide is equally probable, we have

$$P[X \in CG] = 0.5. \qquad \Box$$

Where there is no ambiguity, it is common to suppress the name of the random variable in a probability statement. For example, when the context makes it clear that the random variable is X, we can write $P[x]$ instead of $P[X = x]$ and $P[A]$ instead of $P[X \in A]$.

3.2.2 Probability Distributions

In practice, we almost never know the complete population for any problem of biological interest. Even if we do know the full population, calculations based on exact data tend to be cumbersome. For both of these reasons, it is more convenient to work with a mathematical model of a real population. A mathematical model of a large real population is called a *probability distribution*.

Probability distribution: *a sample space of outcomes along with a means for prescribing their frequencies. The frequencies must all be in the interval* $[0,1]$, *and the sum of the frequencies of all the outcomes must be 1.*

Example 3.2.5. Examples 3.2.1–3.2.3 model the population of DNA in a cell as consisting of 25 % of each of the nucleotides. For this distribution, the values are the letters A, C, G, and T, and the frequencies are all 0.25. This is an example of the *uniform* distribution: a probability distribution with a finite set of possible values, each having the same frequency. $\qquad \Box$

Uniform distribution (discrete): *a probability distribution having n outcomes, each with probability* $1/n$.

So far, we have been assuming that the four nucleotides in DNA are present in equal numbers. This is not generally the case. For humans, the frequencies are approximately 21.5 % each for G and C and 28.5 % each for A and T. The proportions differ for different species, but the proportions of G and C are always equal, as are the proportions of A and T. This happens because a DNA molecule actually consists of two matching strands connected together. Each C nucleotide in one strand corresponds to a G nucleotide in the other, and similarly for A and T.

Example 3.2.6. Molecular biologists define the *CG ratio* for any species to be the combined frequencies of the C and G nucleotides. For humans, the CG ratio is about 0.43, so human DNA can be modeled by the distribution

$$P[C] = 0.215, \qquad P[G] = 0.215, \qquad P[A] = 0.285, \qquad P[T] = 0.285.$$

More generally, if u is the CG ratio for any species, then the DNA for that species can be modeled by the distribution

$$P[C] = 0.5u, \qquad P[G] = 0.5u, \qquad P[A] = 0.5(1 - u), \qquad P[T] = 0.5(1 - u). \qquad \square$$

Example 3.2.6 illustrates that probability distributions for experiments with finitely many outcomes can be defined simply by listing the probabilities for each outcome. The numbers assigned as probabilities must satisfy two mathematical requirements.

Range rule: *For any event A and sample space Ω,*

$$0 \le P[A] \le P[\Omega] = 1. \tag{3.2.1}$$

Sum rule: *For any event $A = \{x_1, x_2, \ldots, x_J\}$,*

$$P[A] = \sum_{x_j \in A} P[x_j] \equiv P[x_1] + P[x_2] + \cdots + P[x_J]. \tag{3.2.2}$$

In the sum rule, the notation $\sum_{x_j \in A}$ simply means that the sum is restricted to those outcomes that are in the event A.

3.2.3 Sequences and Complements

Probabilities for a uniform distribution can be determined simply by counting outcomes, as in Example 3.2.2. Other methods are required for nonuniform distributions.

Example 3.2.7.

Q) Suppose a sequence of three nucleotides is drawn from the distribution of human DNA. What is the probability that the sequence represents glycine?

A) Drawing a glycine codon is equivalent to drawing G in two successive experiments consisting of selection of a single nucleotide. Each of these has probability 0.215. Now suppose we set up a more complicated experiment with 10,000 trials. Since the probability is the theoretical frequency, we can say that we expect to draw G for the first nucleotide in 2,150 of the trials. For these 2,150 trials, the probability of drawing a second G is 0.215, so we expect to see this occur in $0.215 \times 2,150 = 462.25$ of the trials. Theoretically, we expect to get glycine in 462.25 out of every 10,000 trials, so the probability is 462.25/10,000, or 0.0462.[11]

□

Example 3.2.8.

Q) Suppose a sequence of three nucleotides is drawn from the distribution of human DNA. What is the probability that the sequence represents tryptophan?

A) We can follow the same reasoning as in Example 3.2.7, but there is a much simpler calculation. Clearly, the probability of a sequence is always going to be the product of the probabilities. Also, the probability of a given nucleotide does not depend on whether that nucleotide is the first one in the sequence or a later one.[12] Thus, the probability of the sequence TGG is simply the product of the probabilities of drawing T once and G twice.

$$P[TGG] = P[T] \times P[G] \times P[G] = (0.285)(0.215)^2 \approx 0.0132.$$ □

The sequence rule illustrated in Examples 3.2.7 and 3.2.8 works for a sequence of any length.

Sequence rule: *Suppose an experiment is a sequence of J individual experiments. Let p_1 be the probability of event A_1 in experiment 1, p_2 be the probability of event A_2 in experiment 2, and so on. Then the probability that the combined experiment results in the sequence $A_1, A_2, \ldots, A_J$ is $p_1 p_2 \cdots p_J$.*

The complement rules, based on the requirement that the probabilities of all outcomes must sum up to 1, allow us to calculate probabilities of events using the outcomes that are not in the event.

Complement rules: *For any events A and B, from either the same or different experiments,*

$$P[A] + P[A^c] = 1, \qquad P[A \cap B] + P[A^c \cap B] = P[B]. \qquad (3.2.3)$$

Example 3.2.9.

Q) The amino acid isoleucine is represented by the codons ATA, ATC, and ATT. Find the probability that a sequence of three nucleotides from the distribution of human DNA represents isoleucine.

A) We could use the sequence rule to determine the probabilities of each of the three isoleucine codons and add the results. Alternatively, we could use the symbol G^c to indicate any nucleotide other than G. Then isoleucine corresponds to the sequence ATG^c. Using the

---

[11] Data for random variables is subject to demographic stochasticity, with considerable variation among samples. It is customary to report probabilities with four decimal digits, which implies a higher degree of precision than is usually warranted. Reporting any more than four decimal digits, except as a one-digit approximation for a tiny probability, is an example of the "measure it with your hand, mark it with a pencil, cut it with a laser" fallacy.

[12] This is another advantage of working with probability distributions rather than actual populations. Removing an individual from a population changes the probabilities for the remaining population. Sampling does not change a probability distribution. In effect, a probability distribution is a model of an infinite population.

probabilities for human DNA, we have $P[A] = P[T] = 0.285$ and $P[G] = 0.215$. The complement rule gives us $P[G^c] = 1 - 0.215 = 0.785$. Finally, the sequence rule gives us the result:

$$P[ATG^c] = P[A] \times P[T] \times P[G^c] = (0.285^2)(0.785) \approx 0.0638. \qquad \square$$

The complement rules mean that we always have two options for calculating the probability of an event: we can do it directly using the outcomes that comprise the event, or we can do it indirectly by calculating the probability of the complement.

Example 3.2.9 illustrates an important point about the use of algebraic rules. The symbols used in the rules are generic, not specific. Equation (3.2.3) does not apply only to the pair A and A^c. Rather, it applies to any complementary pair of events. The rules are written with specific symbols to make them easier to write, but the specific symbols chosen for the rules have no significance. Every time you use a rule, it is important to remind yourself of the generic interpretation of that rule.

Problems

3.2.1.* Suppose a couple has three children.

(a) By enumerating all possible sequences of outcomes, determine the probability that all three children have the same gender.
(b) Use the sequence rule to find the probability that all three children have the same gender. [Note that the first child could be of either gender.]

3.2.2.*

(a) Determine the probability of rolling a sum of 8 with two standard dice by counting the number of sequences of two die rolls that give the right sum. Write a brief explanation of your reasoning.
(b) Identify the experiment and sample space for part (a).

3.2.3.* Suppose 60 % of Americans have dark hair and 25 % have both light hair and blue eyes. What fraction of Americans have light hair but not blue eyes? Explain your answer as an example of the complement rule.

3.2.4. Find the probability of obtaining a stop codon from a sequence of three nucleotides chosen randomly from human DNA.

3.2.5. Use the result of Problem 3.2.4 to determine the probability that a random DNA sequence of 50 codons does not contain a stop codon. Repeat for 200 codons.

3.2.6. In the "birthday problem" of probability, a group of n people compare their birthdays (month and day, but not year). Ignoring the possibility of a birthday on leap day, the goal is to determine the probability that at least two of the people share a birthday.

(a) Think of the experiment of randomly selecting four people and determining their birthdays, in the order in which the people were selected. How many outcomes are in the sample space of this experiment?
(b) Suppose A is defined as the set of outcomes in which there are no repeats. How many outcomes from the experiment of part (a) are in the set A?
(c) Use the results of parts (a) and (b) to determine the probability that at least two of the four people share a birthday.
(d) Repeat (a)–(c) for the case of 20 people rather than four.

3.2.7. The script GeneFinder.R, available at http://www.springer.com/978-1-4614-7275-9 examines a long sequence of DNA nucleotides for the purpose of identifying sequences without stop codons. The file ChiliLeafCurl Genome.txt contains a complete genome for the chili leaf curl virus. Run the program with this data file to see the output. Then explain the function of each of the statements that is not already annotated.

3.2.8. Use the GeneFinder script of Problem 3.2.7 to identify possible genes in the file Ecoli_plasmid_pEK01102.txt. This file contains the complete sequence of a *plasmid*[13] in *E. coli* KO11 bacteria.[14]

3.3 Discrete Probability Distributions

After studying this section, you should be able to:

- Determine a probability distribution for a finite tree.
- Determine survival probabilities for a finite tree.
- Determine probability distribution functions and cumulative distribution functions for a probability distribution.
- Make connections between survival probability and the cumulative distribution function.
- Compute means and standard deviations for discrete probability distributions.

In Section 3.1, we considered data for sleep gain experienced by a group of individuals, and we made sense of the data in several ways. We counted the numbers of individuals corresponding to each numerical value and prepared histograms to visualize the distribution of the data. We also summarized the data using measures of central tendency and variability.

When real data is unavailable, it is often necessary to create a model to represent a real data set. Such a model is similar to a real data set, but with some key differences. First, we think of the model as representing an infinite population rather than a finite population, as we did in Section 3.2 with an infinite population of DNA nucleotides. With an infinite population, we cannot talk about numbers of individuals, but we can talk about frequencies (in other words, probabilities). Second, we must prescribe the frequencies of each outcome as part of the model, as when we assumed in Section 3.2 that the frequency of both the C and G nucleotides in human DNA is 0.215.

Example 3.3.1. Consider a hypothetical animal species for which individuals have a 60 % chance of surviving each of their first 3 years, but never reach the age of 4. Let x be the age at which an individual dies. Thus, the sample space of the experiment is $\{0, 1, 2, 3\}$, since no individual ever becomes 4 years old. To specify the probability distribution for this hypothetical population, we need to determine $P[X = x]$ for each of the four outcomes. These probabilities must all be positive and less than one, and their sum must be 1.[15]   □

This is a convenient time to review the probability notation introduced in Section 3.2. In Example 3.3.1, we use x to denote a particular (although unspecified) possible outcome of the experiment and X to denote the randomly determined outcome of a particular instance of the experiment. The quantity X is a *random variable*. This term can be very confusing. In algebra and calculus, variables are either independent or dependent, with the dependent ones given in terms of the possible values of the independent ones. The random variable X is not an

---

[13] A plasmid is a DNA strand that appears in cells although it is not part of a chromosome.

[14] This strain of *E. coli* is well studied because it is used in the process of making ethanol from corn.

[15] See the range and sum rules of Section 3.2.

independent variable, because its value is determined by the experiment rather than chosen in advance. Nor is it a dependent variable, because its value is not prescribed uniquely in terms of an independent variable. In fact, the experiments that yield the outcomes are all identical, so there is no independent variable in the usual sense.

Think of the experiment described in Example 3.3.1 as a sequence of simple experiments. Unlike the DNA codon experiment of Section 3.2, the number of simple experiments in the sequence can vary. The overall experiment ends whenever the individual dies. In these cases, it can help to portray the experiment as a type of graph called a *finite tree*. The graph incorporates the probability calculations as well as the structure of the overall experiments in terms of the simple component experiments:

1. There is a single start node, labeled with probability 1.0.
2. The first simple experiment is displayed as a set of edges, one for each of the possible outcomes ("live" and "die," for example), labeled with the probability ascribed to that outcome. The total of the probabilities for this simple experiment must, of course, be 1.
3. Each of the edges that represents an outcome of the first simple experiment ends in a node that represents the state after that outcome. In the experiment of Example 3.3.1, the state after the "live" outcome is "1 year old," while the state after the "die" outcome is "died at the age of zero." In the latter case, the node is called *terminal* because the experiment is now over.
4. The probability associated with each of these new nodes is the probability associated with the simple experiment outcome that leads to the node.
5. Additional experiments appear as sets of edges emanating from the nonterminal nodes. In each case, the total of the probabilities of the set is 1. Each of these edges ends in a new node. In general, the probability associated with a node is the product of the probability on the line that ends in that node and the probability of the preceding node.
6. Eventually, assuming a finite sequence of possible experiments, all paths end in terminal nodes. These nodes represent the possible outcomes of the overall experiment, and the sum of these probabilities must be 1.

Example 3.3.2. Figure 3.3.1 depicts the structure of Example 3.3.1 in terms of simple component experiments, each of which covers survival through 1 year of life. The initial node on the left is labeled with probability 1.0 because the overall survival experiment always starts at age 0. The two edges connected to the initial node represent the first simple experiment of

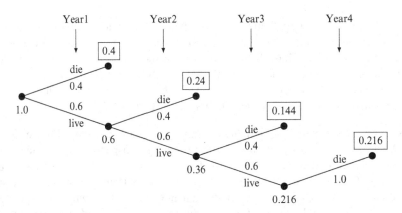

Fig. 3.3.1 A schematic representation of the calculations to determine the probability distribution for Examples 3.3.1 and 3.3.2. The probability of reaching any particular node is the product of the probability on the edge coming from the previous node and the probability at the previous node. The numbers in the boxes are the probability of dying in years 1–4

survival through year 1. These edges are labeled with the corresponding probabilities for the simple experiment. Each of them ends in a node that represents the state of the experiment at the end of year 1 and is labeled with the corresponding probability. The upper node, corresponding to the death of the individual, is terminal.[16] The lower node corresponds to individuals who have survived through year 1 and are now 1 year old.

The second simple survival experiment is depicted by the two edges that exit this lower node to the right. Each edge is labeled with the corresponding probability for the simple experiment; hence, they sum to 1. Each of these edges again ends in a node, the upper one terminal and the lower one leading to the next simple experiment. The probabilities associated with these nodes are the product of the probability at the previous node (which is the probability that this simple experiment will actually occur) and the probability of the outcome on the edge connecting the new node with its predecessor. In this way, the entire experiment is represented by the graph. In general, probabilities on the edges corresponding to each simple experiment must always sum to 1, while the probability on each node is always the product of the probabilities on the preceding node and the edge that connects the two nodes. The experiment ends when the total of the probabilities on terminal nodes is 1. For the example, the probability distribution is defined by the formulas

$$P[X = 0] = 0.4, \quad P[X = 1] = 0.24, \quad P[X = 2] = 0.144, \quad P[X = 3] = 0.216. \qquad \square$$

The information used to define a discrete probability distribution can be used in different ways. Up to now, we have defined the probability distribution by the probabilities of the outcomes. One can also define the probabilities that the random variable will exceed a given outcome or not. For example, in the roll of a fair six-sided die, we can write $P[X > 2] = 2/3$ to indicate that 2/3 of die rolls result in values larger than 2. Similarly, we can write $P[X \leq 2] = 1/3$ to indicate that 1/3 of die rolls result in values of 2 or less.

Example 3.3.3. Consider the scenario of Examples 3.3.1 and 3.3.2. Both of these examples defined the outcome x to be the age at which the individual dies. Biologists are often more interested in knowing the probability that an individual survives through age x than in the probability of dying at age x. The probability of surviving through age x is equivalent to the probability of dying at an age larger than x, which is $P[X > x]$. In this scenario, the probability of surviving year j is the probability associated with the lower of the two nodes after year j in Figure 3.3.1; hence, we obtain the distribution

$$P[X > 0] = 0.6, \quad P[X > 1] = 0.36, \quad \times P[X > 2] = 0.216, \quad P[X > 3] = 0. \qquad \square$$

3.3.1 Distribution Functions

In keeping with the general definition of a probability distribution,[17] a *discrete* probability distribution has a sample space of discrete outcomes. Sometimes there are only finitely many outcomes, as in the distribution of Examples 3.3.1 and 3.3.2. However, in many cases, there are infinitely many outcomes, which makes it impossible to define the distribution by listing the probability for each outcome. In such cases, it is common to define a distribution by means of a formula that can be used to calculate the values.

---

[16] For better clarity, the probabilities at terminal nodes are in boxes.

[17] See Section 3.2.

Example 3.3.4. Consider a model population in which there is no maximum age for individuals and the survival probability is 0.6 for each year of life. The survival probabilities for $X > 0$, $X > 1$, and $X > 2$ are again 0.6, 0.36, and 0.216, or 0.6^1, 0.6^2, and 0.6^3. The pattern is clear: given x as the age at death, the cumulative survival probability (the probability of dying at age greater than x) is $P[X > x] = 0.6^{x+1}$. □

Example 3.3.5. The probability for each outcome in the model population of Example 3.3.4 is a little harder to determine. The key is to focus on the list of the factors rather than the result of the calculations. From Figure 3.3.1, we have

$$P[X = 0] = 0.4, \quad P[X = 1] = (0.6)(0.4), \quad P[X = 2] = (0.6)^2(0.4).$$

As in Example 3.3.4, the pattern is clear. The probability of dying in year x is

$$P[X = x] = (0.6)^x(0.4).$$

This result comes about because of the sequence of events necessary for an individual to die at age x. First the individual must survive x years, and then it must die in the following year. This is represented by a sequence of $x + 1$ experiments for which the probability of following the correct edge in the graph is 0.6 for the first x experiments and 0.4 for the last one. The probability of dying at age x is the product of these individual probabilities. □

There is no important difference between making a list of probability values and using a formula for probabilities. In either case, the probability of each outcome has a unique value, so we can think of it as being determined by a function. The advantage of using a function is that it frees us of the cumbersome notation we have used so far. There are two common ways of using function notation to define a probability distribution; these correspond to the probabilities $P[X = x]$ and $P[X \leq x]$.

Probability distribution function: *a function $f(x)$ that represents the probability that a given experiment has outcome x (in other words, $f(x) \equiv P[X = x]$); a probability distribution function must satisfy the range and sum rules,*

$$0 \leq f(x) \leq 1, \qquad \sum_{x \in \Omega} f(x) = 1. \tag{3.3.1}$$

Cumulative distribution function (given a probability distribution function f): *the function F defined by*

$$F(x) = P[X \leq x] = \sum_{y \leq x} f(y). \tag{3.3.2}$$

For some biological applications, the survival probability, defined as

$$S(x) = P[X > x] = \sum_{y > x} f(y) = 1 - F(x), \tag{3.3.3}$$

is used in place of the cumulative distribution function. Note that the sum of the survival and cumulative distribution functions is always 1, because every outcome not greater than x is less than or equal to x. Which function one uses is a matter of taste and convention: in biology, survival probability is more commonly used, but elsewhere it is the cumulative distribution function.

Example 3.3.6. We have already computed the probability distribution function for the distribution of Examples 3.3.4 and 3.3.5:

$$f(x) = (0.6)^x(0.4).$$

The cumulative distribution function, using the survival probability formula of Example 3.3.4, is

$$F(x) = 1 - 0.6^{x+1}.$$

$\square$

In the definitions of probability distribution functions (pdf) and cumulative distribution functions (cdf), I have deliberately used the same letter, with the lowercase for the pdf and the uppercase for the cdf. In the next section, we will begin a catalog of common probability distributions, so it will be very helpful to use a different letter for each. The notation is much easier to understand if we retain the convention of using the same letter for both functions and distinguishing between the functions by case.

3.3.2 Expected Value, Mean, and Standard Deviation

In Section 3.1, we used the mean ($\bar{x}$) as a measure of central tendency and the standard deviation (s) as a measure of variability for populations of arithmetic data. Mean and standard deviation can be used as measures of central tendency and variability for probability distributions as well, but the definitions need to be modified. To distinguish population values from sample values, we use the symbols μ and σ for the mean and standard deviation of a probability distribution. Derivation of formulas for mean and standard deviation of a probability distribution is most conveniently done by introducing the concept of expected value. We begin with an elementary example in a context outside of biology.

Example 3.3.7. In 2011, the Nebraska Lottery offered a scratch game called "Hot 20." The web page for this game listed the prize amounts, the number of winners for each prize, and the odds of winning each prize. It did not list the number of losing tickets, but that was easy to reconstruct from the information given. Table 3.3.1 presents all the relevant data. Now suppose you were able to buy all of the tickets for this game yourself. It is easy to calculate the value of your purchase as the total of the prize money, which is $543,200. Given the total of 840,000 tickets, this means that the mean ticket value is $543,200/840,000, which is $64\frac{2}{3}$ cents. Unfortunately for your bank account, the purchase price of a ticket is $1.00.[18] No single ticket is worth exactly $64\frac{2}{3}$ cents, so it may seem counterintuitive to call it the "expected" value of a ticket. On the other hand, it represents a fair purchase price in the sense that it would be a good business decision to buy thousands of lottery tickets, provided the price were less than this amount.

All of the results we obtained for the population of tickets can be stated in terms of a probability distribution, where the outcomes are the ticket values in dollars and the random variable X is the value of a randomly selected ticket. Suppose we model the population of tickets as being infinite, with the probabilities as given in the table.[19] Now consider the hypothetical "average"

---

[18] All state lottery authorities hire mathematicians to determine the profiles of the ticket populations, but it is unlikely that any of the tickets are purchased by mathematicians. Those in the know refer to a lottery as "a tax on those who are bad at math."

[19] The distribution is not exactly equivalent to the actual population because the probabilities have been rounded to decimal values; nevertheless, the difference between the actual population and the model distribution is insignificant. Note that we need to round 2/3 down to 0 for one value so that the total probability is *exactly* 1. It is

ticket. This ticket is divided up into portions according to the values of the real tickets. For example, 10 % of the real tickets are worth $2; hence, 10 % of the hypothetical average ticket is valued at the rate of $2 per ticket, for a total value of $0.20. This is the contribution to the value of the hypothetical average ticket that is made by the $2 tickets. Similarly, each outcome contributes to the value of the hypothetical average ticket an amount equal to the product of the outcome and the probability. The sum of all these values is $0.64668. This represents the value of a hypothetical average ticket. In terms of probability, it is the *mean* of the probability distribution and the *expected value* of the random variable X.   □

Table 3.3.1 Computations of mean and expected value for Example 3.3.7

Ticket value	Number of tickets	Total value	Outcome	Probability	Contribution to average
20	5,600	112,000	20	0.00667	0.13340
9	5,600	50,400	9	0.00666	0.05994
4	14,000	56,000	4	0.01667	0.06668
3	19,600	58,800	3	0.02333	0.06999
2	84,000	168,000	2	0.10000	0.20000
1	98,000	98,000	1	0.11667	0.11667
0	613,200	0	0	0.73000	0
Total:	840,000	543,200	Total:	1	0.64668

Example 3.3.7 motivates the definition of *expected value*, which applies not just to the random variable X, but also to functions of X.

Expected value of a function g of a random variable X: *the average value we would expect $g(x)$ to take over infinitely many experiments, usually denoted $Eg(X)$ and calculated as the sum of contributions from each outcome:*

$$Eg(X) = \sum_{x \in \Omega} g(x)P[X = x]. \tag{3.3.4}$$

Example 3.3.8.

Q) Find EX and EX^2 for the distribution of Examples 3.3.1–3.3.3.
A) Using the probabilities from Example 3.3.2, we have

$$EX = (0)(0.4) + (1)(0.24) + (2)(0.144) + (3)(0.216) = 1.176$$

and

$$EX^2 = (0)(0.4) + (1)(0.24) + (4)(0.144) + (9)(0.216) = 2.76.$$

□

---

more important that the probabilities satisfy the definition of a probability distribution than for them to exactly equal the real probabilities.

The mean and standard deviation for discrete probability distributions are defined in terms of expected values.

> - The mean (μ) of a discrete probability distribution is given by
>
> $$\mu = EX. \tag{3.3.5}$$
>
> - The variance (σ^2) of a discrete probability distribution is given by the equivalent formulas
>
> $$\sigma^2 = E[X - \mu]^2 = EX^2 - \mu^2 = EX^2 - (EX)^2. \tag{3.3.6}$$
>
> - The standard deviation (σ) is simply the square root of the variance.

Example 3.3.9. From Example 3.3.8, the mean age at death is

$$\mu = EX = 1.176,$$

the variance is

$$\sigma^2 = EX^2 - (EX)^2 = 2.76 - (1.176)^2 \approx 1.377,$$

and the standard deviation is

$$\sigma = \sqrt{\sigma^2} \approx 1.173.$$

□

Note that the equivalence between the two formulas for the variance is not obvious.[20]

Problems

3.3.1.* Approximately four million women gave birth in the United States in the year 2000, with single births, twins, and triplets occurring 98.51 %, 1.38 %, and 0.11 % of the time, respectively, with a negligible probability of more than three births. Determine the mean and standard deviation for this probability distribution.

   (This problem is continued in Problem 4.3.7.)

3.3.2. Use the definition of expected value for discrete probability distributions to show that

$$E[X - \mu]^2 = EX^2 - \mu^2.$$

3.3.3.* Are some couples prone to having babies of one particular sex? To answer this question from data, it is first necessary to know the theoretical probabilities. Use a tree diagram to determine the probability that a set of four siblings are all of the same sex.

3.3.4. The "law of averages" asserts that an event is more likely if it has not occurred for a long time. Perhaps belief in this bit of folk wisdom is based on confusion of different types of experiments.[21]

---

[20] See Problem 3.3.2.

[21] This is also known as the *gambler's fallacy*.

(a) Suppose Mrs. McCave has had three sons. What is the theoretical probability that her next child will be a girl? [You can use the tree diagram from Problem 3.3.3.]

(b) Can you think of a biological reason why the actual probability of a daughter after three sons might be different than 0.5?[22] Would it be larger or smaller?

(c) Suppose you deal 12 red cards from the top of a well-shuffled deck of standard playing cards. What is the probability that the next card will be black?

(d) What feature distinguishes settings in which the "law of averages" has some value from settings in which it has no value or worse?

3.3.5.* An urn contains three red balls and two black balls. You remove two balls from the urn without replacement. Let X be the number of black balls removed.

(a) Determine the probability distribution function and cumulative distribution function.

(b) Determine the mean, variance, and standard deviation.

(c) Plot a histogram of the probability distribution function.

3.3.6. A certain species of organism has an 80 % chance of surviving each of the first 2 years of life, a 40 % chance of surviving a third year, and a 0 % chance of surviving a fourth year. Let X be the age at death.

(a) Determine the probability distribution function, cumulative distribution function, and survival function.

(b) Determine the mean, variance, and standard deviation.

(c) Plot a histogram of the probability distribution function.

3.3.7. Find the expected value for a scratch game available in your state or a nearby state.

3.3.8. Some species of birds try to have a second nest if the first one fails early enough. Suppose $g(x)$ is the fraction of birds that build a second nest if the first one is raided on day x, where $g(x) = 1$ for $x < 4$, $g(4) = 0.8$, $g(5) = 0.6$, $g(6) = 0.4$, $g(7) = 0.2$, and $g(x) = 0$ for $x > 7$. Given that 0.1 is the probability of a successful raid on any given day and that a nest can only be destroyed once, determine the expected value of the function $g(X)$. What is the biological meaning of the value $E(g(X))$? [Assume that the birds never build a third nest, in which case you do not need to know if second nests are raided.]

3.3.9. Suppose $\Omega = \{1, 2, 3, 4\}$ with probabilities in the ratios $1 : 2 : 3 : 4$. Determine the probability distribution function, mean, and standard deviation. [Hint: Think about the properties that probability distribution functions must satisfy. If $P[X = 1] = p$, then only one value is possible for p.]

3.4 The Binomial Distribution

After studying this section, you should be able to:

- Identify the context in which a binomial distribution is appropriate.
- Use the probability distribution function for binomial distribution calculations.
- Use the formulas for mean, variance, and standard deviation for a binomial distribution.
- Discuss the significance of the parameters in a binomial distribution.

The fundamentals of genetics were first published by the Austrian monk Gregor Mendel in 1866 and only became well known in 1900.[23] Mendel's key experiment began with two

---

[22] The implications of this question were of great interest to the children's author "Dr. Seuss" [16].

[23] *On the Origin of Species* was published in 1859. Not only was Charles Darwin ignorant of the DNA *mechanism* of inheritance—he was also ignorant of the basic *facts* of inheritance!

strains of pea plants, one whose seeds were always smooth and one whose seeds were always wrinkled.[24] He bred one variety with the other, being careful to fertilize the ovules of each flower with pollen from the opposite strain. All of the seeds produced by both strains of plants were smooth. Next, Mendel planted this first generation of offspring seeds and used them to produce a second generation. The result was that approximately 3/4 of the next generation of seeds were smooth and 1/4 were wrinkled. Mendel worked out his conceptual model of genetics from this experiment. This conceptual model consists of two assumptions.

1. The smoothness trait in peas is determined by a pair of *genes*, one from the pollen and one from the ovule. The genes come in two varieties, smooth and wrinkled. A pair of smooth genes produces a smooth seed and a pair of wrinkled genes produces a wrinkled seed.
2. An individual with one gene of each type produces smooth seeds; we say that the smooth gene is *dominant* and the wrinkled gene is *recessive*.

The assumptions of Mendel's model completely explain the data. If a population always produces wrinkled seeds, then all of the individuals in it must have pairs of wrinkled genes. While not as obvious, the same conclusion holds for a population that always produces smooth seeds. If there are some wrinkled genes hiding inside smooth seeds, then at least some of the seeds produced in the next generation might have two wrinkled genes. Hence, two pure strains are the starting point for the experiment. When we produce a first generation of seeds using one gene from a strain that always produces smooth seeds and one gene from a strain that always produces wrinkled seeds, then all of the first generation seeds will have one of each kind of gene; hence, they will all be smooth. Plants grown from these seeds have one of each kind of gene, so the pollen grain and the ovule each have a 50 % chance of carrying the wrinkled gene. Both of these events have to occur to produce a wrinkled seed. By the sequence rule, the probability of a wrinkled seed is 1/4.

In 1936, the statistician and geneticist Ronald A. Fisher analyzed Mendel's published data. From this analysis, Fisher came to the conclusion that the data had been adjusted (intentionally or unconsciously) to give better support to Mendel's theory.[25] To the beginning student of probability, this is a curious statement. Shouldn't Mendel have found *all* of the first generation offspring to be smooth and *exactly* 1/4 of the second generation to be wrinkled? Clearly the answer to the first question is yes, but the second question is problematic, and it was this data that Fisher examined. Let us take a look over Fisher's shoulder. What should we expect from Mendel's second generation data?

Let's begin by considering a single trial of the experiment. If we produce only one seed, it will be either smooth or wrinkled, not 25 % wrinkled. With one seed, "all" will be smooth or "all" will be wrinkled. Clearly, having a probability of 1/4 for a wrinkled seed is not the same as having *exactly* 1/4 of the seeds be wrinkled. The key to understanding the difference is in a correct interpretation of what it means to have a probability of 1/4. It means that the theoretical frequency of wrinkled seeds is 1/4 for the model population, which is infinite. The frequency in an actual population will vary because of demographic stochasticity.[26] In a finite population of seeds, the probability of containing any particular number of wrinkled seeds must be calculated using the basic probability rules.

---

[24] The peas also differed in color, but the extra complication is not necessary for our mathematical modeling.

[25] Mendel did, in fact, clearly state that he used only a portion of the data for his calculations. Careful control of data to minimize bias is a fairly recent innovation in science. A more complete discussion of Mendel's data and Fisher's analysis appears in [2].

[26] See Section 2.1.

3.4.1 Bernoulli Trials and Binomial Distributions

Each seed produced in the second generation is a little experiment of its own, with two possible outcomes: wrinkled and smooth. If we want to count numbers of wrinkled seeds in a large number of trials, it is convenient for individual trials to use the outcome 1 to represent a wrinkled seed and 0 to represent a smooth seed. The individual experiments for each seed are called *Bernoulli trials*.

Bernoulli trial: *an experiment in which the sample space is $\Omega = \{0, 1\}$, with $P[1] = p$ and $P[0] = q \equiv 1 - p$.*

Bernoulli trials serve as the basic component of the *binomial* distribution.[27]

Example 3.4.1.

Q) Suppose we produce three second generation seeds. What is the probability distribution for the number of wrinkled seeds in this experiment?

A) The graph of Figure 3.4.1 displays a sequence of three Bernoulli trials with "success" probability 1/4. Each pair of edges corresponds to one Bernoulli trial, with the upper edge indicating a wrinkled seed and the lower edge indicating a smooth seed. For ease of reading, only the probabilities are labeled in the second and third trials. Each of the nodes on the far right is a terminal node, because the experiment always has exactly three Bernoulli trials. Each terminal node is labeled with the sequence of outcomes and the overall probability, which is the product of the probabilities on the edges leading to the terminal node. The outcome for the three-seed experiment is the number of wrinkled seeds, regardless of order. Thus, we obtain the probability distribution function

$$f(0) = \frac{27}{64}, \qquad f(1) = \frac{27}{64}, \qquad f(2) = \frac{9}{64}, \qquad f(3) = \frac{1}{64}.$$

□

Check Your Understanding 3.4.1:
Repeat Example 3.4.1 to determine the probability distribution function for a sequence of three Bernoulli trials, each with $p = 1/3$ rather than $p = 1/4$.

The distribution in Example 3.4.1 is the *binomial distribution*. This distribution arises in any context where a random variable X represents the number of successes (outcomes of 1) in some number of Bernoulli trials with equal probability of success for each. Two parameters are needed: the number of Bernoulli trials and the success probability for each.

Binomial distribution: *the discrete probability distribution having outcomes k that represent the number of successes in a set of n Bernoulli trials, each with the same success probability p; we denote this distribution by the probability distribution function $b_{n,p}(k)$.*[28]

---

[27] Bernoulli trials are also the basic components for the geometric and negative binomial distributions.

[28] There is no standard notation for the binomial distribution. Most authors use the letter "b" or some variation of it. The number of trials is almost always n and the success probability for one Bernoulli trial is almost always

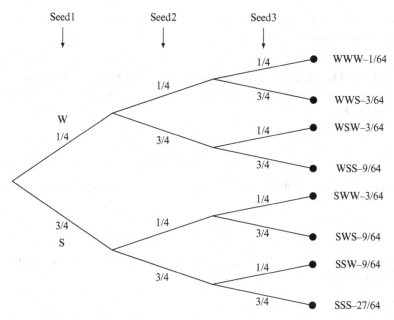

Fig. 3.4.1 Schematic calculation of the probability distribution $b_{3,0.25}$ for the number of wrinkled seeds out of three

Example 3.4.2. Using the results from Example 3.4.1, we can list all of the values of the binomial distribution $b_{3,0.25}$:

$$b_{3,0.25}(0) = \frac{27}{64}, \qquad b_{3,0.25}(1) = \frac{27}{64}, \qquad b_{3,0.25}(2) = \frac{9}{64}, \qquad b_{3,0.25}(3) = \frac{1}{64}.$$

The expected value of the distribution (Equation (3.3.4)) is

$$EX = \sum_{k \in \Omega} k\, b_{3,0.25}(k) = (0)\left(\frac{27}{64}\right) + (1)\left(\frac{27}{64}\right) + (2)\left(\frac{9}{64}\right) + (3)\left(\frac{1}{64}\right) = \frac{48}{64} = \frac{3}{4}.$$

□

The expected value in Example 3.4.2 is equal to the product of the number of trials (3) and the success probability for each trial (0.25). This is intuitively correct. The product np cannot be the *actual* number of wrinkled seeds in this experiment, but it is nevertheless the *expected* number of wrinkled seeds. If we do the experiment with four seeds instead of three, then the expected value of the distribution is 1. In this case, the actual number of wrinkled seeds *might* be the expected value np, but it does not have to be.

The brute force method of Example 3.4.1 is reasonable only when the number of Bernoulli trials is small. Suppose we want to determine the probability distribution of wrinkled seeds from a population of 24. Clearly we need a more sophisticated method for doing the calculations. Such a method is available, thanks to the branch of mathematics known as *combinatorics*. The

---

p. The outcomes are usually labeled as either x or k. Many authors put the parameters inside the parentheses along with the outcome. I prefer to use the parameters as subscripts to emphasize that they are fixed parameters of the distribution function, whereas k is the independent variable in the function.

development of the result is outside the scope of our treatment, but it can be found in any standard probability or statistics text.

> **Theorem 3.4.1 (The Binomial Distribution Formula and Properties).** *The probability of k successes in n Bernoulli trials, each with the same success probability p, is given by the binomial distribution formula*
>
> $$b_{n,p}(k) = \frac{n!}{k!(n-k)!} \left(\frac{p}{q}\right)^k q^n, \quad q = 1-p. \tag{3.4.1}$$
>
> *The mean and variance of the binomial distribution with n trials and success probability p are*
>
> $$\mu = np, \quad \sigma^2 = npq. \tag{3.4.2}$$

As in our first example, the mean of the binomial distribution agrees with our intuition that the expected number of successes should be np.

Example 3.4.3.

Q) Find the probabilities of getting k wrinkled seeds out of 24, for $0 \le k \le 13$.
A) For $n = 24$ and $p = .25$, we have the binomial distribution

$$b_{24,0.25}(k) = \frac{24!}{k!(24-k)!} \left(\frac{0.25}{0.75}\right)^k 0.75^{24} = \frac{(24!)(0.75^{24})}{3^k k!(24-k)!} \approx \frac{6.2255 \times 10^{20}}{3^k k!(24-k)!}.$$

For example,

$$b_{24,\,0.25}(6) = \frac{6.2255 \times 10^{20}}{3^6 6! 18!} \approx 0.1853.$$

The full set of results is given in Table 3.4.1 and displayed in Figure 3.4.2a. $\square$

> **Check Your Understanding 3.4.2:**
> Verify any two of the results in Table 3.4.1.

There is no simple formula for the cumulative distribution function for the binomial distribution. This makes some calculations difficult.

Example 3.4.4.

Q) What is the probability of obtaining 10 or more wrinkled seeds in 24 trials?
A) The event $k \ge 10$ is the complement of the event $k \le 9$. Using the complement rule, we can write the desired probability in terms of the cumulative distribution:

$$P[K \ge 10] = 1 - P[K \le 9] = 1 - B_{24,\,0.25}(9),$$

Table 3.4.1 Probabilities for the number of wrinkled peas out of 24

k	0	1	2	3	4	5	6
$b_{24,.25}(k)$	0.0010	0.0080	0.0308	0.0752	0.1316	0.1755	0.1853
k	7	8	9	10	11	12	13
$b_{24,.25}(k)$	0.1588	0.1125	0.0667	0.0333	0.0141	0.0051	0.0016

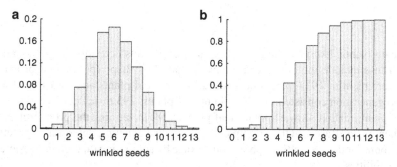

Fig. 3.4.2 (a) The probability distribution function $b_{24,\,0.25}$ and (b) the cumulative distribution $B_{24,\,0.25}$

where $B_{n,p}(k)$ is defined to be the probability of having k or fewer successes in n trials with success probability p. There is no special formula for this calculation. The best we can do is add the relevant probabilities from Table 3.4.1, with the result that the probability of 10 or more successes in 24 trials is 0.0546. The cumulative distribution is displayed in Figure 3.4.2b.

□

Human intuition is a great tool for mathematical modeling, but it can be misleading in probability. The result of Example 3.4.4 is surprising if you think about its significance. Suppose 20 other biologists did the same experiment as Mendel at the same time, and each reported his/her work to the others. Each would have collected 24 seeds from the first generation plants. According to the probability calculations, it is very likely that at least one would have found at least 10 wrinkled peas out of 24. Based on his data, this biologist would have proposed an incorrect theory of inheritance and faced ridicule by his colleagues. Yet the unusual result would be due to neither an error in Mendelian genetics nor experimental error by the biologist. As Example 3.4.4 shows, it is a predictable consequence of demographic stochasticity. The difficulty for the individual biologist doing experimental work is that she might unwittingly collect an unusual data set that lies far from the population mean in spite of flawless experimental design and technique. This point cannot be overemphasized:

A random sample is not necessarily a representative sample.

The only cure for this problem is to use a larger sample. This issue is more broadly addressed in Sections 4.1 and 4.3.

Check Your Understanding Answers

1.

$$b_{3,1/3}(0) = \frac{8}{27}, \qquad b_{3,1/3}(1) = \frac{12}{27}, \qquad b_{3,1/3}(2) = \frac{6}{27}, \qquad b_{3,1/3}(3) = \frac{1}{27}.$$

Problems

3.4.1. Tomato plants, like pea plants, have a number of traits determined by a small number of genes. These include the height (tall or dwarf) and leaf shape (potato leaf or cut leaf). A genetics experiment in 1931 was done with tall potato-leaf plants and dwarf cut-leaf plants. The second generation consisted of 926 tall cut-leaf plants, 293 dwarf cut-leaf plants, 288 tall potato-leaf plants, and 104 dwarf potato-leaf plants [10]. Determine the probabilities for each of these phenotypes. Based on the principles of Mendelian genetics, determine a genetic model that can account for these results. Support your choice by computing theoretical expected values for the probabilities.

3.4.2. Suppose the probability that a bird nest will be raided by a predator in any given day is 0.1. Assume that the fledglings survive if a nest is kept safe for 20 days. Determine the probability that the fledglings survive. Express this probability in terms of the binomial distribution.

3.4.3.* Suppose a predator succeeds in 30 % of its attacks on prey. How likely is it that it will need at least three tries to achieve its first success? What about five or more tries? State the probability of needing y or more tries in terms of the binomial distribution.

3.4.4. Suppose pairs of birds have a 25 % chance of a successful nest. Given 10 pairs of birds, find the complete set of probabilities for the number of successful nests. Prepare a histogram and find the mean and standard deviation.

3.4.5. According to human genetics professor Daniel Geschwind, "Six out of the past 12 presidents [being left-handed] is statistically significant, and probably means something" [6, 11, 13].[29] Suppose the probability of left-handedness is 0.15.[30] Find the probability distribution for the number of left-handed people in a group of 12 randomly chosen people. Prepare a histogram and find the mean and standard deviation. How many standard deviations away from the mean is the outcome $X = 6$? To what extent do you think these calculations support Geschwind's view?[31]
(This problem is continued in Problem 4.4.8.)

3.4.6. The chi-square test in statistics was developed by Karl Pearson using data obtained by his colleague Walter Frank Raphael Weldon, who reported the results from 26,306 runs of an experiment consisting of rolls of a set of 12 dice [12].[32] The data is reproduced in Table 3.4.2.

(a) Determine the probability distribution for the number of successes in 12 die rolls, with "5" and "6" counted as successes. Prepare histograms of this distribution and Weldon's corresponding data. Do the dice appear to be fair?

---

[29] For a more scholarly discussion, see [9].

[30] This estimate is probably a little high, but not much. It is hard to know how many people are naturally left-handed, since some natural left-handers, such as the author's sister, were "trained" in school to be right-handed. Irrational prejudice against left-handers dates back many centuries; for example, the Latin word for "left-handed" is the original source of the English word "sinister."

[31] Even if 6 of 12 is highly unusual, it does not mean that the result is significant. While any one coincidence discovered in a group of 12 people is unusual, there was no particular reason to look for the specific coincidence of left-handedness. The number of possible coincidences that could be discovered is probably quite large, so perhaps there is nothing unusual in discovering one.

[32] Why the dice were rolled 26,306 times, rather than some larger or smaller number, is lost to posterity. According to Weldon, 7,006 rolls were done by a clerk deemed "reliable and accurate," but Weldon did the other 19,300 rolls himself. There is no evidence that Weldon's experiments inspired the invention of the game "Yahtzee."

(b) Use a stopwatch to estimate how long it takes to do 10 runs of the experiment (rolling 12 dice and recording the results as a list of tick marks). Estimate the time Weldon personally spent doing 19,300 experiment runs. Can you imagine a professional scientist rolling a set of dice 19,300 times for a scientific experiment?[33]

(This problem is continued in Example 4.4.3.)

Table 3.4.2 The frequency of N successes in 26,306 rolls of 12 dice, where a success means a roll of "5" or "6"

N	0	1	2	3	4	5	6	7	8	9	10	11	12
X	185	1,149	3,265	5,475	6,114	5,194	3,067	1,331	403	105	14	4	0

3.4.7. The geometric distribution is used for experiments in which Bernoulli trials with success probability p are performed until the first success, with X being the number of failures prior to that first success.

(a) Use the binomial distribution formula and the sequence rule to determine the probability distribution function for the geometric distribution.
(b) Determine the probability distribution function for the number of codons in a strand of non-coding DNA before the first stop codon. [Use the result of Problem 3.2.4 as the probability that one randomly selected codon is a stop codon.]

3.4.8. A predator must capture 10 prey to grow large enough to reproduce. The probability of capturing prey on a given day is 0.1.

(a) Find a formula for the probability that it takes n days for the predator to reproduce. [Hint: Think of this scenario as a sequence of two experiments, in which the second experiment is a success on the nth day. What is the first experiment in the sequence?]
(b) Write a computer program to calculate the probabilities for n from 10 to 300. Note that you may not be able to calculate factorials of large numbers; however, a ratio of factorials is equivalent to a product of consecutive numbers, which is easier for a computer to calculate. Test your program by having it compute the sum of all the probabilities. Note that this sum should be the probability that the predator finds 10 meals in 300 or fewer days.
(c) Based on straightforward reasoning, what should the mean be?
(d) Use your program to calculate the mean and standard deviation for the distribution. Note that your list of probability values does not go beyond $n = 300$. For the purpose of expected value calculations, assume that the full probability for $n > 300$ is actually for $n = 301$. With this adjustment, the mean should be very close to your result for part (c).

3.5 Continuous Probability Distributions

After studying this section, you should be able to:

- Discuss the concept of a continuous probability distribution, including the probability density function.
- Use a cumulative distribution function to compute the probability for a continuous random variable to fall in a given interval.

---

[33] Nowadays, a professional scientist would have his/her graduate student roll the dice 19,300 times.

- Compute means and standard deviations for continuous probability distributions.

In Section 3.3, we defined discrete probability distributions using a probability distribution function f, defined by

$$f(x) = P[X = x].$$

We would like to define continuous probability distributions in the same way, but there is a problem.

Example 3.5.1. The probability of obtaining exactly 200 heads in 400 tosses of a fair coin can be computed using the binomial distribution:

$$b_{400,0.5}(200) = \frac{400!}{(200!)^2} \left(\frac{1}{2}\right)^{200} \left(\frac{1}{2}\right)^{200} = \frac{(400)!}{4^{200}(200!)^2} \approx 0.040.$$

Similarly, the probability of obtaining exactly 800 heads in 1,600 tosses is 0.020. In general, it can be shown that the probability of getting heads in *exactly* one-half of a large number of Bernoulli trials is approximately

$$b_{2m,\,0.5}(m) \approx \frac{1}{\sqrt{m\pi}},$$

where m is the number of heads and $2m$ is the number of trials. Thus, the probability of exactly 50 % heads approaches 0 as the number of trials increases. □

Example 3.5.1 illustrates the general principle that as the number of possible outcomes increases, even the most common of them becomes very unlikely. When that number becomes infinite, we lose the ability to calculate the probability of an individual outcome altogether.

Example 3.5.2. A dozen large eggs in the United States must weigh a minimum of 24 oz. Assuming that a dozen large eggs always weighs between 24 and 24.2 oz, what is the probability that a dozen large eggs weighs *exactly* 24.1 oz?

The answer is that the probability is 0. One argument for this is to note that the sample space that consists of the interval $24 \leq x \leq 24.2$ contains infinitely many values, of which 24.1 is only one. With equally probable outcomes, the probability of the outcome 24.1 would be $1/\infty = 0$. Even if 24.1 is the *most likely* outcome, it remains only one outcome out of infinitely many, and the probability is still 0. □

Example 3.5.2 points out the fundamental difficulty of continuous probability distributions.

It is not possible to define a continuous probability distribution with an algebraic formula that prescribes the probabilities of the individual outcomes.

Nevertheless, we can define and work with continuous probability distributions by shifting our focus from a probability distribution function to a cumulative probability distribution function.

3.5.1 Cumulative Distribution Functions

In our study of discrete probability distributions, we had two alternative ways to define a probability distribution: probability distribution functions, which prescribe probabilities of specific outcomes, and cumulative distribution functions, which prescribe probabilities that a random variable does not exceed a specific outcome. Only the latter works for continuous probability distributions.

Example 3.5.3. Table 3.5.1 shows data for the length of petals in a species of iris.[34] Figure 3.5.1b shows the relative frequencies of the data as a histogram. For comparison, Figure 3.5.1a shows the relative frequencies of the same data, but with bins of width 0.2 cm. These relative frequencies correspond to experimental probabilities of obtaining a particular measurement from a randomly chosen flower. The corresponding cumulative probabilities are shown in Figure 3.5.1c,d. In the relative frequency plots, the total probability (always exactly 1) is the sum of the bar heights. With twice as many bars in Figure 3.5.1b as compared to Figure 3.5.1a, the bar heights are only half as large. Thus, the scales on the vertical axes are different. This difference does not occur for the cumulative frequency plots because each bar height represents the total probability up to that point, which is independent of the way in which outcomes are grouped into bins.   □

Table 3.5.1 Frequencies of petal length in centimeters of *Iris versicolor* flowers

Length	3.0	3.1	3.2	3.3	3.4	3.5	3.6	3.7	3.8	3.9	4.0
Frequency	1	0	0	2	0	2	1	1	1	3	5
Length	4.1	4.2	4.3	4.4	4.5	4.6	4.7	4.8	4.9	5.0	5.1
Frequency	3	4	2	4	7	3	5	2	2	1	1

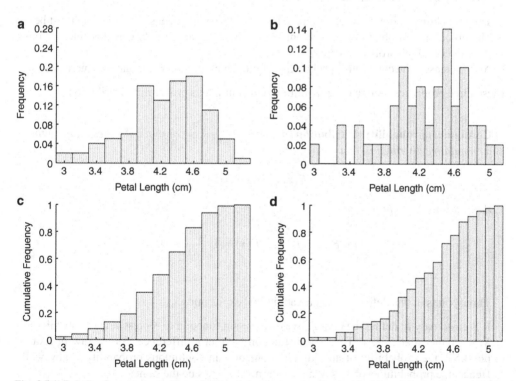

Fig. 3.5.1 Relative frequencies and cumulative relative frequencies for the petal length in centimeters of *Iris versicolor* flowers (**a**) and (**b**): frequencies with bin widths of 0.2 cm and 0.1 cm; (**c**) and (**d**): cumulative frequencies with bin widths as in (**a**) and (**b**)

---

[34] This data is part of a larger set in a classic paper by Ronald A. Fisher [3].

Check Your Understanding 3.5.1:
Use the data in Table 3.5.1 to make a table of relative frequencies and cumulative frequencies with bin widths of 0.2 cm. Remember to round half of the frequency for an odd value up to the nearest 0.2 and half down to the nearest 0.2. Check your table by comparing it to Figure 3.5.1a,c.

The data in Table 3.5.1 is given to the nearest millimeter, which makes it essentially discrete. In principle, however, petal lengths could be reported to any desired degree of precision, which would allow us to plot histograms with progressively narrower bins. As Figure 3.5.1a,b shows, the relative frequencies of each bin become smaller as the bins narrow. However, Figure 3.5.1c,d shows that the cumulative frequencies do not become smaller as the bins narrow. In principle, we could make the bins arbitrarily narrow, and the cumulative frequencies would remain meaningful.

Example 3.5.3 and the subsequent discussion show that cumulative probability distribution functions make sense for a continuous probability distribution; the challenge is to find a suitable definition. The definition used for discrete probability distributions (Equation (3.3.2)) does not work because it relies on the probability distribution function. The key is to understand three basic properties of cumulative probability. The cumulative probability for any numerical value in a data set, such as that of Table 3.5.1, is the sum of the height of its corresponding bar in the relative frequency histogram and all the bars to the left. Thus:

1. The probability of obtaining an outcome less than the smallest possible outcome must be 0.
2. The probability of obtaining an outcome less than or equal to the largest possible outcome (the sum of all the probabilities) must be 1.
3. As x increases, the cumulative probability of outcomes less than or equal to x increases.

These three properties motivate the definition of a continuous probability distribution.

Continuous probability distribution: *the interval $(-\infty, \infty)$ of real outcomes along with a cumulative distribution function*

$$F(x) = P[X \leq x]$$

that satisfies the requirements

$$\lim_{x \to -\infty} F(x) = 0, \ \lim_{x \to \infty} F(x) = 1, \ \frac{dF}{dx} \geq 0.$$

Three points in this definition require additional explanation.

1. It is customary to define the cumulative distribution function to represent the probability of $X \leq x$ rather than $X < x$. This is a distinction without a difference. As we saw in Example 3.5.2, the probability of an individual outcome in continuous probability is always 0. Hence, the probabilities of $X \leq x$ and $X < x$ must always be the same.
2. The derivative of the cumulative distribution function is the rate at which probability is accumulated as the interval widens. Normally this will be strictly positive, but it could be 0 for an interval of impossible outcomes. For example, if the distribution represents the amount of time that elapses before some event occurs, then $F(0) = 0$ and dF/dx is 0 on the interval $(-\infty, 0)$.

3. For the sake of generality, the definition assumes that all real numbers are possible outcomes, but the definition could be revised for smaller intervals. For example, if outcomes are restricted to the interval $(0, \infty)$, then the property $F(-\infty) = 0$ can be replaced by $F(0) = 0$.

A cumulative distribution function can be used to calculate the probability for any interval.

Example 3.5.4. Consider the continuous probability distribution defined by the outcomes $(0, \infty)$ and the cumulative distribution function $F(x) = 1 - e^{-x}$, as seen in Figure 3.5.2. Note that this function has the required properties $F(0) = 0$, $\lim_{x \to \infty} F = 1$, and $F' = e^{-x} \geq 0$. We can easily use F to calculate probabilities of intervals starting at 0:

$$P[0 \leq X \leq 1] = F(1) = 1 - e^{-1} \approx 0.63, \quad P[0 \leq X \leq 2] = F(2) = 1 - e^{-2} \approx 0.86.$$

If we want to calculate the probability for an interval that does not start at 0, we can look at the difference between two intervals that do start at 0. For example, the interval $1 < X \leq 2$ includes all of those outcomes that are in the interval $0 \leq X \leq 2$ but not in the interval $0 \leq X \leq 1$. Using the complement rule (Equation (3.2.3)), the probabilities of $0 \leq X \leq 1$ and $1 < X \leq 2$ must add up to the probability of $0 \leq X \leq 2$. Thus,

$$P[1 < X \leq 2] = F(2) - F(1) \approx 0.23.$$

As in the first note in the discussion of the definition, the exclusion of 1 and the inclusion of 2 from the interval $1 < X \leq 2$ is a matter of custom and has no practical significance. □

Check Your Understanding 3.5.2:
Calculate $P[2 < X \leq 3]$ for the distribution of Example 3.5.4.

3.5.2 Probability Density Functions

A discrete distribution can always be obtained from continuous data by partitioning the continuous interval of outcomes into a discrete collection of subintervals, just as in Example 3.5.1. By examining a sequence of discrete approximations with more and more subintervals, we can

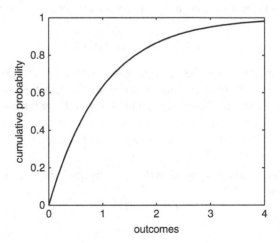

Fig. 3.5.2 The cumulative distribution function $F(x) = 1 - e^{-x}$ with outcomes $(0, \infty)$

discover another approach to continuous probability. First, however, we need to make an important modification to our histogram plots. The numerical values on the vertical axes of the histograms in Figure 3.5.1a,b represent the frequencies of ranges of outcomes. Thus, cutting the interval width in half cuts the probabilities approximately in half as well, which means that the range of values used for the vertical axis changes. This property is what prevents us from generalizing the probability distribution function to continuous distributions.

The fundamental insight needed to proceed is this: since both the probability and the width are cut in half, the ratio of probability to width stays the same. This observation suggests that we could get a better comparison of histograms if we changed the definition of the vertical coordinate; instead of using the probability for the given bin, we could divide the probability by the interval width. This change would not alter the appearance of the histogram plot, because the relative heights are unaffected in division by a fixed number. However, it would keep the numbers marked on the vertical axis from decreasing.

Example 3.5.5. Figure 3.5.3a,b,c shows histograms for the probability distribution with outcomes $(0, \infty)$ and the cumulative distribution function $F(x) = 1 - e^{-x}$. The heights of the first three bars in Figure 3.5.3a were calculated in Example 3.5.4 and Check Your Understanding 1, and the others are obtained by similar calculations. The three plots of Figure 3.5.3a,b,c have interval widths of 1, 0.5, and 0.125 respectively, with corresponding y coordinates of 1, 2, and 8 times the frequency. With this change, the scales on the y axis are the same for all three plots.

□

As a consequence of the change to plotting frequency divided by bin width on the vertical axis, the probability is no longer represented by the height of the bars. Nevertheless, we can still visualize it in the histogram because we get frequency if we multiply the vertical coordinate (frequency/width) by the horizontal coordinate (width). This means that the probability for a given interval of outcomes is represented by the *area* under the curve within that interval. This association of probability with area means that the areas, rather than the heights, add up to 1. Each bar in Figure 3.5.3a corresponds to two bars in Figure 3.5.3b, each of roughly the same height and half the width of the original. Thus, the area corresponding to a particular interval of outcomes is preserved as the bins in the histogram are narrowed. The same statements can be made to compare Figure 3.5.3b,c, but with a factor of 4 rather than 2. If we continue the refinement process, the plots eventually approach a smooth curve, as shown in Figure 3.5.3d. The area under this curve is 1, and the probability of an outcome within any particular interval is the area encompassed by that interval. The curve is called a **probability density function**.

Because probability is the area under a probability density function f, we can use the integral formula

$$P[a < X \leq b] = \int_a^b f(x)\,dx \tag{3.5.1}$$

to compute probabilities. Probability density functions are therefore an alternative to cumulative distribution functions for defining a continuous probability distribution. In general, the probability density function f and cumulative distribution function F are related by the equation

$$\int_{-\infty}^x f(\bar{x})\,d\bar{x} = P[X \leq b] = F(x). \tag{3.5.2}$$

Differentiating (3.5.2) reveals the relationship between the probability density function and the cumulative distribution function:

$$F'(x) = f(x). \tag{3.5.3}$$

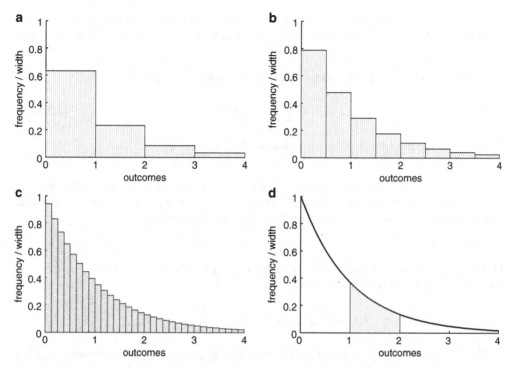

Fig. 3.5.3 Histograms for the cumulative distribution function $F(x) = 1 - e^{-x}$ with (**a**) $\Delta x = 1$, (**b**) $\Delta x = 0.5$, (**c**) $\Delta x = 0.125$, with heights computed as frequency divided by width, and (**d**) the probability density function $f(x) = e^{-x}$, showing the area corresponding to $P[1 < X \le 2]$

Hence, we can recast the properties of cumulative distribution functions as corresponding properties of probability density functions:

$$f(x) \ge 0, \qquad \int_{-\infty}^{\infty} f(x)\,dx = 1. \tag{3.5.4}$$

Example 3.5.6. Let $F(x)$ be the cumulative distribution function $F(x) = 1 - e^{-x}$. From (3.5.3), the probability density function is

$$f(x) = F'(x) = e^{-x}.$$

We can calculate probabilities by integrating this probability density function; for example,

$$P[1 < X \le 2] = \int_1^2 e^{-x}\,dx.$$

$\square$

3.5.3 Expected Value, Mean, and Variance

In our study of discrete probability distributions, we used the concept of expected value to define the mean (Equation (3.3.5)) and variance (Equation (3.3.6)) of a distribution as

$$\mu = EX, \qquad \sigma^2 = EX^2 - (EX)^2. \tag{3.5.5}$$

These definitions for mean and variance carry over to continuous distributions once we have a continuous analog of the formula for the expected value for the discrete case (Equation (3.3.4)).

The expected value of a function g of a continuous random variable X is given by

$$Eg(X) = \int_{-\infty}^{\infty} g(x)f(x)\,dx, \qquad (3.5.6)$$

where f is the probability density function for the distribution.

Note that the integral in the expected value formula is improper, so it may not converge. While this is mathematically possible, it will not occur for $g = x$ or $g = x^2$ (the functions needed to find mean and variance) and any useful probability density function f.

Example 3.5.7.

Q) Find the mean and variance for the distribution given by $F(x) = 1 - e^{-x}$.
A) Given the probability density function $f(x) = e^{-x}$, we have

$$EX = \int_0^{\infty} xe^{-x}\,dx, \qquad EX^2 = \int_0^{\infty} x^2 e^{-x}\,dx.$$

We could compute these integrals using integration by parts, a method commonly taught in the second semester of calculus. However, we can avoid these calculations by making use of the well-known integral formula

$$\int_0^{\infty} x^n e^{-x}\,dx = n!, \quad n = 0, 1, \ldots.$$

Thus,

$$\mu = EX = 1, \quad \sigma^2 = EX^2 - (EX)^2 = 2 - 1 = 1.$$

The mean, the variance, and the standard deviation (the square root of the variance) are all 1.

$\square$

Check Your Understanding Answers

2. 0.086

Problems

The data sets for Problems 3.5.2–3.5.6 are from [5] and can be found in both csv and xlsx form at http://www.math.unl.edu/~gledder1/MLS/, http://www.springer.com/978-1-4614-7275-9.

3.5.1.* A fair coin is flipped $2n$ times.

(a) What is the probability that there have been exactly n heads?
(b) What is the probability of exactly 10 heads in 20 coin flips?
(c) What is the probability of exactly 50 heads in 100 coin flips?

(d) Conclude that it is rare for an experimental probability to be exactly equal to the theoretical probability. We will revisit this issue when we study the probability distributions of sample means.

3.5.2.* The file SOLDIERS.* contains measurements of the chest circumferences of 5,732 Scottish soldiers, to the nearest inch.[35]

(a) Prepare a table of relative frequencies for chest size rounded to the nearest 2 in. Round half of the frequency for each odd inch up and half down.
(b) Plot histograms of the original relative frequencies and the relative frequencies from (a). Note the values on the vertical axes.
(c) Plot histograms for the cumulative relative frequencies. Again note the values on the vertical axes.
(d) Plot the histograms of the relative frequencies again, but this time divide the frequencies by the bin width. Again note the values on the vertical axes.

(This problem is continued in Problem 3.6.6.)

3.5.3.* The file CUCKOOS.* contains measurements of the lengths, to the nearest 0.5 mm, of the eggs of a species of cuckoo.

(a) Prepare a table of relative frequencies for the egg length rounded to the nearest millimeter. Round half of the frequency for each odd half-millimeter up and half down.
(b) Plot histograms of the original relative frequencies and the relative frequencies from (a). Note the values on the vertical axes.
(c) Plot histograms for the cumulative relative frequencies. Again note the values on the vertical axes.
(d) Plot the histograms of the relative frequencies again, but this time divide the frequencies by the bin width. Again note the values on the vertical axes.

(This problem is continued in Problem 3.6.7.)

3.5.4. The file BUTTER.* contains measurements of the butterfat content of the milk from several breeds of cows, measured in hundredths of a percent.

(a) Prepare a table of relative frequencies for the butterfat content of all breeds, rounded to the nearest 0.2.
(b) Repeat part (a) but with rounding to the nearest 0.1.
(c) Repeat part (a) but with rounding to the nearest 0.05.
(d) Plot histograms of the relative frequencies in (a), (b), and (c). Note the values on the vertical axes.
(e) Plot histograms for the cumulative relative frequencies. Again note the values on the vertical axes.
(f) Plot the histograms of the relative frequencies again, but this time divide the frequencies by the bin width. Again note the values on the vertical axes.

(This problem is continued in Problems 3.6.12 and 3.6.13.)

---

[35] This data set was originally published in the *Edinburgh Medical and Surgical Journal* in 1817. Since then, it has been used as an example for the development of statistical methods. The full history of the data set is recounted in [17].

3.5.5. The file IRISES.* contains measurements of the lengths and widths of petals and sepals of iris flowers of three different species.

(a) Prepare a table of relative frequencies for the sepal length of *Iris setosa*.
(b) Prepare a table of relative frequencies for the sepal lengths rounded to the nearest 0.2 mm. Round half of the frequency for each odd half-millimeter up and half down.
(c) Plot histograms of the relative frequencies in (a) and (b). Note the values on the vertical axes.
(d) Plot histograms for the cumulative relative frequencies. Again note the values on the vertical axes.
(e) Plot the histograms of the relative frequencies again, but this time divide the frequencies by the bin width. Again note the values on the vertical axes.

(This problem is continued in Problem 3.6.10.)

3.5.6. Repeat Problem 3.5.5 using the data for the sepal length of *Iris virginica*.
(This problem is continued in Problem 3.6.11.)

3.5.7. Let $F(x) = 1 - e^{-0.2x}$ for $x \geq 0$ and $F(x) = 0$ for $x < 0$.

(a) Verify that F satisfies the requirements for a cumulative distribution function.
(b) Find the corresponding probability density function.
(c) Find the probability $P[4 \leq X \leq 6]$.
(d) Plot the probability density function along with lines and shading to mark the probability in part (c).
(e) Find the mean and standard deviation of this distribution, using the integral formula

$$\int_0^\infty x^n e^{-x} dx = n!, \quad n = 0, 1, \ldots$$

along with substitution in the definite integral.
(f) Find the probability $P[X \leq \mu]$.
(g) Find the probabilities $P[\mu - \sigma \leq X \leq \mu]$ and $P[\mu \leq X \leq \mu + \sigma]$.
(h) Prepare histograms similar to those of Figure 3.5.3a–c, using bin widths of 4, 2, 1, and 0.5.

3.6 The Normal Distribution

After studying this section, you should be able to:

- Identify contexts in which a normal distribution might be appropriate.
- Use a probability density function and cumulative distribution function for calculations.
- Discuss the significance of the parameters in a normal distribution.

If you measure the masses of 100 seeds from a seed packet, group them into appropriately chosen intervals, and prepare a histogram of the data, you will almost certainly see a "bell-shaped curve." This sort of curve is characteristic of what is called the *normal distribution*. In addition to fitting many data sets, the normal distribution has some mathematical properties that make it relatively easy to interpret, as will be seen in the following discussion. We begin with the mathematical definition.[36]

---

[36] Note the use of $\bar{x}$ inside the definite integral rather than x. The symbol for the integration variable must be distinct from that for the independent variable. Any symbol other than x can be chosen; the advantage of $\bar{x}$ is that it serves as a reminder that we are integrating over values of x.

Normal distribution: *the continuous probability distribution with outcomes* $-\infty < x < \infty$
and probability density function

$$N'_{\mu,\sigma}(x) = \frac{1}{\sqrt{2\pi\sigma^2}} \exp\left(-\frac{(x-\mu)^2}{2\sigma^2}\right), \qquad (3.6.1)$$

where μ *is any real number and* $\sigma > 0$. *The cumulative distribution function is then given
by*

$$N_{\mu,\sigma}(x) = P[X \le x] = \int_{-\infty}^{x} N'_{\mu,\sigma}(\bar{x})\, d\bar{x}. \qquad (3.6.2)$$

The normal distribution definition raises several questions, including these.

1. How do we calculate normal distribution probabilities without a simple formula for the cumulative distribution function N?
2. Is it just a coincidence that the parameters in the normal distribution formula have the symbols we have been using for the mean and standard deviation?
3. The normal distribution has sample space $(-\infty, \infty)$. Does this mean we cannot use it for random variables that must always be positive?

Question 1 has two fully satisfactory answers. In the pre-computer era, people referenced tables of normal distribution values. These were readily available and not difficult to use. Now, any good mathematical software has a built-in function for their calculation, as do any handheld calculators designed for scientific use.[37] Questions 2 and 3 are more easily addressed after an intuition-building example.

Example 3.6.1. In Section 3.5, we examined a data set consisting of the petal lengths of iris flowers. The data has mean $\mu = 4.26$ and standard deviation $\sigma = 0.47$. Could this data fit a normal distribution? Figure 3.6.1 shows a histogram of the data set along with a plot of the probability density function $N'_{4.26,0.47}$. The data does not seem to fit the distribution very well, but this could be because the sample size is so small. If the correct distribution *is* normal, it appears plausible that the mean and standard deviation of the distribution are the same as that for the data.[38] □

3.6.1 The Standard Normal Distribution

Complete answers to Questions 2 and 3 are easier to give after we have examined the *standard normal distribution*.

---

[37] In essence, we now calculate normal distribution values the same way we calculate values for exponential and trigonometric functions.

[38] The question of whether a data set is approximately normal is addressed in Section 4.2.

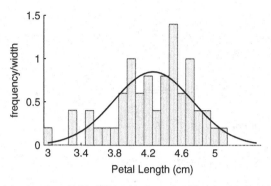

Fig. 3.6.1 A histogram of the data from Table 3.5.1 with the probability density function $N'_{4.26,0.47}$

Standard normal distribution: *the continuous probability distribution with outcomes* $-\infty < z < \infty$, *probability density function*

$$N'(z) = \frac{1}{\sqrt{2\pi}} \exp\left(-\frac{z^2}{2}\right),\tag{3.6.3}$$

and cumulative distribution function

$$N(z) = P[Z \le z] = \int_{-\infty}^{z} N'(\bar{z})\,d\bar{z}.\tag{3.6.4}$$

Note that the standard normal distribution is just the normal distribution with $\mu = 0$ and $\sigma = 1$. These values can be omitted from the notation because of the standard practice of using the random variable Z and outcomes z for the standard normal distribution rather than the symbols X and x used for a general normal distribution.

We noted in the answer to Question 1 above that normal distribution values have been calculated from tables. One would expect difficulties in creating tables for distributions with two continuous parameters. This is generally true, but normal distributions have an amazing mathematical property (see Problem 3.6.1):

Theorem 3.6.1 (Relationship of Normally Distributed Random Variables). *Let X be a normally distributed random variable with parameters μ and σ. Then the random variable*

$$Z = \frac{X - \mu}{\sigma}\tag{3.6.5}$$

is normally distributed with mean 0 and standard deviation 1.

Theorem 3.6.1 explains why it is possible to calculate values for a normal distribution using tables. Only one table—that for the standard normal distribution—is necessary.

Example 3.6.2. Suppose we want to calculate the probability that a random variable from the distribution $N_{10,2}$ is less than 8. There are two ways we can do this using mathematical software. First, we can use the specific cumulative distribution function:

$$P[X < 8] = N_{10,2}(8) \approx 0.1587.$$

Second, we can rewrite the statement $X < 8$ in terms of the standard normal random variable Z. Using $z = (x - \mu)/\sigma = (8 - 10)/2 = -1$, we have

$$P[X < 8] = P[Z < -1] = N(-1) \approx 0.1587.$$

These probabilities are represented as areas in Figure 3.6.2. Note that the areas are the same because the scale on the horizontal axis in Figure 3.6.2a is twice as large as that in Figure 3.6.2b, while the scale on the vertical axis is only half as large.   □

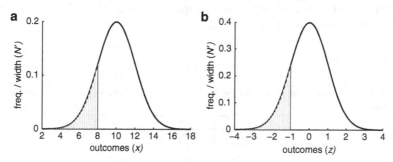

Fig. 3.6.2 The probabilities (**a**) $N_{10,2}(8)$ and (**b**) $N(-1)$

We are now ready to address Question 2 from the beginning of the section. From Section 3.5, the mean of a continuous probability distribution F is

$$EX = \int_{-\infty}^{\infty} xF'(x)\,dx;$$

hence, the mean of the standard normal distribution is

$$EZ = \int_{-\infty}^{\infty} \frac{z}{\sqrt{2\pi}} \exp\left(-\frac{z^2}{2}\right)\,dz.$$

We can evaluate this integral without actually integrating it if we notice that the integrand is an *odd* function; that is, it satisfies the property $f(-z) = -f(z)$. Every contribution to the integral from a point $z > 0$ is exactly balanced by a negative contribution from the corresponding point $-z < 0$. Thus, the mean of the random variable Z is 0. From (3.6.5), this means that the mean of any random variable X from a normal distribution with parameter μ is the parameter μ. Hence, it is not coincidental that the symbol for the first parameter in the normal distribution is the same as the symbol used for the mean of a distribution. Similarly, the parameter σ is the standard deviation, although the computation of the fact $EZ^2 = 1$ is beyond the scope of our presentation. The results are summarized in Theorem 3.6.2.

Theorem 3.6.2 (Mean and Standard Deviation of the Normal Distribution). *The normal distribution $N_{\mu,\sigma}$ has mean μ and standard deviation σ.*

Question 3 from the beginning of the section can now be addressed in terms of the standard normal distribution. Again, we first consider a motivating example.

Example 3.6.3.

Q) What is the probability that a normally distributed random variable X is more than n standard deviations less than or greater than the mean?

A) By Theorem 3.6.1, the answer to this question is independent of the numerical values of μ and σ; hence, it is sufficient to consider only the standard normal distribution. We have $P[Z < -n] = N(-n)$. The results are shown in Table 3.6.1. Out of a population of one million values drawn from the standard normal distribution, the expected number of values less than -5 is about $1/3$.[39] As a general rule, values more than 5 standard deviations away from the mean of a normal distribution are extremely rare. Even 4 standard deviations away from the mean is highly unlikely; on the average, only about three values out of 100,000 are that far from the mean. By symmetry, $P[Z > z] = P[Z < -z]$.

□

Table 3.6.1 Probabilities of outcomes more than n standard deviations below the mean for a normally distributed random variable

z	-1	-2	-3	-4	-5	-6
$P[Z < z]$	0.1587	0.0228	0.0013	0.00003	3×10^{-7}	10^{-9}

Example 3.6.3 helps us understand when we can use the normal distribution for data that can only be positive, such as measurements of iris petal lengths. The problem is that the normal distribution, regardless of the mean and standard deviation, will have a nonzero probability of values less than 0. As always, we must remember that we are using distributions as *models* for data sets rather than as *descriptions* of data sets. In cases where the standard deviation is less than 20% of the mean, the outcome $x = 0$ lies 5 or more standard deviations away from the mean. As shown in Example 3.6.3, the probability of a value corresponding to a negative measurement, which would be at least 5 standard deviations from the mean, is extremely small; hence, we should not shy away from the normal distribution in such cases. When the standard deviation is a larger percentage of the mean, we should be more cautious. Certainly the normal distribution should not be used if the standard deviation is more than one-third of the mean.

Since the probabilities in the normal distribution are given in terms of the number of standard deviations away from the mean, we can do a general calculation of the probabilities for a collection of intervals delimited by multiples of the standard deviation. Figure 3.6.3 illustrates the probabilities for intervals whose width is one-half of the standard deviation.

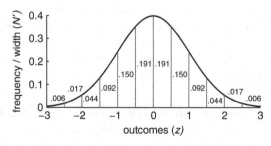

Fig. 3.6.3 The probability distribution function for the standard normal distribution, showing the probabilities associated with adjacent multiples of $\sigma = 0.5$

---

[39] Think of drawing a value from the normal distribution as a Bernoulli trial. For one million Bernoulli trials, each with probability 3×10^{-7}, the expected number of successes is the product 0.3.

3.6.2 Standard Probability Intervals

The strong connection between the standard deviation and probability values allows us to define intervals corresponding to specific probabilities.[40]

Example 3.6.4.

Q) Find the probability that a normally distributed random variable lies within 1.96 standard deviations of the mean.

A) We do not need to work with the random variable X to answer this question because the z value of 1.96 is given. By the complement rule and the symmetry of the normal distribution,

$$P[-1.96 < Z \le 1.96] = 1 - P[Z < -1.96] - P[Z > 1.96] = 1 - 2P[Z < -1.96]$$
$$= 1 - 2N(-1.96) = 0.9500.$$

$\square$

Check Your Understanding 3.6.1:
Find the probability that a normally distributed random variable lies within 2.58 standard deviations of the mean.

Example 3.6.4 shows that 95 % of measurements for a normally distributed random variable are within 1.96 standard deviations of the mean; that is, in the interval $(\mu - 1.96\sigma, \mu + 1.96\sigma)$. Similarly, 99 % of measurements lie within 2.58 standard deviations of the mean. The probabilities 0.99 and 0.95 are an accepted part of statistics practice, although the specific choices of 95 % and 99 % are arbitrary. One could just as easily focus on 96 % or 98 %.[41]

Problems

The data sets for Problems 3.6.6–3.6.13 are from [5] and can be found in both csv and xlsx form at http://www.math.unl.edu/~gledder1/MLS/, http://www.springer.com/978-1-4614-7275-9.

3.6.1. Derive Theorem 3.6.1.

3.6.2.* Suppose X is normally distributed with mean 10 and standard deviation 2.

(a) Find the probability $P[X > 13]$.
(b) Find the probability $P[11 < X \le 13]$.

3.6.3. Suppose X is normally distributed with mean 7.1 and standard deviation 1.6. Find $P[X > 9]$.

3.6.4. Suppose X is normally distributed with mean 4.7 and standard deviation 0.5. Find $P[4 < X \le 5]$.

3.6.5. Suppose $\ln X$ is normally distributed with mean -1 and standard deviation 1.

---

[40] These are *not* confidence intervals, which are discussed in Section 4.3.

[41] In general, statistics can be described as "probability augmented by arbitrary standards." The advantage of such standards is that they provide a unified language to use in comparing results, but the disadvantage is that the standards can acquire a perceived significance out of proportion to their actual value.

(a) Find the probability $P[X < 1]$. [Change the statement to a statement about $\ln X$ and then use the normal distribution.]
(b) Find the probability $P[X > 0.5]$.
(c) Find the symmetric interval corresponding to 95 % of measurements for $\ln X$.
(d) Find the corresponding 95 % interval for X. [In other words, find the range of X values that correspond to the 95 % interval for $\ln X$. This interval is not symmetric.]

3.6.6.* (Continued from Problem 3.5.2.)
Prepare a histogram of the chest measurements of Scottish soldiers in the file SOLDIERS.*. Superimpose the probability density function for the corresponding normal distribution with $\mu = \bar{x}$ and $\sigma = s$. Does the normal distribution seem to be a reasonable model for this data set?

3.6.7. (Continued from Problem 3.5.3.)
Repeat Problem 3.6.6 with the lengths of cuckoo eggs in the file CUCKOOS.*.

3.6.8. Repeat Problem 3.6.6 with the litter size of pigs in the file PIGS.*.

3.6.9. (Continued from Problem 3.1.7.)

(a) Repeat Problem 3.6.6 with the pulse rates of native Peruvians in the file PULSE.*.
(b) Based on part (a), what do you think is the "normal" range of resting pulse rates in this population?
(c) Do an Internet search for information about how normal ranges of physiological parameters are defined and report what you learn.

(This problem is continued in Problem 4.2.5.)

3.6.10. (Continued from Problem 3.5.5.)
Repeat Problem 3.6.6 with the sepal lengths of *Iris setosa* in the file IRISES.*.
(This problem is continued in Problem 4.2.7.)

3.6.11. (Continued from Problem 3.5.6.)
Repeat Problem 3.6.6 with the sepal lengths of *Iris virginica* in the file IRISES.*.
(This problem is continued in Problem 4.2.7.)

3.6.12. (Continued from Problem 3.5.4.)
The file BUTTER.* contains data for the amount of butterfat in the milk of several breeds of cows.

(a) Prepare three relative frequency tables, with bin width 0.2: (1) Jersey cows, (2) Ayrshire cows, (3) Jersey and Ayrshire cows.
(b) Prepare histograms for the three relative frequency tables.
(c) On a single graph, plot the normal probability density functions using (1) the mean and standard deviation for the Jersey data and (2) the mean and standard deviation for the Ayrshire data.
(d) On a single graph, plot (1) the normal probability density function using the mean and standard deviation of the combined Jersey and Ayrshire data and (2) the probability density function obtained by averaging the probability density functions from part (c).
(e) Discuss the differences in the two functions in the plot from part (d), referring to the part (c) graphs.

(f) Compare the graphs in part (d) with the histograms in part (b). Discuss the issue of whether it appears more reasonable to combine the data for these two breeds or to keep them distinct.

(This problem is continued in Problem 4.2.4.)

3.6.13. (Continued from Problem 3.5.4.)
Repeat Problem 3.6.12, but with the Guernsey data instead of the Ayrshire data.
(This problem is continued in Problem 4.2.4.)

3.6.14. Suppose two theoretical populations of normally distributed data are combined into a single data set, as considered in Problems 3.6.12 and 3.6.13. In this problem, we explore the relationship between the shape of the resulting distribution and the number of standard deviations by which the means differ. To keep things from getting too complicated, we consider only the case where both populations have the same standard deviation.

(a) Plot the probability density function f for a random variable consisting of an equal mix of data from the distributions $N_{9,2}$ and $N_{11,2}$. Use the viewing window $2 \leq x \leq 18$, $0 \leq f \leq 0.2$ to facilitate comparison with the probability density function for $N_{10,2}$ that appears in Figure 3.6.2.
(b) Repeat part (a) for a random variable consisting of an equal mix of data from the distributions $N_{8,2}$ and $N_{12,2}$;
(c) Repeat part (a) for a random variable consisting of an equal mix of data from the distributions $N_{7,2}$ and $N_{13,2}$;
(d) Repeat part (a) for a random variable consisting of an equal mix of data from the distributions $N_{6,2}$ and $N_{14,2}$.
(e) How does the shape of a distribution consisting of a sum of two normal distributions depend on the relationship between the standard deviation and the difference between the means?
(f) Until now, we have been assuming that a data set drawn from two different distributions has a probability density function that is the average of that for the two individual distributions. To test this assumption, use computer software to create a data set consisting of 10,000 values chosen from each of $N_{6,2}$ and $N_{14,2}$. Prepare a histogram and compare it with the probability density function from part (d).

(This problem is continued in Problem 4.2.10.)

3.7 The Poisson and Exponential Distributions

After studying this section, you should be able to:

- Identify the contexts in which the Poisson and exponential distributions are appropriate.
- Use the probability distribution function for Poisson distribution calculations.
- Use the cumulative distribution function for exponential distribution calculations.
- Use the formulas for mean, variance, and standard deviation for the Poisson and exponential distributions.
- Discuss the significance of the parameters in the Poisson and exponential distributions.

Consider an event that recurs at random intervals, such as radioactive decay in a sample of uranium or the consumption of flies by a hungry frog. We might want to know the probability that k events happen in a given amount of time, or we might want to know the probability that the next event happens before time t. There are probability distributions for each of these cases; the simplest are the Poisson distribution, which models the probabilities for the number of events occurring in a fixed amount of time, and the exponential distribution, which models probabilities for the amount of time before the next occurrence of an event.

3.7.1 The Poisson Distribution

Example 3.7.1. Suppose you put a frog and some flies in a cage and set a video camera to record the scene. In 10 h, you record the frog eating 120 flies. This means that the *average* rate of fly consumption by the frog is 12 flies per hour. Now suppose you repeat the experiment again the next day with the same frog. We might reasonably ask, "What is the probability that the frog will eat exactly 6 flies in the first half-hour?" Because of demographic stochasticity, we know that an average of 12 per hour is not the same thing as *exactly* 12 in every hour. We need a probability distribution that can address questions about numbers of events over a period of time. □

We cannot use the binomial distribution for Example 3.6.1, because it is meant specifically for settings with a specified number of Bernoulli trials. Although Example 3.6.1 is about the probability of a given number of successes, those successes are distributed over continuous time rather than accumulated through a sequence of individual trials. Example 3.6.1 calls for a *Poisson*[42] *distribution.*

> **Poisson distribution**: *the discrete probability distribution for a random variable consisting of the number of occurrences of an event in a time interval of duration t, given that the events occur with mean rate λ.*

Both the binomial and Poisson distributions compute the probability of a given number of successes. In the binomial distribution, the k successes occur over n discrete trials. In the Poisson distribution, the k successes occur over a continuous time interval of duration t. For the binomial distribution, we need a parameter p to indicate the success probability for each discrete trial; for the Poisson distribution, we need a parameter λ to indicate the expected rate of successes per unit time.

Example 3.7.2. In basketball, each player plays some number of minutes while the clock is running and takes some number of free throws with the clock stopped. The free throws are Bernoulli trials because each is an experiment with outcomes of success or failure. Based on past performance, we can calculate each player's success rate in successes per attempt.[43] Given a success rate p, we can calculate the probability of k successes in a given number of attempts by using a binomial distribution. However, a binomial distribution cannot be used to determine the probability of a player scoring k points or making k baskets in t minutes of playing time, because (1) each minute of playing time could have a variable number of points/baskets, and (2) minutes of playing time are not single experiments. To calculate the probability of a player scoring k points in t minutes, we need a Poisson distribution[44]. The mean rate λ needed for the distribution should be the player's total points divided by the total number of minutes played. □

---

[42] For those of us who don't speak French, an acceptable pronunciation in English is "pwa-SOHN."

[43] Basketball statisticians report this as a percentage rather than a probability.

[44] Technically, we should only consider the distribution of baskets as suitable for a Poisson distribution. Since points can come in two or threes, they are not technically Bernoulli trials.

Check Your Understanding 3.7.1:
Suppose we are interested in the probability distribution for the number of points scored by a given player per game. Is this a binomial distribution, a Poisson distribution, or neither?

Here is the Poisson distribution formula, which we present without proof.

Theorem 3.7.1 (The Poisson Distribution Formula and Properties). *The probability of k events occurring in time t, given an average rate of λ events per unit time, is given by the Poisson distribution formula*

$$f_\mu(k) = \frac{\mu^k}{k!} e^{-\mu}, \qquad where\ \mu = \lambda t. \tag{3.7.1}$$

The quantity μ is the variance of the Poisson distribution as well as the mean.

Example 3.7.3. We can use (3.7.1) to answer the question posed in Example 3.7.1. We want the probabilities for each number of successes for one half-hour with a mean rate of 12 per hour. Thus, the mean is $\mu = 6$ and the distribution is

$$f_6(k) = \frac{6^k}{k!} e^{-6}.$$

Specifically, the probability of 6 successes is

$$f_6(6) = \frac{6^6 e^{-6}}{6!} \approx 0.16.$$

A portion of this probability distribution is given in Table 3.7.1 and displayed in Figure 3.7.1a.

□

Some interesting properties can be observed in the data of Table 3.7.1.

1. It may come as a surprise that the probability of 5 successes for f_6 is as high as the probability of 6 successes. Shouldn't the probability of hitting the mean be higher than the probability for being one less than the mean? To understand the result, note that the average success rate corresponds to one success every 5 min. Suppose there have been 5 successes in the first 25 min. With a mean rate of one every 5 min, it seems reasonable that the probability of getting the next success in the next 5 min should be about 1/2. This suggests that the probabilities for 5 successes and for 6 successes should be about the same, as it turns out they are.

2. A comparison of the results in Table 3.7.1 with those in Table 3.4.1 for the binomial distribution with the same mean suggests that the Poisson distribution has more variability than the binomial distribution. Indeed, this is the case. The mean and variance of the Poisson distribution are the same, while the variance of the binomial distribution is always less than the mean. The number of successes in a time interval is less predictable than the number of successes in a given number of trials. This makes sense in the basketball example. We should expect more variation in points per minute than in points per free throw attempt because there is variability in both the number of attempts per minute and the number of baskets per attempt.

Table 3.7.1 Probabilities for the number of flies captured in one half-hour given a mean rate of 12 per hour

k	0	1	2	3	4	5	6
$f_6(k)$	0.0025	0.0149	0.0446	0.0892	0.1339	0.1606	0.1606
k	7	8	9	10	11	12	13
$f_6(k)$	0.1377	0.1033	0.0688	0.0413	0.0225	0.0113	0.0052

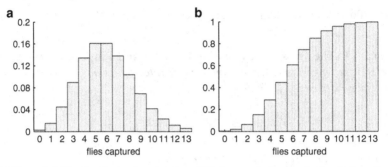

Fig. 3.7.1 (a) The probability distribution function f_6 and (b) the cumulative distribution F_6 from Example 3.7.1

As with the binomial distribution, the formula for the mean of the Poisson distribution agrees with our intuition. If events occur at a rate of *exactly* λ events per unit time, then λt is the number of events that will occur in time t (provided λt is an integer). If the events occur *randomly* with *average* rate λ, we should still expect to see λt events in time t, *on the average*.

A curious property of the Poisson distribution is that the mathematical formula requires only the mean. The specific rate and time do not matter.

Example 3.7.4. Consider three different scenarios:

1. An experiment with a mean success rate of 0.5 per hour runs for 6 h.
2. An experiment with a mean success rate of 1 per hour runs for 3 h.
3. An experiment with a mean success rate of 2 per hour runs for 1.5 h.

All three of these scenarios have the same Poisson distribution, f_3. For example, in each case, the probability of at least 3 successes is

$$P[X \geq 3] = 1 - P[X \leq 2] = 1 - P[0] - P[1] - P[2] = 1 - \left(\frac{3^0}{0!} + \frac{3^1}{1!} + \frac{3^2}{2!} \right) e^{-3}$$

$$= 1 - \frac{17}{2} e^{-3} = 0.577.$$

$\square$

3.7.2 The Exponential Distribution

Like the Poisson distribution, the exponential distribution deals with events that occur randomly in time. Both assume that the individual events are independent of each other, such as in radioactive decay. Whereas the number of events in a given amount of time follows a Poisson distribution, the times between events are exponentially distributed. As the exponential distribution is continuous, we begin with a brief summary of the principles of continuous distributions.[45]

---

[45] See Section 3.5.

1. Continuous probability distributions can be defined by a cumulative distribution function

$$F(x) = P[X \leq x].$$ (3.7.2)

2. The probability density function is the derivative F'.
3. The probability of an outcome falling in a particular interval[46] can be determined using either the cumulative distribution function or the probability density function.

$$P[a < X \leq b] = F(b) - F(a) = \int_a^b F'(x)\,dx,$$ (3.7.3)

4. The expected value of any function $g(X)$ is determined by integrating the product of g with the probability density function:

$$Eg(X) = \int_{-\infty}^{\infty} g(x)F'(x)\,dx.$$ (3.7.4)

5. The mean, variance, and standard deviation are determined using the same expected value formulas as for discrete probability distributions:

$$\mu = EX, \qquad \sigma^2 = EX^2 - (EX)^2.$$ (3.7.5)

The exponential distribution is conveniently defined by a cumulative distribution function.

Exponential distribution: *the continuous probability distribution with outcomes $0 < t < \infty$ and defined by the cumulative distribution function*

$$E_\lambda(t) = 1 - e^{-\lambda t},$$ (3.7.6)

where λ is the mean rate per unit time at which the events occur.

Example 3.7.5.

Q) Suppose a mosquito trap collects an average of 2 mosquitoes per minute. Find the probability that the first capture occurs after 1 min.
A) This question could be answered with either the exponential or the Poisson distribution, which shows the connection between them. Assuming that the individual capture events are independent, the times are exponentially distributed with mean rate $\lambda = 2$. This distribution of event times is illustrated in Figure 3.7.2. If T is the time of the next capture, then the probability that the first capture occurs after the first minute has passed is

$$P[T > 1] = 1 - P[T \leq 1] = 1 - E_2(1) = 1 - (1 - e^{-2}) = e^{-2} \approx 0.135.$$

Alternatively, if X is the number of captures in 1 min, then we can use the Poisson distribution to calculate the probability that no captures occur in the first minute, which is

$$P[X = 0] = f_2(0) = \frac{2^0}{0!}e^{-2} = e^{-2} \approx 0.135.$$

---

[46] We use $<$ at the lower bound of the interval and $\leq$ at the upper bound because this is the custom. In practice, it is a distinction without a difference, since $P[X = x] = \int_x^x f(s)\,ds = 0$.

These results are clearly the same, as the statement that the first capture occurs after the first minute is equivalent to the statement that no captures occur in the first minute.

□

We can obtain the basic properties of the exponential distribution from the formulas provided in the summary. The probability density function is the derivative $E' = \lambda e^{-\lambda t}$. Then we have

$$ET = \int_0^\infty \lambda \times t e^{-\lambda t}\, dt, \quad ET^2 = \int_0^\infty \lambda t^2 e^{-\lambda t}\, dt.$$

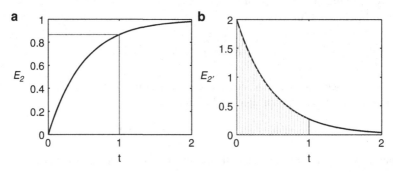

Fig. 3.7.2 (**a**) The cumulative distribution E_2 and (**b**) the probability distribution function E_2' from Example 3.7.5, both showing the probability that the first event occurs in the first time unit

These integrals can be computed using integration by parts or by the method of undetermined coefficients.[47] The results are summarized here.

Theorem 3.7.2 (Exponential Distribution Properties). *The exponential distribution has probability density function*

$$E_\lambda'(t) = \lambda e^{-\lambda t}, \tag{3.7.7}$$

and its mean, variance, and standard deviation are given by

$$\mu = \frac{1}{\lambda}, \quad \sigma^2 = \frac{1}{\lambda^2}, \quad \sigma = \frac{1}{\lambda}. \tag{3.7.8}$$

3.7.3 Memory in Probability Distributions

Our human intuition suggests that events distributed over time become more likely as we wait longer for them to occur. While true under some specialized circumstances, this is not generally the case. As an example, consider the radioactive decay of uranium atoms, which occurs spontaneously without any preparation. We can show by calculation that the probability that a particular atom decays in a period of time equal to one half-life is exactly 0.5, no matter how long the atom has existed prior to the start of the experiment.

---

[47] See Problem 3.7.14.

Example 3.7.6.

Q) Suppose a radioactive element has a half-life of 50,000 s (about 14 h). Consider a (hypothetical) scientific experiment in which we count out 1,000 atoms of this element and then start a clock. What is the probability that the first decay in the sample occurs in the first 100 s?

A) We expect 500 of the atoms to decay in 50,000 s, so the mean rate of decay is $\lambda = 500/50,000 = 0.01$. The decay times are exponentially distributed, so we have

$$P[T \leq 100] = E_{0.01}(100) = 1 - e^{-1} = 0.632.$$

This may be counterintuitive. On average, we expect one decay every 100 s ($1/\lambda$), but the probability that the first decay occurs in less than 100 s is more than 0.5.

Q) Suppose there is no decay in the first 100 s. What is the probability that the first decay occurs in the next 100 s?

A) At the end of 100 s, we still have 1,000 atoms. The mean rate is still 0.01, so we still have the distribution $E_{0.01}$. The clock is started again for the second 100 s. Hence, the answer is the same as for the first 100 s: 0.632.

$\square$

In Example 3.7.6, having already "waited" for 100 s without a decay makes no difference to the probability of its happening in the next 100 s. Specifically, the probability of the first decay occurring within 100 s of "right now" is always 0.632, no matter what time is on the clock. The probability of the next decay occurring within 100 s will not change until the first decay occurs, and then only because reducing the sample size by one changes the mean rate.[48]

Example 3.7.6 illustrates that the exponential distribution is memoryless. We will approach this point in another way with the discussion of conditional probability in Section 4.6. It is a difficult property to understand, because the world of our everyday experience is mostly occupied with events whose timing depends on historical information. For example, the probability that you will complete your purchase at the grocery store depends on how long ago you got in line. You might finish in the next 10 seconds if the clerk started your order some time ago, but you cannot finish in the next 10 seconds if the clerk is just starting your order now. Similarly, the probability that you will recover from a cold in the next 5 minutes depends on when you first got sick. You can't recover in 1 hour if you only began experiencing symptoms a few minutes ago. Situations where memory is irrelevant, such as radioactive decay, are more common in nature than in our everyday experience.

The exponential distribution is much easier to work with than other distributions of interevent times; hence, it is common to use it even in cases where memory ought to matter. This is sometimes bad modeling practice, but not always.

Example 3.7.7. Many models in ecology assume lifespans are exponentially distributed, but this is often a significant modeling error. With the assumption of exponentially distributed lifespans, probability and optimization models predict a probability of reaching maturity on the order of 1/8, whereas for most species the probability of reaching maturity is far less than 1/1,000 [8]. $\square$

Example 3.7.8. Epidemic models often assume that the time required to recover from a disease is exponentially distributed, even though time to recovery is often highly dependent on the amount of time since becoming sick. This sounds like a serious error, but in many cases it makes little difference. At any given moment, there are individuals with various durations of illness. As long as the distribution of illness durations stays roughly the same over time, there is no harm in assuming that recovery times are exponentially distributed. $\square$

---

[48] In most cases, the sample size is so large that individual decays do not measurably change the mean rate.

Problems

The data sets for Problems 3.7.12–3.7.17 are from [5] and can be found in both csv and xlsx form at http://www.math.unl.edu/~gledder1/MLS/, http://www.springer.com/978-1-4614-7275-9.

3.7.1. Consider a population of parasitoids and hosts.[49] Suppose parasitoids oviposit an average of once per day and each host is vulnerable for a period of 6 days, and assume that the parasitoid population is 1/3 the size of the host population.

(a) Find the probability that a given host avoids attack.
(b) Find the probability that a given host is attacked exactly once.

3.7.2.* (a) Suppose an event occurs an average of once per hour. Find the probability that the event occurs once in a 1-h period.
(b) Suppose an event occurs an average of twice per hour. Find the probability that the event occurs twice in a 1-h period.
(c) Suppose an event occurs an average of n times per hour. Find the probability that the event occurs n times in a 1-h period for the values $n = 3, 4, 5$.
(d) Is there a pattern in the results? Note that the case $n = 6$ was done in Example 3.7.3.

3.7.3.* Suppose X is exponentially distributed with mean rate $\lambda = 4$. Find the probability that the next occurrence of the event occurs within 1 time unit.

3.7.4. Suppose X is exponentially distributed with mean rate λ.

(a) What is the probability that the value of X is greater than the mean of the distribution?
(b) What is the probability that the value of X is more than one standard deviation greater than the mean?
(c) What is the probability that the value of X is more than one standard deviation less than the mean?

3.7.5. Use computer software to generate 1,000 data points from the Poisson distribution F_{10}. Find the mean and variance of the data and plot a histogram.

3.7.6. Table 3.7.2 presents data for the counts of malaria parasites in erythrocytes. These data were collected to test the hypothesis that malaria parasites attack certain cells preferentially [20]. If the infection rate is the same for infected and uninfected erythrocytes, then the distribution should be of Poisson type.

(a) Determine the mean and variance for the data.
(b) Plot the data as a histogram of relative frequencies.
(c) Plot the probability distribution function for the Poisson distribution corresponding to the mean of the data.
(d) To what extent do your results appear to support the claim that malaria parasites prefer either uninfected cells or previously infected cells?

(This problem is continued in Problem 4.5.9.)

3.7.7.* Table 3.7.3 presents the frequencies of spiders found under boards [7].

(a) Determine the mean and variance for the data.
(b) Plot the data as a histogram of relative frequencies.
(c) Plot the probability distribution function for the Poisson distribution corresponding to the mean of the data.
(d) Does the Poisson distribution seem to be a reasonable model for this data set?

(This problem is continued in Problem 4.5.10.)

---

[49] Parasitoids are animals that lay their eggs inside a living organism. The parasitoid larvae feed on the host.

Table 3.7.2 Numbers of malaria parasites in erythrocytes

Number	0	1	2	3	4
Frequency	40,000	8,621	1,259	99	21

Table 3.7.3 Frequencies of a given number of spiders found under boards.

Number	0	1	2	3
Frequency	159	64	13	4

3.7.8. Table 3.7.4 presents the frequencies of weevil eggs on beans.

(a) Determine the mean and variance for the data.
(b) Plot the data as a histogram of relative frequencies.
(c) Plot the probability distribution function for the Poisson distribution corresponding to the mean of the data.
(d) Assuming that spatial distributions are usually of Poisson type, suggest a biological reason why the distribution of weevil eggs on beans should be so different. [Hint: Think about why a deviation from a strictly random distribution would be advantageous, and therefore subject to natural selection.]

Table 3.7.4 Frequencies of a given number of weevil eggs found on beans [7]

Number	0	1	2	3
Frequency	5	68	88	32

3.7.9. Table 3.7.5 contains data for the numbers of insects in the order Hemiptera in 33 traps [4].

(a) Determine the mean and variance for the numbers of Hemiptera.
(b) Plot the data as a relative frequency histogram.
(c) Plot the probability distribution function for the Poisson distribution corresponding to the mean of the data.
(d) Does the Poisson distribution seem to be a reasonable model for this data set?

Table 3.7.5 Numbers of insects in the order Hemiptera found in 33 traps

Number	0	1	2	3	4
Frequency	6	8	12	4	3

3.7.10. Table 3.7.6 contains data for the numbers of insects in the family Staphylinidae in 33 traps. Repeat Problem 3.7.9 using this data set.
(This problem is continued in Problem 3.7.16.)

3.7.11. Table 3.7.7 contains data for the numbers of plants found in portions of a field site. Repeat Problem 3.7.9 using this data set.

Table 3.7.6 Numbers of insects in the family Staphylinidae found in 33 traps [4]

Number	0	1	2	3	4	5	6
Frequency X	10	9	5	5	1	2	1

Table 3.7.7 Numbers of *Eryngium maritimum* plants per quadrat square [1]

Number	0	1	2	3	4	5	6	7
Frequency X	16	41	49	20	14	5	1	1

3.7.12. The file YEAST.* contains data collected for the numbers of yeast cells detected on a microscope slide by William Sealy Gosset [18].

(a) Repeat Problem 3.7.9 for subset (A), using an appropriate bin size for the histogram.
(b) Repeat Problem 3.7.9 for subset (B), using an appropriate bin size for the histogram.
(c) Repeat Problem 3.7.9 for subset (C), using an appropriate bin size for the histogram.
(d) Repeat Problem 3.7.9 for subset (D), using an appropriate bin size for the histogram.
(e) Does the Poisson distribution seem to be a reasonable model for these data sets?

3.7.13. (a) Prepare a histogram of the relative frequencies of data for the waiting times between firings of motor cortex neurons of an unstimulated monkey, recorded in the file WAITING.*. [Note that there are no headings in this file.] Superimpose the probability density function for the corresponding exponential distribution. Is the exponential distribution a reasonable model for this data set? What about the normal distribution?

(b) Repeat part (a) with the data for time intervals between successive pulses along a nerve fiber, recorded in the file NERVE.*.

(This problem is continued in Problem 4.2.8.)

3.7.14. Use the definition of expected value to calculate the mean and standard deviation for an exponential distribution with parameter λ.

3.7.15. The Poisson distribution can be used to approximate the binomial distribution when the success probability for each Bernoulli trial is small and the number of trials is large.[50] The error in using the Poisson distribution f_{np} to approximate the binomial distribution $b_{n,p}$ is given approximately by[51]

$$b_{n,p}(k) - f_{np}(k) \approx \frac{k - (k - np)^2}{2n},$$

$$\max(|b_{n,p} - fnp|) \approx \frac{P}{2} fnp(k_m)$$

where k_m is the integer part of $\mu + 2/3$. To test this claim, consider a family of binomial distributions having $np = 3$.

(a) Compute $f_3(k)$ for $0 \le k \le 10$. These probabilities are the approximation for all binomial distributions in the family.
(b) Compute $b_{30,0.1}$ for $0 \le k \le 10$. Find the maximum error in the approximation and compare with the estimate $0.05 f_3(3)$.

---

[50] This was a very important technique in the period before modern computers. It is still used in software implementations of the binomial distribution for extreme cases such as that of part (d).

[51] See PoissonApproximation.pdf at http://www.math.unl.edu/~gledder1/MLS for details.

(c) Repeat part (b) with $n = 100$ and the appropriate value of p.

(d) There are approximately 6×10^9 base pairs in the human genome, and it has been estimated that the genome duplication process, which occurs in cell division, has an error probability of about 1.1×10^{-8} per base pair [14]. Estimate the maximum error in using the result of part (a) to estimate the probabilities of k errors in a genome duplication process.

The *negative binomial distribution* is sometimes used for data collected in a scenario that seems to call for a Poisson distribution, but for which the data shows much more variability than a Poisson distribution. The probability distribution function for the negative binomial distribution with mean μ and variance σ^2 is given by

$$f_{\mu,r}(k) = \frac{\Gamma(k+r)}{k!\,\Gamma(r)} \left(\frac{r}{\mu+r} \right)^r \left(\frac{\mu}{\mu+r} \right)^k,$$

where the dispersion parameter $r > 0$ is taken to be

$$r = \frac{\mu^2}{\sigma^2 - \mu} > 0$$

and Γ is the gamma function, one of the standard special functions available in mathematical computer software. Note that $r \to \infty$ as $\sigma^2 \to \mu$. In fact, the limiting case $f_{\mu,\infty}$ is the Poisson distribution.

3.7.16. (a) Find the appropriate value of r for the data of Table 3.7.6.

(b) Compute the probability values for outcomes $k = 0, 1, \ldots 6$ using the appropriate negative binomial distribution function.

(c) Compare the negative binomial probabilities with the measured relative frequencies. Is the negative binomial distribution a reasonable model for this data set?

3.7.17. The file TICKS.* contains data for the number of ticks found on each individual in a flock of 82 sheep.

(a) Determine the mean and variance for the data.

(b) Plot the data as a histogram of relative frequencies using an appropriate bin size.

(c) Plot the probability density function for the negative binomial distribution corresponding to the mean and variance of the data.

(d) To what extent do your results support the claim that the number of ticks found fits a negative binomial distribution?

References

1. Blackman GE. A study by statistical methods of the distribution of species in grassland associations. *Annals of Botany* **49**, 749–777 (1935). Cited in *Goodness-of-Fit Techniques*, eds. RB DAgostino and MA Stephens. Marcel Dekker, New York (1986)

2. Fairbanks DJ and B Rytting. Mendelian controversies: A botanical and historical review. *American Journal of Botany* **88**, 737–752 (2001)

3. Fisher RA. The use of multiple measurements in taxonomic problems. *Annals of Eugenics*, **7**, 179–184 (1936)

4. Gilchrist W. *Statistical Modeling*. John Wiley and Sons, New York (1984)

5. Hand DJ, F Daly, K McConway, D Lunn, and E Ostrowski. *Handbook of Small Data Sets*. CRC Press, Boca Raton, FL (1993)

6. James SD. Four Out of Five Recent Presidents Are Southpaws. ABC News (2008-02-22). http://abcnews.go.com/politics/story?id=4326568CitedSep2012

7. Kulasekeva KB and DW Tonkyn. A new discrete distribution, with applications to survival, dispersal, and dispersion. *Communications in Statistics (Simulation and Computation)*, **21**, 499–518 (1992)

8. Ledder G, JD Logan, and A Joern. Dynamic energy budget models with size-dependent hazard rates. *J Math Biol*, **48**, 605–622 (2004)

9. Llaurens V, M Raymond, and C Faurie. Why are some people left-handed? An evolutionary perspective. *Philosophical Transactions of the Royal Society of London, B: Biological Sciences*, **364**, 881–894 (1999)

10. MacArthur JW. Linkage studies with the tomato, III, Fifteen factors in six groups. *Transactions of the Royal Canadian Institute*, **18**, 1–19 (1931)

11. Macrae F. As two lefties vie for the American presidency... why are so many U.S. premiers left-handed? The Daily Mail (2008-10-24). http://www.dailymail.co.uk/sciencetech/article-1080401/As-lefties-vie-American-presidency--U-S-premiers-left-handed.htmlCitedSep2012

12. Pearson K. On the criterion that a given system of deviations from the probable in the case of a correlated system of variables is such that it can be reasonably supposed to have arisen from random sampling. *Philosophical Magazine*, **5**, 157–175 (1900)

13. Pilkington E. Revealed: The leftist plot to control the White House. The Guardian (2008-10-24). http://www.guardian.co.uk/world/2008/oct/24/barack-obama-mccain-white-house-left-handedCitedSep2012

14. Roach JC, G Glusman, AFA Smit, CD Huff, R Hubley, PT Shannon, L Rowen, KP Pant, N Goodman, M Bamshad, J Shendure, R Drmanac, LB Jorde, L Hood, and DJ Galas. Analysis of genetic inheritance in a family quartet by whole-genome sequencing. *Science*, **328**, 636–639 (2010)

15. Senn SJ and W Richardson. The first *t*-test. *Statistics in Medicine*, **13**, 785–803 (1994)

16. Seuss, Dr. (1961), Too Many Dave's. In *The Sneetches and Other Stories*. Random House, New York (1961)

17. Stigler SM. *The History of Statistics: The Measurement of Uncertainty Before 1900*, p. 208. Harvard University Press (1986)

18. "Student". On the error of counting with a haemocytometer. *Biometrika*, **5**, 351–360 (1906)

19. "Student". The probable error of a mean. *Biometrika*, **6**, 1–25 (1908)

20. Wang CC, Multiple invasion of erythrocyte by malaria parasites. *Transactions of the Royal Society of Tropical Medicine and Hygiene*. **64**, 268–270 (1970)

Chapter 4
Working with Probability

We now build on our background knowledge of probability distributions by examining other topics in probability. Three practical problems motivate the seven sections of this chapter.

Do people who take a particular medication have lower cholesterol than those who don't? Are coffee drinkers more or less likely than others to have some particular medical condition? These are the sorts of questions that are addressed by inferential statistics. This subject is introduced in Section 4.1 with a simple example and developed in detail in Section 4.3, where we consider probability distributions of random variables that represent sample means, and Section 4.4, which presents the key results for assessing inferences. Section 4.2 introduces the issue of how to know if a distribution is approximately normal. This material appears ahead of the section on probability distributions of samples so that it can serve as an extra tool for understanding the central limit theorem.

Probability distributions always involve one or more parameters. So far, we have assumed these parameters to be known. But how do we obtain those parameter values experimentally? We consider the mathematical issues associated with determining unknown parameter values in Section 4.5. Two examples are given to illustrate the ideas: the problems of finding the success probability for a binomial distribution and estimating the unknown size of a finite population.

Sometimes one is interested in knowing how the probability of some event changes when additional information is provided. This is the subject of conditional probability. We develop the principal ideas of conditional probability in Section 4.6 and then apply these ideas to the interpretation of medical tests in Section 4.7.

A table showing connections among problems in this chapter and Chapter 3 can be found in the introduction to Part II.

4.1 An Introduction to Statistical Inference

After studying this section, you should be able to:

- Explore the use of data to search for systematic differences between a subpopulation and a population.
- Understand that we must look at all values less than or equal to a value x to judge whether that value is unusually small.

G. Ledder, *Mathematics for the Life Sciences: Calculus, Modeling, Probability, and Dynamical Systems*, 207
Springer Undergraduate Texts in Mathematics and Technology, DOI 10.1007/978-1-4614-7276-6_4,
© Springer Science+Business Media, LLC 2013

- Understand that random samples are not necessarily representative, but larger samples are more likely to be so.

This section is the first of three that deal with the use of probability to assess inferences from data, a subject with a well-deserved reputation for being hard to understand. To address this difficulty, we begin with a simple example couched in a narrative that builds on the reader's everyday experience rather than biological knowledge. The key ideas are introduced in this informal setting before their detailed development in Section 4.3 on distributions of samples and Section 4.4 on probability of inferences.

Three hundred out of the 6,000 students at Lincoln College (5 %) are members of the Alpha Beta Epsilon (ABE) honorary fraternity. ABE leaders claim with pride that their members are active in all campus activities. One such activity is the weekly foreign art film screening, which is held in a room that seats only 50 students. Different films appeal to different students, but there are always enough to fill the room. While watching the film one week, freshman Shelly Sharp notices that there is only one ABE member in attendance. Shelly is a reporter for the college newspaper, and this gives her an idea. Only one ABE member out of 50 students is only 2 %. Should she write an exposé about ABE not living up to its image? She does not want to risk embarrassing herself, so she needs to be sure she understands the significance of the data before writing the article.

The first step in analyzing Shelly's data is to identify it as a random variable for a probability experiment. The actual experiment here consists in selecting 50 individuals from a collection of 6,000; the random variable X is the number of those individuals who are ABE members. We do not know whether ABE members are more or less likely to attend the film screenings than other students. For the purpose of probability calculations, we assume that there is no significant difference between ABE students and the larger population,[1] but we hope that our analysis will either confirm or reject this hypothesis.

Before doing the actual calculation for Shelly's data, let's consider a simpler experiment. Suppose we select just two students. Our selection could include zero, one, or two ABE members; if one, it could be either the first or second person chosen. To calculate the probability that we select one ABE member, we start with the fact that the probability of getting an ABE member with the first selection is 300/6,000, or 0.05. Assuming this happens, the first individual is now at the film screening and cannot be selected again. The remaining population of 5,999 contains only 299 ABE members, so the probability that the second individual is an ABE member is 299/5,999, or 0.04984. Of course the probability that the second individual is *not* an ABE member is then 5,700/5,999, or 0.95016. By the sequence rule, the probability of selecting an ABE member followed by a non-member is the product of these individual probabilities, or 0.04751. Now suppose the first individual is not an ABE member, which has probability 0.95. The probability that the second individual is an ABE member is then 300/5,999 = 0.05001, which is different from that for the second individual in the previous case. The probability of selecting a member after a non-member is 0.95 times 0.05001, which is 0.04751. This is the same as the probability of selecting one member before one non-member, which makes sense— once we know which two people are selected, it is equally likely for either to have been selected first. Taken together, the probability of getting one member and one non-member is 0.09502.

---

[1] Of course there has to be *some* difference between ABE members and other students, or else we wouldn't be able to count them. Until proven otherwise, the difference is assumed to be some distinctive but meaningless marking.

This calculation for the actual experiment of selecting two individuals from a population of 6,000 is rather complicated because selected individuals are removed from the population. However, the change in the population caused by this removal is rather small because the population is large. The calculation would be much easier if we used a model of an infinite population. Then each selection of a student is a Bernoulli trial with a success rate of 0.05. Using the binomial distribution, we can calculate the probability of one success in two tries as $b_{2,0.05}(1) = 0.095$. The error made in modeling the population as infinite is 0.00002 out of approximately 0.1, which is about 0.02 %. Such a small error is trivial, given that calculations in statistics are done with experimental data.

Returning to our actual example, using a model of an infinite population makes the calculation of probabilities for a sample of 50 much easier. The error in the approximation will be larger, so we should take some care to justify the approximation. To do so, consider that replacement only affects the sample if it results in selecting the same person twice, and we can calculate the probability that this does not happen. The probability of getting 50 distinct individuals is easily calculated by thinking of a sequence of 50 experiments of choosing one individual.[2] The probability that the first individual was not previously chosen is 1. The probability that the second individual was not previously chosen is then 5,999/6,000. The probability that the third individual is also distinct is 5,998/6,000. The pattern continues, with the probability that the fiftieth is distinct being 5,951/6,000. All of these events must happen to achieve a fully distinct set of 50, so the sequence rule applies, with the result that samples of 50 individuals from a population of 6,000 will contain 50 distinct individuals 81.5 % of the time. Most samples of 50 individuals selected with replacement will not contain any repeats, and will therefore be equivalent to the same sample selected without replacement. Nearly all of the other 19 % will contain only one repeat, which will have very little impact on the overall probabilities for the samples. This calculation serves as an adequate validation for using the model of an infinite population to analyze Shelly's data.

Check Your Understanding 4.1.1:
Verify the calculation that the probability of 50 distinct individuals is 0.815.

We are now ready to do the statistical analysis. The probability of getting k ABE members in a sample of 50 individuals, given a 5 % chance of any individual being an ABE member, is $b_{50,0.05}(k)$, which can be calculated using the binomial distribution formula[3]

$$b_{n,p}(k) = \frac{n!}{k!(n-k)!} \left(\frac{p}{q} \right)^k q^n, \qquad q = 1 - p. \tag{4.1.1}$$

The results for k from 0 to 6 are given in Table 4.1.1, along with the cumulative probabilities $B_{50,0.05}(k)$, which represent the probability of getting k or fewer successes.

Table 4.1.1 Probabilities for the number of ABE members in a sample of 50

k	0	1	2	3	4	5	6
$b_{50,0.05}(k)$	0.0769	0.2025	0.2611	0.2199	0.1360	0.0658	0.0260
$B_{50,0.05}(k)$	0.0769	0.2794	0.5405	0.7604	0.8964	0.9622	0.9882

---

[2] See Problem 3.2.6.

[3] See Section 3.4.

Check Your Understanding 4.1.2:
Verify the results in Table 4.1.1.

The obvious (but incorrect) answer to Shelly's question about whether having only 1 ABE member is unusual is that there will be only 1 ABE member about 20 % of the time. This statement is correct, but misleading. The probability of having 2 ABE members is only 26 %, which is not much different from the probability of having 1, yet Shelly would not have thought it unusual if there had been 2. The probability of having 2 *or fewer* ABE members is 54 %, and this number is more helpful in judging whether 2 is unusual than the probability of having *exactly* 2. We'll see a more striking example of this later. The important conclusion for Shelly to draw is this: if ABE members are no more or less likely to attend than anyone else, we should expect to see 1 or fewer 28 % of the time. This is small enough to arouse suspicion, but it is certainly not small enough to be considered unusual.

Check Your Understanding 4.1.3:
Use the data of Table 4.1.1 to determine how unusual it would be to find 5 ABE members at the screening.

The analysis of Shelly's problem offers an important lesson about interpreting statistical data.

The standard for a result to be considered unusual is much stricter than the standard needed to arouse suspicion that a result is unusual.

Just how unlikely an event must be to classify it as unusual is a matter of judgment, but it certainly should be less than 28 % likely.

While the data clearly do not justify the claim that ABE is a fraud, there is ample justification for doing a more thorough experiment. We therefore consider an updated problem.

Shelly returns to the film screening for 3 more weeks. She counts 2 ABE members in the second week, 2 in the third, and 0 in the fourth. Does this data provide stronger evidence that ABE members are less interested in the film screenings than the college average? If so, is the evidence strong enough to warrant a newspaper article about ABE not living up to its image?

There are two different ways to look at Shelly's new data. We could consider her weekly counts as separate data points for something.[4] For now, we choose the alternative, which is to consider the 200 individuals as one collection of Bernoulli trials. The problem is then to find the probability that there will be 5 or fewer successes in 200 Bernoulli trials with a success probability of 0.05 for each. We have

$$B_{200,0.05}(5) = b_{200,0.05}(0) + b_{200,0.05}(1) + b_{200,0.05}(2) + b_{200,0.05}(3)$$
$$+ b_{200,0.05}(4) + b_{200,0.05}(5) \approx 0.062.$$

Table 4.1.2 shows the results for k from 0 to 10.

---

[4] See Example 4.3.3.

Table 4.1.2 Probabilities for the number of ABE members in a sample of 200

k	0	1	2	3	4	5
$b_{200,0.05}(k)$	0.000035	0.00037	0.00193	0.00671	0.01740	0.03590
$B_{200,0.05}(k)$	0.000035	0.00040	0.00233	0.00905	0.02645	0.06234

k	6	7	8	9	10
$b_{200,0.05}(k)$	0.06140	0.08956	0.11372	0.12769	0.12836
$B_{200,0.05}(k)$	0.12374	0.21330	0.32702	0.45471	0.58307

> **Check Your Understanding 4.1.4:**
> Verify the results in Table 4.1.2.

In the first problem, Shelly found an average of 2 % instead of the expected 5 %, and we concluded that a random average will be this low 28 % of the time. With a sample of 200 rather than 50, an average of 2.5 % is expected to happen only 6 % of the time. Even the probability of 7 or fewer out of 200, corresponding to an average of 3.5 %, is noticeably less than the probability of having 2 % or less in a sample of 50. This calculation offers another important lesson about statistical data.

> Larger samples are more likely than smaller ones to match the population from which the sample was collected.

Larger samples also provide stronger evidence of the need to judge whether a low count is unusual by considering the probability of getting that count *or less*. For a sample of 200, we expect to have 9 ABE members about 13 % of the time. Although that is less likely than having 1 ABE member out of 50, it is clearly far from unusual, as an average of 4.5 % is very close to the expected average. On the other hand, the probability of a result of 9 or less is 45 %, which clearly indicates that a result of 9 is not unusual.

If ABE members were as likely to attend the film screenings as the average student, there would only be a 6 % chance of getting a result of 5 or less. Shelly now has much stronger evidence that ABE members are underrepresented at the film screenings. Whether the new evidence is strong *enough* is a matter of opinion. As Shelly's editor, I would say that the evidence is already strong enough to publish the story. As a scientist, I would say that 6 % is still not quite unusual enough to be conclusive. It is common in biology to consider 5 % to be a standard cutoff for judging a result to be significant enough to count as unusual, but in some research areas the standard cutoff for significance is taken to be 1 %, and in a few it is even less. For example, I would not want to convict someone of a crime on the basis of data that could be obtained randomly in 1 % of experiments; this would imply an acceptance of false convictions in as many as 1 % of cases.[5] The important thing to keep in mind is that results that deviate significantly from those expected can and do occur simply by chance. Suppose we were able to choose a group of Lincoln College students that is truly average in attendance at the film screenings. We can expect to have a total of 5 or fewer attendees from this group over 4 consecutive screenings about 6 % of the time. If we continue to count group members at screenings for a full year, we will have done the 4-week experiment 13 times, and it is quite likely that there will be at least one count that is 5 or less.[6] This will likely happen in spite of the fact that

---

[5] Standard cutoff values are arbitrary. There is no particular reason why the cutoff used in most fields should be 5 % rather than 4 %, and yet most papers on statistical analysis make definitive judgments based on this arbitrary choice. In a sense, statistics is a way of codifying judgments based on probability; in many cases it is better scientific practice to base these judgments on context rather than fixed arbitrary rules.

[6] See Problem 4.1.2.

our group of 300 does not differ from the full population in film attendance, simply because random processes occasionally produce results that are far from average.

Check Your Understanding Answers

3. The probability of $X \geq 5$ is $1 - P[X \leq 4] = 0.1036$.

Problems

4.1.1. (a) Write a computer program that randomly selects N numbers from the $b_{n,0.05}$ distribution and determines the probabilities that the selected number is 0, 1, and so on up to 10. Test your program with $n = 50$ and $N = 100$. Theoretically, the probabilities should be those of Table 4.1.1. Do not worry about whether your probabilities match those in the table. Instead, repeat the test several times. Your results for each k should be higher than the table sometimes and lower other times.

(b) Check your program more thoroughly by running it once with $N = 1,000,000$. This time your results should be very close to the theoretical results.

4.1.2. This problem uses the computer simulation from Problem 4.1.1 with $n = 200$.

(a) Suppose you were to visit the film screening at Lincoln College for 4 weeks and count the total number of ABE members in attendance. Assuming that ABE members are as likely to attend as anyone else, this experiment is equivalent to that of drawing a value randomly from the $b_{200,0.05}$ distribution. Run your program once with $N = 13$. How many times is the result equal to 5 or less?

(b) Repeat part (a) several times to see how much variation there is in the answer. The text asserts that this seemingly unusual result is likely to occur at least once in 13 trials. Does this seem to be the case?

(c) Can we really be sure that Shelly's observation is due to the characteristics of ABE members and not mere chance? In more general terms, are random samples always representative?

4.1.3. This problem uses the computer simulation from Problem 4.1.1.

(a) Suppose you were to visit the film screening at Lincoln College for a full year and that ABE members are as likely to attend as anyone else. Your experiment is equivalent to that of drawing 52 values randomly from the $b_{50,0.05}$ distribution. Run your program once with $N = 52$. How many times is the result equal to 1 or less? This represents the number of times that you would have thought ABE members to be underrepresented.

(b) Repeat part (a) several times to see how much variation there is in the answer. In 10 runs, representing 10 years of data, record the largest and smallest values for the number of weeks in which there are 1 or fewer ABE members present along with the average number of weeks.

(c) Repeat part (a) again, but this time do 10 runs of 208 values each.

(d) If possible, compare your results for parts (b) and (c) with classmates.

(e) Discuss the effect of sample size on the extent to which observations are likely to match theoretical probabilities.

4.1.4. With a success probability of 0.05 for each Bernoulli trial, we determined in the text that the probability of getting 5 or fewer successes out of 200, which represents 2.5 % (half the expected value), is 0.0623, which is too large to meet the 5 % standard for a result to be considered unusual. How does the probability of getting 2.5 % or less given an expected value of 5 % depend on the size of the sample? To address this question, determine the probability of getting k or fewer successes out of $40k$ trials, with individual success probability $p = 0.05$, for $1 \leq k \leq 10$. How large a sample is needed to reduce the probability to the usual 5 % standard for biology? How large a sample is needed to reduce it to the 1 % standard that is used in some disciplines? What does this say about the difficulty of drawing conclusions from small random samples?

4.1.5. With a healthy amount of scientific skepticism, we might wonder if the theoretical results for the binomial distribution, which we are using to model something that is not really a collection of Bernoulli trials, actually give a good approximation of reality. We addressed this question theoretically in the text, and will do so in more detail in Problem 4.1.6. Here we address it using a simulation.

(a) Write a computer simulation that represents the process of selecting the students for the film screening and counting the number who are ABE members:

 1. Set up a vector of 6,000 components, each initially 0.
 2. Repeat until the sum of values in the vector is 50:
     a. Randomly choose an integer Y from 1 to 6,000.
     b. Set element Y of the vector to 1.
 3. The random variable X is the sum of entries in the first 300 components of the vector.

(b) Explain how the program makes sure that exactly 50 students are selected.
(c) Run the simulation 100 times. The results should be roughly similar to Table 4.1.1.
(d) Run the simulation 10,000 times and use the results to estimate the probabilities of having k ABE members at a film screening. Compare the results to Table 4.1.1.

4.1.6.* The results in Table 4.1.1 used a binomial model to approximate the actual scenario of randomly choosing 50 individuals out of a population of 6,000. We can check some of the entries in the table by calculating the actual probabilities. Each of these should be very close to the binomial approximation.

(a) Calculate the actual probability that there are no ABE members in a collection of 50 individuals from a population of 6,000. Compare with the binomial approximation. [Think of the selection as being a sequence of 50 individual selection events, each of which must yield a non-member. For the first selection, there are 5,700 non-members out of a total population of 6,000, so the probability is 5,700/6,000. With one person already chosen, the probability that the second individual is a non-member is 5,699/5,999. The same pattern continues for all 50 selections.]
(b) Repeat part (a) for the probability of 1 ABE member. [Think of the selection as consisting of several selection events. First, we select 49 non-members using the same calculations as in part (a). Second, we select 1 ABE member from the entire population of 5,951 individuals not yet chosen. This yields the probability of choosing 49 non-members and one member with the member being the last one chosen. It would have been equally likely that the member could have been chosen in any other position, so we must multiply the result of the sequence calculation by 50. Note that it is not necessary to do the entire multiplication. Compared to the calculation of part (a), we have merely replaced a numerator of 5,651 by a numerator of 300 in the last factor and multiplied by 50.]
(c) Repeat part (b) for the probability of 2 ABE members. [This time we start by selecting 48 non-members from the original population, then we select 2 ABE members from the remaining population, and then we multiply by the number of possible placings. This latter is a bit tricky. There are 50 places where we could put the first member and 49 places for the second; however, this will count each of the possible pairs twice, so the number of possible

placings is $50 \times 49/2$. Again, the calculation is less work if we consider only changes from the part (b) calculation.]

(d) Repeat part (c) for the probability of 3 ABE members. [Follow the same pattern. When calculating the number of possible placings, be careful determining how many times each triplet will have been counted.]

4.2 Tests on Probability Distributions

After studying this section, you should be able to:

- Understand the empirical distribution function.
- Understand the graphical approach to testing normality.
- Use the Cramer–von Mises statistic to assess the likelihood that a data set is from a normal distribution.
- Understand the issues involved in rejecting or retaining outliers.

In Section 4.3, we will need to judge how well a normal distribution fits a data set. The purpose of this section is to develop some tools for addressing that question. In addition, we also examine the issue of whether an anomalous value should be rejected from a data set.

4.2.1 A Graphical Test for Distribution Type

Most tests for distribution type make use of a construction called the *empirical distribution function (EDF)*, which is calculated for a given data set and then compared to what would be expected for a given distribution.[7]

Empirical distribution function for a set of observations $X_1 < X_2 < \cdots < X_n$: *the piecewise constant distribution given by*

$$F_n(x) = \begin{cases} (i-0.5)/n & x \in \{X_i\} \\ J(x)/n & \text{otherwise} \end{cases}, \tag{4.2.1}$$

where $J(x)$ is the number of observations less than x. Note that $0 \le F_n(x) \le 1$. Note also that the observations must be sorted in increasing order.

Example 4.2.1. Consider the sleep data from Section 3.1, reproduced in ascending order as Table 4.2.1. The connection between data and individual patients is lost with this change, but

Table 4.2.1 Hours of sleep gained on D. and L. hyoscine hydrobromide

Number	1	2	3	4	5	6	7	8	9	10
D. hyoscine hydrobromide, X	−1.6	−1.2	−1.0	−0.2	0.0	0.7	0.8	2.0	3.4	3.7
L. hyoscine hydrobromide, X	−0.1	0.1	0.8	1.1	1.6	1.9	3.4	4.4	4.6	5.5

---

[7] Most authors use the formula $J(x)/n$ for all x and change the definition of J from "less than" to "less than or equal to." This has the drawback of requiring an explanation for why the formula involving i is the one that is actually used in statistical calculations. The distinction between the definitions is otherwise meaningless.

we are considering only the nonpsychotic patients and need to have the data in ascending order. The EDF for L-hyoscine hydrobromide has the value 0 for all $x < -0.1$, 0.05 for $x = -0.1$, 0.1 for $-0.1 < x < 0.1$, 0.15 for $x = 0.1$, and so on, ending with 0.95 for $x = 5.5$ and 1 for $x > 5.5$.

□

Now suppose the data points come from some distribution with cumulative distribution function $F(x)$. Then the probability of a randomly chosen value being less than or equal to X_i is

$$P[X \le X_i] = F(X_i).$$

This means that the empirical distribution function F_n ought to approximate F; indeed, it can be shown that $F_n \to F$ in the limit $n \to \infty$. Eventually, this idea will be used to devise a statistical measure of how well the data set fits the given distribution type, but for now we just consider its graphical rendering.

Suppose we plot a graph of the points $(F_n(X_i), F(X_i))$. A perfect fit to the given distribution would result in all the points being on the line $y = x$. If the fit is terrible, the points will lie far from the line.

Example 4.2.2. To test whether the L-hyoscine hydrobromide data is normally distributed, we first need to choose parameters for the distribution. From the data, we have mean $\bar{x} = 2.33$ and standard deviation $s = 2.00$; hence our best guess is $\mu = 2.33$ and $\sigma = 2.00$. With these values, we can calculate each value of the theoretical distribution by

$$Z_i = \frac{X_i - 2.33}{2.00}, \qquad F(X_i) = N(Z_i),$$

where N is the standard normal distribution. Table 4.2.2 presents the results of these calculations and Figure 4.2.1 shows the corresponding plot. The points don't appear to be very close to the line. The large jump between the sixth and seventh points is the gap between the F values of 0.415 and 0.703, which corresponds to the gap between 1.9 and 3.4 in the original data. □

Table 4.2.2 Values of F and F_n for Example 4.2.2

i	1	2	3	4	5	6	7	8	9	10
$F_n(i)$	0.05	0.15	0.25	0.35	0.45	0.55	0.65	0.75	0.85	0.95
$F(i)$	0.112	0.133	0.222	0.270	0.358	0.415	0.703	0.849	0.872	0.943

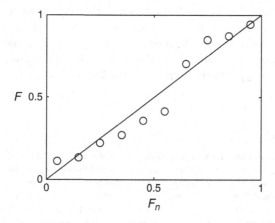

Fig. 4.2.1 EDF plot for the fit of the L-hyoscine hydrobromide data to a normal distribution, with parameters taken from the data

Check Your Understanding 4.2.1:
Prepare a plot similar to that of Figure 4.2.1 for the D-hyoscine hydrobromide data.

4.2.2 The Cramer–von Mises Test

In Section 2.3, we measured the quality of fit for a model to a data set by calculating the sum of square discrepancies. This idea makes sense as a method of quantifying the discrepancies in an EDF plot as well, but there are some subtleties that lead to modifications in the formula. Let $F_i = F(X_i)$ and $F_{n,i} = (i - 0.5)/n$ for data values $X_1 \leq X_2 \leq \cdots \leq X_n$. The residual sum of squares is then

$$RSS = \sum (F_i - F_{n,i})^2 \, ,$$

where the sum is taken over all points. This is a good starting point, but there are two modifications needed to correct for cases where the number of data points is small. For historical reasons, these modifications are applied with two different formulas:

$$W^2 = \sum (F_i - F_{n,i})^2 + \frac{1}{12n}, \quad \hat{W}^2 = \left(1 + \frac{1}{2n}\right) W^2. \tag{4.2.2}$$

The quantity $\hat{W}^2$ is called the **Cramer–von Mises[8] statistic.**[9]

Example 4.2.3. Using the values from Table 4.2.2, we obtain a residual sum of squares of 0.0514, resulting in $W^2 = 0.0598$ and $\hat{W}^2 = 0.0628$. □

Check Your Understanding 4.2.2:
Compute $\hat{W}^2$ for the D-hyoscine hydrobromide data.

Values of the Cramer–von Mises statistic are generally less than 0.1 for data that looks anything like the predicted distribution. The corrections to the residual sum of squares have increased the value by about 0.01 in this case, with $n = 10$. With $n \geq 100$, the correction is generally no larger than 0.001, which is negligible. Experiments with simulated data show that values of $\hat{W}^2$ for small subsets of a large data set are consistent with the value for the whole set. This is a significant fact. If we had 100 values in the L-hyoscine hydrobromide data set, the residual sum of squares would be about the same. With 10 times as many points, the individual square discrepancies would be only about one-tenth as large. Hence, the actual discrepancies we see on the graph would be about one-third as large.[10] This makes interpretation of an EDF plot difficult. The points appear to be farther from the line for a small data set, but this does not

---

[8] Pronounced "MY-zees."

[9] There is a dizzying number of tests for goodness of fit to a distribution type. The most commonly used is probably the Anderson–Darling test, which is similar to the Cramer–von Mises test except that it weights discrepancies in the tails of the distributions higher than discrepancies near the mean. Most experts consider Anderson–Darling to be slightly better, but the difference seems too small to compensate for the extra complexity, given that statistical tests merely produce numerical values that must be interpreted in context. From a modeler's point of view, there is an inherent advantage to using methods that are easy to understand. See [11] for a more complete treatment of normality testing.

[10] This is roughly the square root of one-tenth.

mean that the normal distribution is a poor fit for the smaller data set. This is why a quantitative test based on EDF is better than a visual test.

It remains for us to decide what a particular value of $\hat{W}^2$ means. Like other questions involving interpretations of statistics, there is no definitive answer. The best we can do is to associate confidence levels with the statistic; these are given in Table 4.2.3 [1, 10, 11]. The interpretation of the values in the table is similar to that for a table of normal distribution values themselves. For example, only 5 % of samples drawn from a theoretical normal distribution yield a $\hat{W}^2$ value of 0.126 or greater. Values larger than 0.2 for an actual data set are a near-certain indication that the underlying distribution is not normal.

Table 4.2.3 Values of the Cramer-von Mises test statistic $\hat{W}^2$ for confidence levels corresponding to rejection of the hypothesized distribution type

Confidence level	0.75	0.80	0.85	0.90	0.95	0.974	0.99
$\hat{W}^2$ for rejection	0.074	0.081	0.091	0.104	0.126	0.148	0.179

Normality tests cannot confirm a distribution type; they can either provide strong evidence that a distribution is not the tested type or weak evidence that the tested type cannot be ruled out. The $\hat{W}^2$ value for the L-hyoscine hydrobromide data set falls in the range where no conclusion can be drawn. Given the mathematical advantages of the normal distribution,[11] it is reasonable to assume that data is normally distributed unless the test calls for rejection of the hypothesis of normality.

Smaller values of $\hat{W}^2$ indicate greater consistency with a normal distribution, but only up to a point. Real data is a random sample, so F_n should not be a perfect match for F. Values of the test statistic that are less than 0.02 are suspicious. Such values will only rarely be seen with real data.

4.2.3 Outliers

Suppose a scientific experiment yields data that appears to be normally distributed, but with one point that is not close to the others. Such a point is referred to as an *outlier*. The scientist has to decide which is more likely, that the outlier represents an error, in which case the value should be discarded, or that the outlier is simply part of a random sample and should be retained. Like other statistical issues, this one is hard to resolve. We saw in Table 3.6.1 that results more than three standard deviations from the mean of a normal distribution are quite rare. Yet such results are bound to occur if the sample is large enough. Statisticians have proposed a variety of criteria for use in judging whether an outlier should be retained or discarded.[12] We present here the simplest method, which is to reject any points identified as outliers by *Chauvenet's*[13] *criterion.*[14]

---

[11] See Section 4.4.

[12] See [12] for a treatment that uses the best methods and distinguishes a variety of subcases.

[13] Pronounced "show-veh-NAY."

[14] Whereas the Cramer–von Mises test is at least arguably as good as any other test for distribution type, Chauvenet's test is no longer used. Our purpose here is to explore the issue of outliers; in this context, simplicity is better than correctness. Most statisticians favor the Grubbs test, which appears in Problems 4.2.12–4.2.14.

Chauvenet outlier: *any data value Z from a normal distribution for which*

$$N(-|Z|) < \frac{1}{2n},$$

where n is the number of data points in the sample

Example 4.2.4. The data for the petal lengths of 50 iris flowers in Table 3.5.1 has mean 4.26, standard deviation 0.47, smallest value 3.0, and largest value 5.1. Of the two extreme values, 3.0 is further from the mean, with $Z = (3.0 - 4.26)/0.47 = -2.68$. The probability of a result this far from the mean is

$$N(-|Z|) = N(-2.68) = 0.0037.$$

With 50 measurements, the critical value is $1/100 = 0.01$. For this data set, the value 3.0 is a Chauvenet outlier. By trial and error, we can find the critical Z value for a Chauvenet outlier for this data set from the equation

$$N(-|Z|) = 0.01.$$

The result is $|Z| = 2.33$, which corresponds to X values of $\mu \pm Z\sigma = 4.26 \pm 1.10$. Thus, any measurement smaller than 3.16 or larger than 5.36 is a Chauvenet outlier. □

Should the length 3.0 be rejected from the iris data set? There is no clear answer to this question. If it happened by chance that one of the flowers was unusually small, then the value should be retained. If the flower is in some sense abnormal, then the value should be rejected. The numerical value of $Z = -2.68$ does not indicate the cause of the anomaly. Chauvenet's criterion says that the value should be rejected, but the criterion is an arbitrary rule of thumb. Other criteria may yield a different qualitative answer. Most scientists believe that outliers should be retained unless there is a clear scientific reason for their exclusion. Based on a glance at the frequency table, I would retain the measurement.

Check Your Understanding Answers

1. See Figure 4.2.2.
2. $\hat{W}^2 = 0.0501$.

Problems

The data sets for Problems 4.2.1–4.2.8 and 4.2.12–4.2.13 are from [2] and can be found in both csv and xlsx form at http://www.math.unl.edu/~gledder1/MLS/, http://www.springer.com/978-1-4614-7275-9.

4.2.1.* (Continued from Problem 3.1.3.)

(a) Prepare an EDF plot for the dopamine data for nonpsychotic patients from the file DOPAMINE.*.
(b) Compute the modified Cramer–von Mises statistic $\hat{W}^2$ for the data of part (a).
(c) Repeat parts (a) and (b) for the psychotic patients.
(d) What conclusions can we draw from parts (a) to (c)?

(This problem is continued in Problems 4.4.2 and 4.5.1.)

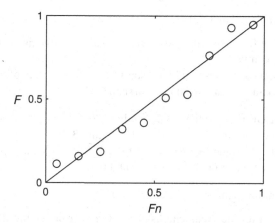

Fig. 4.2.2 EDF plot for the fit of the D-hyoscine hydrobromide data to a normal distribution, with parameters taken from the data

4.2.2. (Continued from Problem 3.1.4.)

(a) Prepare an EDF plot for the fecundity of the control group of fruit flies from the file FRUIT-FLY.*.

(b) Compute the modified Cramer–von Mises statistic $\hat{W}^2$ for the data of part (a).

(c) Repeat parts (a) and (b) for the fruit flies bred for DDT resistance.

(d) Repeat parts (a) and (b) for the fruit flies bred for DDT susceptibility.

(e) What conclusions can we draw from parts (a) to (d)?

(This problem is continued in Problems 4.4.3 and 4.5.2.)

4.2.3. (Continued from Problem 3.1.5.)

(a) Prepare an EDF plot for the weight gain of rats fed a low-protein beef diet from the file WEIGHT.*.

(b) Compute the modified Cramer–von Mises statistic $\hat{W}^2$ for the data of part (a).

(c) Repeat parts (a) and (b) for the rats fed a high-protein beef diet.

(d) Repeat parts (a) and (b) for the rats fed a low-protein cereal diet.

(e) What conclusions can we draw from parts (a) to (d)?

(f) Repeat parts (a) and (b) for the combined beef diet data. Can we say whether we should consider this data to be from a single normal distribution rather than two?

(This problem is continued in Problems 4.4.4 and 4.5.3.)

4.2.4. (Continued from Problems 3.6.12 and 3.6.13.)

(a) Prepare an EDF plot for the butterfat content of milk from Jersey cows in the file BUT-TER.*.

(b) Compute the modified Cramer–von Mises statistic $\hat{W}^2$ for the data of part (a).

(c) Repeat parts (a) and (b) for the Guernsey cows.

(d) Repeat parts (a) and (b) for the Ayrshire cows.

(e) What conclusions can we draw from parts (a) to (d)?

(f) Repeat parts (a) and (b) for the combined Jersey and Guernsey data. Can we say whether we should consider this data to be from a single normal distribution rather than two?

(g) Repeat part (f) for the combined Jersey and Ayrshire data.

(This problem is continued in Problems 4.4.5 and 4.5.4.)

4.2.5. (Continued from Problem 3.6.9.)

(a) Prepare an EDF plot for the resting pulse rates of native Peruvians from the file PULSE.*.
(b) What is unusual about this data?
(c) Compute the modified Cramer–von Mises statistic $\hat{W}^2$ for the data of part (a).
(d) What conclusions can we draw from parts (a) and (c)?

4.2.6. (a) Prepare an EDF plot for the petal lengths of *Iris versicolor* from the file IRISES.*.
Note that a histogram of this data was plotted in Figure 3.5.1.
(b) Compute the modified Cramer–von Mises statistic $\hat{W}^2$ for the data of part (a).
(c) What conclusions can we draw from parts (a) and (b)?

4.2.7. (Continued from Problems 3.6.10 and 3.6.11.)

(a) Prepare an EDF plot for the sepal lengths of *Iris setosa* from the file IRISES.*.
(b) Compute the modified Cramer–von Mises statistic $\hat{W}^2$ for the data of part (a).
(c) Repeat parts (a) and (b) for the sepal lengths of *Iris virginica*.
(d) What conclusions can we draw from parts (a) to (c)?

(This problem is continued in Problem 4.5.5.)

4.2.8. (Continued from Problem 3.7.13.)

(a) Prepare an EDF plot for the waiting times between firings of motor cortex neurons of an
unstimulated monkey from the file WAITING.*.
(b) Compute the modified Cramer–von Mises statistic $\hat{W}^2$ for the data of part (a).
(c) What conclusions can we draw from parts (a) and (b)?

The best way to build intuition about matters of statistics is to do a lot of computer experiments with simulated data. Problems 4.2.9–4.2.11 experiment with the Cramer–von Mises normality test.

4.2.9.* A lot of data sets that don't look like they are normally distributed produce low enough $\hat{W}^2$ values that we cannot reject the hypothesis of normality. What about data that clearly fits a different distribution? Here we consider data from a uniform distribution.

(a) The data set consisting of the integers from 1 to 25 is uniformly spread, although it is not random. Prepare an EDF plot and compute the modified Cramer–von Mises statistic for this data.
(b) Repeat part (a) for the integers 1–50 and for the integers 1–100. What features do you see in the plots? What seems to be happening to $\hat{W}^2$?
(c) Use a computer program to generate a set of 50 numbers from a uniform distribution. Find the average value of the modified Cramer–von Mises statistic for this data over five trials. How does the result compare to that obtained with the integers 1–50?
(d) Run one trial for a set of 100 numbers from a uniform distribution. How does the result compare to part (c) and to the integers from 1 to 100?
(e) Compare EDF plots for single runs of 50 consecutive integers and 100 consecutive integers. Is the result consistent with parts (b)–(d)?

4.2.10. (Continued from Problem 3.6.14.)
In Problem 3.6.14, we created bimodal data sets by combining two normal distributions. We saw that the resulting distribution resembles a normal distribution if the means of the components are close enough, but not when the means are far apart. This suggests a computer experiment for the Cramer–von Mises test.

(a) Write a computer program to generate a data set that contains 50 numbers from each of two normal distributions, one with $\mu = 7$ and $\sigma = 2$, and the other with $\mu = 13$ and $\sigma = 2$. Use the program to create an EDF plot.

(b) Embed the program in a loop that saves the calculations for $\hat{W}^2$ for 100 runs and then computes the average. [Do not put the plot command inside the loop!]

(c) What conclusions can you draw from parts (a) and (b) in conjunction with Problem 3.6.14c?

(d) Repeat parts (a) and (b) with $\mu = 8$ and $\mu = 12$.

(e) What conclusions can you draw from part (e) in conjunction with Problem 3.6.14b?

4.2.11. In this problem, we consider the effect of contaminated data. For each part of the problem, create a data set with 100 numbers drawn from the standard normal distribution and 20 numbers drawn from the given contaminating distribution. Prepare one EDF plot. Also do a run of 100 trials (without the plot) and obtain an average and standard deviation for $\hat{W}^2$.

(a) The contaminating distribution is $N_{0,0.1}$.

(b) The contaminating distribution is $N_{1,0.1}$.

(c) Plot a theoretical probability density function for each of parts (a) and (b) by taking a weighted average of the pdfs for the component distributions.

(d) Both cases have the same amount of the contaminating distribution and the standard deviations of the contaminant are small, but the contaminant is near the mean in one case and away from the mean in the other. Describe the effects these contaminants have on the EDF plot and the Cramer–von Mises test statistic.

Most statistics references recommend the *Grubbs test* for outliers, provided the data appears to be normally distributed. For this test, we calculate G, defined as the maximum value of $|Z|$ for the data set. The corresponding data value is considered to be an outlier if G is greater than a critical value that depends on the number of measurements. The critical value also depends on one's arbitrary choice of significance level. Table 4.2.4 lists the critical values for the 95 % and 99 % significance levels. As with Chauvenet's test, we should consider the Grubbs test to be just one piece of information used to make the final decision, along with information about the population and the sampling method.

Table 4.2.4 Critical values for the maximum $|Z|$ in Grubbs' test, given n data points, and either 95% significance or 99% significance

n	$\alpha = .05$	$\alpha = .01$	n	$\alpha = .05$	$\alpha = .01$	n	$\alpha = .05$	$\alpha = .01$
10	2.29	2.48	30	2.91	3.24	120	3.45	3.82
12	2.41	2.64	40	3.04	3.38	160	3.54	3.91
14	2.51	2.76	50	3.13	3.48	200	3.61	3.98
16	2.59	2.85	60	3.20	3.56	300	3.72	4.09
18	2.65	2.93	70	3.26	3.62	400	3.80	4.17
20	2.71	3.00	80	3.31	3.67	500	3.86	4.23
25	2.82	3.14	100	3.38	3.75	600	3.91	4.27

4.2.12. Use the Grubbs test with $\alpha = 0.05$ to identify (approximately) the largest and smallest measurements that would not be considered as outliers for the specified data sets.

(a) *Iris versicolor* petal length, from Table 3.5.1.

(b) *Iris versicolor* sepal length, from IRISES.* (See Problem 3.5.5.)

(c) Resting pulse rate of native Peruvians, from PULSE.* (See Problem 3.6.9.)

4.2.13. (a) Compute the value of Z for each measurement in the data file BUTTER.*. Note that each column has its own mean and standard deviation.

(b) Which of the measurements are Chauvenet outliers?

(c) Which of the measurements would be rejected by the Grubbs test at the 95 % level?

4.2.14. (a) Suppose you randomly generate ten values from a normal distribution. What is the probability that at least one of these values would be rejected by the Grubbs test with $\alpha = 0.05$?

(b) Repeat part (a) for a sample of 100 values.

(c) Repeat parts (a) and (b) using Chauvenet's test.

4.3 Probability Distributions of Samples

After studying this section, you should be able to:

- Determine the theoretical mean and standard deviation for distributions of sample means.
- Prepare histograms for distributions of means of samples of size n from a given underlying distribution.
- Discuss the properties of distributions of sample means as a function of sample size.

In many scientific experiments, we want to determine the characteristics of some population but are unable to study all the individuals. We want to know how tall adult Americans are, but we cannot measure all American adults. We want to determine how many people have HIV, but we can get records only for people who have volunteered to be tested. We want to determine how much oxygen a particular Olympic swimmer uses in 200- m training swims, but our equipment is only available at one pool. [15]

In practice, we must determine the mean $\bar{x}$ and standard deviation s for a *sample* and then approximate the mean μ and standard deviation σ of the *population* by

$$\mu \approx \bar{x}, \qquad \sigma \approx s.$$

However, conclusions drawn from samples only apply to the full population if the sample is *representative*, which means that its characteristics match the population to a desired degree of accuracy. This poses a difficulty, because no procedure can guarantee that a sample is representative. As an alternative, scientists try to collect a *random sample*, in which each member of the population has an equal chance of being selected. There are two issues that must be faced when doing this. First, a truly random sample is not necessarily representative. We saw in Section 4.1 that random samples can sometimes be noticeably different from the underlying population. This issue is explored in more detail in this section. Second, it is not always possible to obtain a sample that is truly random. Medical tests, for example, are conducted by institutions using subjects who are from a specific geographic region. This poses a difficulty: Can a random sample drawn from a subpopulation be representative of the population at large? [16]

4.3.1 Sums and Means of Two Random Variables

Suppose X_1 and X_2 are random variables drawn from probability distributions with means μ_1 and μ_2 and standard deviations σ_1 and σ_2. Let Y and X be the sum and mean of X_1 and X_2,

---

[15] The "population" in this case consists of the different 200- m swims for the single athlete.

[16] This issue is addressed in Section 4.4.

respectively; thus, $Y = X_1 + X_2$ and $X = Y/2$. Because they depend on random variables, both Y and X are also random variables. As such, there are probability distributions for Y and X whose characteristics depend on those of the distributions for the components X_1 and X_2. The characteristics of the distributions for the sum and mean can be easily computed, with the results summarized here.[17]

Theorem 4.3.1 (Mean and Variance for a Sum and Mean of Two Random Variables). *Suppose X_1 and X_2 are independent random variables drawn from distributions with means μ_1 and μ_2 and variances σ_1^2 and σ_2^2. Then the random variable $Y = X_1 + X_2$ has mean $\mu_1 + \mu_2$ and variance $\sigma_1^2 + \sigma_2^2$. Furthermore, the distribution of the mean $X = Y/2$ has a mean and standard deviation that are one-half of those of the sum Y.*

This theorem provides the basic tool needed to characterize the probability distribution of sample means.

4.3.2 General Characteristics of Samples

Suppose we have a population that is large enough to model with a probability distribution. This means that removing individuals to create a sample does not affect the distribution of individuals remaining in the population. Hence, a sample of size n from the population corresponds to a set of n independent identically distributed random variables $X_1, X_2, \ldots, X_n$. We can aggregate the results by computing a sum Y and a mean X, defined by

$$Y = \sum_{i=1}^{n} X_i, \qquad X = \frac{Y}{n}. \qquad (4.3.1)$$

These aggregate quantities are random variables as well; our primary interest is in determining the shapes, means, and standard deviations of their probability distributions. We begin our study with a motivating example.

Example 4.3.1. Single die rolls have a uniform distribution. However, a roll of two dice is much more likely to produce a mean of 3.5 (a sum of 7) than a mean of 1 (a sum of 2). In fact, we can calculate these means without much difficulty. A mean of 1 requires a 1 on both dice. This is a sequence of two experiments, each with probability 1/6; hence, the probability of a mean of 1 is 1/36. In contrast, a mean of 3.5 is equivalent to a sum of 7. No matter what the number on the first die, there is always one roll of the second die that makes the sum 7; hence, the probability of a mean of 3.5 is 1/6. Clearly the distribution of sample means is not uniform. Figure 4.3.1 illustrates the distribution of means for samples of two die rolls and samples of four die rolls. □

Observe that the possible values for the means of two die rolls are $1, 1.5, 2, \ldots, 6$, while the possible means of four die rolls are $1, 1.25, 1.5, \ldots, 6$. Means for samples from a discrete distribution are always discrete, but the number of discrete outcomes increases as the sample size increases. Given that probability is based on *models*, there is no harm in thinking of the means of large samples as having a continuous distribution rather than a discrete one. One important characteristic of distributions of sample means is their shape; we will see that they are always approximately normal for large sample sizes. For example, the underlying distribution in Example 4.3.1 is uniform, but the histogram in Figure 4.3.1b shows that a sample size of 4 is enough for the distribution of sample means to look approximately normal.

---

[17] See Problem 4.3.1.

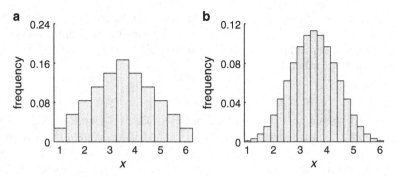

Fig. 4.3.1 Histograms for means of samples of (**a**) two and (**b**) four die rolls

4.3.3 *Means and Standard Deviations of Sample Sums and Means*

Repeated application of Theorem 4.3.1 gives us simple formulas for the mean μ_Y and variance σ_Y^2 of the sum, and we can calculate the standard deviation from the latter:

$$\mu_Y = \sum_{i=1}^{n} \mu = n\mu\,, \qquad \sigma_Y^2 = \sum_{i=1}^{n} \sigma^2 = n\sigma^2, \qquad \sigma_Y = \sqrt{\sigma_Y^2} = \sqrt{n}\,\sigma\,. \qquad (4.3.2)$$

The mean X is merely Y/n, so its mean and standard deviation can be computed immediately from those for Y, simply by dividing by n. This result is sufficiently important to be given as a separate theorem.

> **Theorem 4.3.2 (Mean and Standard Deviation for a Distribution of Sample Means).**
> *Let X be the mean of n independent random variables, each drawn from a distribution having mean μ and standard deviation σ. Then the mean μ_X and standard deviation σ_X for the random variable X are given by*
>
> $$\mu_X = \mu\,, \qquad \sigma_X = \frac{\sigma}{\sqrt{n}}\,. \qquad (4.3.3)$$

Theorem 4.3.2 allows us to calculate theoretical means and standard deviations for sample means, given only that we know the mean and standard deviation for the population from which the samples are drawn.

Example 4.3.2.

Q) Suppose we draw a sample of four individuals from the Poisson distribution having mean $\mu = 4$ and define X to be the mean of the sample. What are the mean and standard deviation for the random variable X?

A) From Theorem 3.7.1, we know that $\sigma^2 = \mu$ for the Poisson distribution; hence, the standard deviation is $\sigma = \sqrt{\mu} = 2$. The random variable X, which is the mean of a sample of four individuals, therefore has mean and standard deviation

$$\mu_X = \mu = 4\,, \qquad \sigma_X = \frac{\sigma}{\sqrt{4}} = \frac{2}{2} = 1\,.$$

Note that the distribution of sample means cannot be a Poisson distribution because $\sigma_X^2 \neq \mu_4 X$. $\qquad \square$

Because sampling is such a common occurrence, it is important to be clear about the conceptual framework for discussing samples. A total of n individual measurements are drawn from a base population having some distribution type with mean μ and standard deviation σ. The sample mean is a random variable X, which has its own distribution with mean and standard deviation given by Theorem 4.3.2 and a shape that is different from the underlying distribution.

4.3.4 Distributions of Sample Means

We cannot make a general statement about the shape of distributions of sample means. However, in the case of the binomial distribution, we can express the distribution of sample means in terms of the binomial distribution with all the Bernoulli trials combined, as we did in Section 4.1.

Example 4.3.3. The scenario of Section 4.1 involved individual experiments consisting of a count of marked individuals in a sample of 50. Since 5 % of the population consisted of marked individuals, we used the binomial distribution $b_{50,\,0.05}$ to model the probability; that is, we used $b_{50,\,0.05}(k)$ as the probability of obtaining k marked individuals in the experiment. Then we considered four successive counts as equivalent to a single experiment with a sample of 200 and obtained the probability distribution $b_{200,\,0.05}$. Instead, suppose we think of this new experiment as having two layers. The lower layer is the original experiment consisting of 50 Bernoulli trials, which we use four times to obtain random variables X_1, X_2, X_3, and X_4. The upper layer is the combination of these four random variables into a sample, and we can think of it as generating a random sum Y and a random average X. Because $Y = X_1 + X_2 + X_3 + X_4$, its distribution is $b_{200,\,0.05}$. The distribution for the mean X is then $0.25b_{200,\,0.05}$. The underlying distribution $b_{50,\,0.05}$ has mean $\mu = np = 2.5$ and standard deviation $\sigma = \sqrt{np(1-p)} = 1.541$. By Theorem 4.3.2, the distribution for the sample mean X has mean $\mu_X = 2.5$ and standard deviation $\sigma_X = 1.541/\sqrt{4} = 0.77$. Although the distribution of sample means is still binomial, the shapes of the distributions $b_{50,\,0.05}$ and $0.25b_{200,\,0.05}$ are different. We can quantify the difference in these distribution shapes by calculating values of the modified Cramer–von Mises statistic. Samples of 100 values from $b_{50,\,0.05}$ have an average of $\hat{W}^2 \approx 0.45$, while samples of the same size from $0.25b_{200,\,0.05}$ average $\hat{W}^2 \approx 0.16$. $\qquad\square$

> **Check Your Understanding 4.3.1:**
> Identify the distribution for the mean X of random variables $X_1, \cdots, X_{10}$, each drawn from $b_{50,0.05}$. Compute the mean and standard deviation of X. Write a program to determine $\hat{W}^2$ for a sample of 50 values of the distribution for X. Run enough experiments to see convincing evidence that the distribution of means of ten numbers from $b_{50,0.05}$ has a smaller value of $\hat{W}^2$ than the distribution of means of four numbers from the same underlying distribution.

A detailed look at histograms of sample means is particularly revealing if the underlying distribution is highly skewed. The exponential distribution works well in this role.

Example 4.3.4. Suppose we draw four individuals from the exponential distribution $E_{0.5}$.[18] The underlying distribution has mean $\mu = 2$ and standard deviation $\sigma = 2$. If we draw samples of size 4, 16, and 64, each will have a distribution with mean $\mu_X = 2$, while the standard deviations will gradually decrease, with $\sigma_X = 1$, $\sigma_X = 0.5$, and $\sigma_X = 0.25$. To visualize these sample

---

[18] See Section 3.7.

distributions numerically, we can use a random number generator to draw a sample of size n and compute the mean, which will be X for the sampling experiment. We can embed this calculation in a loop that runs the same experiment 10,000 times, which is enough for a good approximation of the distribution of sample means. We can then sort the 10,000 sample means into bins and produce a histogram. The result is Figure 4.3.2. Figure 4.3.2a shows the highly skewed underlying distribution, in which very small values are more common than larger values. Figure 4.3.2b shows the distribution of means of samples of size 4. Very small values are no longer common. They still arise in the underlying distribution, but most often in conjunction with three other values that are not very small. The first bin, with means less than 0.25, occurs only when the sum of the four underlying values is less than 1, and this is rare. The distribution is still skewed; for example, a sample mean of 1.8 is much more common than a sample mean of 2.2, even though both are 0.2 standard deviation away from the distribution mean of 2.0. Proceeding through the figures, we see that larger samples produce distributions that are progressively less skewed and also narrower.   □

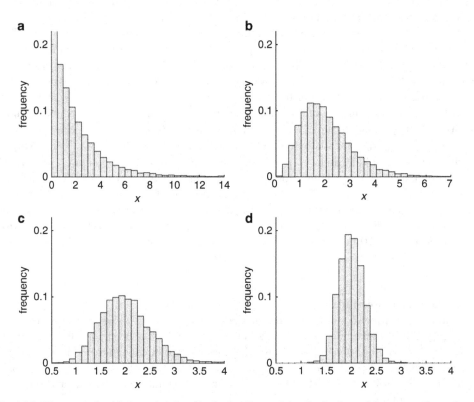

Fig. 4.3.2 Histograms for (**a**) the underlying distribution $E_{0.5}$ and the distributions of the means of samples of size (**b**) 4, (**c**) 16, and (**d**) 64

Figure 4.3.3 shows the histogram for the distribution of the means of samples of size 64 along with the normal distribution having the same mean and standard deviation. The figure clearly shows the similarity of the distribution of sample means to the corresponding normal distribution. We can also document the progression toward a normal distribution by examining values of the Cramer–von Mises statistic for the four distributions in Example 4.3.4, which are approximately 0.84 for $E_{0.5}$, 0.24 for samples of size 4, 0.10 for samples of size 16, and 0.07 for samples of size 64. Data drawn from a normal distribution by a random number generator yields an average Cramer–von Mises statistic of 0.06, so samples of size 64 from $E_{0.5}$ are effectively normal. The value of 0.10 is small enough to justify using the normal distribution for samples of size 16 as well.

The evidence from Example 4.3.4, as seen in the results of the Cramer–von Mises normality test and Figure 4.3.3, suggests that means of large samples are normally distributed. This is confirmed by a key theorem of probability, which we state here without proof.

> **Theorem 4.3.3 (Central Limit Theorem).** *Suppose X is a random variable obtained as a mean of n independent identically distributed random variables. In the limit as n →* ∞, *the distribution of X approaches a normal distribution with mean μ and standard deviation* $\sigma/\sqrt{n}$, *where μ and σ are the mean and standard deviation of the distribution from which the independent random variables are drawn.*

The central limit theorem is a valuable tool to use for drawing inferences about populations, as we will see in Section 4.4.

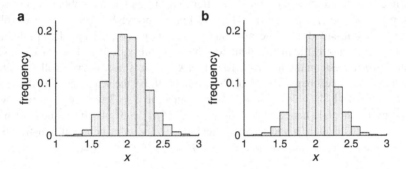

Fig. 4.3.3 A comparison of histograms: (**a**) the distribution of means of samples of size 64 and (**b**) the normal distribution having the same mean and standard deviation

Check Your Understanding Answers

1. The distribution is $0.1b_{500,\,0.05}$. It has mean 2.5 and standard deviation 0.49. A typical value for $\hat{W}^2$ for a sample of 100 values from this distribution is 0.10.

Problems

4.3.1. Prove Theorem 4.3.1. You may assume without proof that $E(X_1X_2) = E(X_1)E(X_2)$.

4.3.2. Suppose we draw a sample of 100 individuals from the probability distribution indicated below. Determine the mean and standard deviation of the means for each of these samples.

(a) The probability distribution for the number of successful attacks out of 5 tries given a success probability of 0.3 for each attack.
(b) The probability distribution for the number of successful predator attacks in a 2-h period, given a mean attack rate of 3 per hour.

4.3.3.* Consider samples drawn from an underlying binomial distribution $b_{20,0.1}$ and let X be the mean of a sample of size n.

(a) Determine the means and standard deviations of the random variable X for the cases where $n = 4$, $n = 16$, $n = 64$.
(b) Draw 10,000 values of X for each of the three cases and plot histograms. Compare with the underlying distribution and with the corresponding normal distributions.
(c) Compute the average Cramer–von Mises statistic over 10,000 sets of 100 values of X for each of the three cases.
(d) Summarize the results of these experiments.

4.3.4. Repeat Problem 4.3.3 using the Poisson distribution f_1 as the underlying distribution.

4.3.5. Repeat Problem 4.3.3 using the exponential distribution $E_{0.2}$ as the underlying distribution. Also compare with Example 4.3.4.

4.3.6. Repeat Problem 4.3.3 using the *geometric distribution* g_p, which is the distribution for the experiment in which X is the number of Bernoulli trials taken to obtain a success, given a success probability of p for each trial. For example, suppose X is the number of rolls taken to roll a 1 on a four-sided die. The distribution of X is given by $g_{0.25}$. The probabilities of this distribution are not difficult to calculate. The probability of $X = 1$ is 1/4 because $X = 1$ corresponds to a success on the first roll. To obtain $X = 2$, it is necessary to roll a number other than 1 on the first roll before rolling a 1 on the second roll. By the sequence rule, the probability is $P[X = 2] = (3/4)(1/4) = 3/16$. Similarly, the outcome $X = 3$ requires two rolls without a 1 followed by a roll of 1; hence, $P[X = 3] = (3/4)^2(1/4) = 3^2/4^3$. It can be shown that the distribution g_p has mean $1/p$ and standard deviation $\sqrt{1-p}/p$; hence, the distribution $g_{0.25}$ has mean $\mu = 4$ and standard deviation $\sigma = \sqrt{0.75}/0.25 = 2\sqrt{3} \approx 3.464$.

4.3.7. (Continued from Problem 3.3.1.)
Assume the distribution of babies for a successful pregnancy is given by $p(1) = 0.9851$, $p(2) = 0.0138$, and $p(3) = 0.0011$. Repeat Problem 4.3.3 with this distribution, but with samples of size 100, 400, 2,500, and 10,000.

4.3.8. In a news story on climate change that ran on National Public Radio's "All Things Considered" on August 6, 2012, the reporter said, "One standard deviation is like rolling snake eyes three times in a row."[19]

(a) Determine the mean and standard deviation for a single die roll and use it to determine the mean and standard deviation of a sum of six rolls.
(b) How many standard deviations below the mean is a total score of 6?
(c) What total score comes closest to being one standard deviation below the mean?
(d) If the distribution of the sum of six dice were actually normal, what would be the probability of rolling a sum less than 6? What would be the probability of a sum of 5 or less? What is the relationship between these two probabilities in the actual experiment of rolling six dice?
(e) Repeat part (d) to compare the probability of a sum less than 21 with a sum of 20 or less, using a normal distribution model.
(f) Use an experiment similar to that of Problem 4.3.3c to assess how close the distribution of the sum of six die rolls is to being normal.

---

[19] The reporter did not identify the source of this claim. It certainly was not in the research paper that inspired the climate change news story.

4.4 Inferences About Populations

After studying this section, you should be able to:

- Determine the probability that a specific sample mean could be obtained from a general population of quantitative data.
- Determine the probability that a specific number of successes could be obtained in Bernoulli trials with a general population.

In Section 4.3, we raised the question of whether a sample drawn randomly from a subpopulation can be representative of the full population. We identified two issues—whether random samples are necessarily representative and whether a subpopulation can represent a full population—and addressed the first of these. The answer to the second issue depends on whether there are systematic differences between the subpopulation and the full population.

Example 4.4.1. In chemistry, we take a sample of pure hydrochloric acid by pouring some out of a single bottle. If we mix the contents before pouring, we are getting a random sample from the subpopulation of all molecules of hydrochloric acid that happen to be in that one bottle. Nevertheless, the sample is representative of all molecules of hydrochloric acid, because there is no systematic difference among bottles of pure hydrochloric acid. □

In contrast with the usual situation in chemistry, a truly random sampling procedure in biology is often impossible.

Example 4.4.2. If we are studying bison, we might have to restrict our sample to the subpopulation of bison living in Custer State Park in South Dakota. We need to employ a well-designed procedure to obtain a random sample of the subpopulation, and then we need to be concerned about the possibility of systematic differences between bison who live in Custer State Park and those who live elsewhere. □

Whether the difference between two data sets is "significant" is one of the central questions of data analysis. This question is only partly mathematical, because the answer depends on what we mean by a significant difference. This issue belongs to the subject of statistics, which we can think of as "drawing conclusions from probability calculations." Here, we are concerned only with the probability calculations themselves. We consider two scenarios, one in which the experiment yields quantitative results and one in which the experiment yields a result of "positive" or "negative."[20]

4.4.1 Inferences About Sample Means

Assume that we have a set of n measurements $X_1, X_2, \ldots, X_n$, each drawn randomly from a large population of mean μ and standard deviation σ. If X is the mean of the individual measurements, then it is also a random variable, but it comes from the distribution of means of samples of size n. While the distribution of sample means is different from the underlying distribution, we do know a lot about it. From Theorem 4.3.2, we know that the mean is μ and the standard deviation is $\sigma/\sqrt{n}$. If the sample size is large enough, we know from the central limit theorem that the distribution of sample means is approximately normal.[21]

---

[20] For example, does an individual have a heart attack in a given period or not?

[21] See Theorem 4.3.3.

Problem 3.1.3 uses a set of data for dopamine concentration in people identified as either psychotic or not psychotic. The mean concentration for the psychotic subjects is higher, but is it higher enough to justify the conclusion that high dopamine concentration and psychosis are correlated?[22] This is a very common type of question in biology. A biologist collects data for a general population or control group and also for a test population that differs from the general population in some way. Invariably, the means will be at least a little bit different. If they seem noticeably different, we want to know if that difference could arise simply by chance. If not, then we can conclude that the characteristic that identifies the test population is correlated with the characteristic being measured.[23]

To answer questions about the consistency of data with an assumed population, we measure a single sample of size n from the test population. Let $\bar{x}$ be the mean obtained for the specific sample. If the test population is representative of the full population, then the distribution of means for test population samples is the same as that of the full population. In particular, the expected value for test population means is μ, the mean of the underlying distribution for the full population. But suppose $\bar{x}$ is noticeably less than (or greater than) μ. This provides evidence that the test population and the full population are significantly different. To determine the strength of this evidence, we need to know how unusual the value of $\bar{x}$ would be if it actually came from the full population. So the mathematical question is:

What is the probability that the mean (X) of a random sample of size n from the full population is *at least as far*[24] below (above) the expected value (μ) as the measured mean ($\bar{x}$) of the test sample?

In mathematical notation, the problem is to determine $P[X \leq \bar{x}]$. For large enough n, we know that the distribution for X is approximately normal, with mean μ and standard deviation $\sigma_X = \sigma/\sqrt{n}$.[25] Thus,[26]

$$P[X \leq \bar{x}] = N_{\mu,\sigma_X}(\bar{x}) = N(z), \quad z = \frac{\bar{x} - \mu}{\sigma_X}, \quad \sigma_X = \frac{\sigma}{\sqrt{n}},$$

---

[22] See Problem 4.4.2.

[23] It is a common fallacy to interpret a correlation in terms of causation; for example, we might think that a correlation between high dopamine levels and psychosis means that too much dopamine makes a person psychotic. This may be true, but demonstrating a correlation is not enough to support the claim. It is also possible that being psychotic raises dopamine levels, or that some other condition is responsible for both the psychosis and the high level of dopamine. A good rule of thumb is that causation should not be inferred from correlation unless only one of the possibilities of "A causes B" and "B causes A" is plausible. For example, if people who eat carrots have better night vision than people who don't, we can safely conclude that something in the carrots is responsible for the improved vision, as it seems clearly implausible that being able to see well at night makes a person hungry for carrots.

[24] See Section 4.1 for a discussion of why this detail is necessary.

[25] See Theorem 4.3.2.

[26] See Theorem 3.6.1.

where N is the cumulative distribution function for the standard normal distribution. If the sample mean is unusually large, then the corresponding result is

$$P[X \geq \bar{x}] = 1 - N_{\mu,\sigma_X}(\bar{x}) = 1 - N(z) = N(-z), \quad z = \frac{\bar{x} - \mu}{\sigma_X}, \quad \sigma_X = \frac{\sigma}{\sqrt{n}}.$$

These formulas can be combined with the observation that the argument of N in both cases is negative, given $z < 0$ for values less than the mean and $z > 0$ for values greater than the mean. Hence, we can summarize the key result in one statement:

Theorem 4.4.1 (Probability of Obtaining a Given Mean from Large Samples of a Given Population). *Let $\bar{x}$ be the mean of a sample of size n drawn from an underlying distribution with mean μ and standard deviation σ. In the limit as $n \to \infty$, the probability of obtaining a mean at least as far from μ as $\bar{x}$ is $N(z)$, where*

$$z = \frac{-|\bar{x} - \mu|}{\sigma_X}, \quad \sigma_X = \frac{\sigma}{\sqrt{n}}. \tag{4.4.1}$$

Note that Theorem 4.4.1 provides a theoretical result. In practice, n is finite, so the use of the normal distribution does not yield an exact answer. As we saw in Section 4.3, the convergence of the distribution of sample means to the normal distribution is fast; for all but the most skewed underlying distribution, a sample of size 50 is large enough for the distribution of sample means to be effectively normal.

Example 4.4.3.

Q)   Karl Pearson's landmark paper introducing the chi-square test used a data set consisting of frequencies for the number of successes, defined as rolls of "5" or "6," in rolls of 12 dice [7].[27] The data is reproduced in Table 4.4.1. Were the dice fair?

A)   Of course we cannot answer the question as posed. What we can do is look at the distribution of sample means for rolls of fair dice and determine the probability that the data could have come from that distribution. The individual experiments consisted of 12 Bernoulli trials, each with success probability 1/3; hence, the underlying distribution is $b_{12,1/3}$. This distribution has mean $\mu = 4$ and standard deviation $\sigma = 1.633$.[28] The sample mean $\bar{x}$ is the expected value determined from the frequency table, which is $\bar{x} = 4.052$. The distribution of sample means is normal, with mean $\mu_X = 4$ and standard deviation $\sigma_X = \sigma/\sqrt{26306} = 0.010$. From (4.4.1), we obtain $z = -5.2$, which corresponds to a probability of $N(-5.2) = 10^{-7}$. Thus, the probability of a mean of 4.052 or larger for fair dice is vanishingly small. There are two possible conclusions: either the dice were not fair or the data is inaccurate.[29]   □

Check Your Understanding 4.4.1:
Verify the calculations in Example 4.4.3.

---

[27] See Problem 3.4.6 for background and commentary.

[28] See Theorem 3.4.1.

[29] Pearson drew the same conclusion using his more sophisticated chi-square test. He also suggested an explanation: large numbers were more likely than small numbers because the pips on the dice reduced the weight of each face; hence, the faces with larger numbers were lighter than those with smaller numbers. See [3] for a further discussion.

Table 4.4.1 The frequency of N successes in 26306 rolls of 12 dice, where a success means a roll of '5' or '6'

N	0	1	2	3	4	5	6	7	8	9	10	11	12
X	185	1149	3265	5475	6114	5194	3067	1331	403	105	14	4	0

4.4.2 Inferences About Sample Proportions

A lot of biological data, such as height cholesterol levels, and numbers of offspring, is quantitative. However, other important data is qualitative in nature, such as whether or not a person has had a heart attack. Suppose the fraction of heart attack patients in some group appears to be significantly less than the overall heart attack rate for the population at large. We want to know the probability that the measured heart attack rate for the small group could occur in a representative sample drawn from the larger population. This is a question of drawing inferences about proportions rather than means.

Think of each measurement as a Bernoulli trial. A given patient has either had a heart attack (outcome $X_j = 1$) or not (outcome $X_j = 0$). The total number of heart attacks in a collection of n individuals is governed by a binomial distribution with probability distribution function $b_{n,p}$. We assume that the probability p for success in each Bernoulli trial is known for the general population. Then we can compute $b_{n,p}(k)$, which is the probability of k "successes" in n Bernoulli trials, each of which has probability p of success. If there are $\bar{k}$ heart attacks in the sample of size n, we want to know the probability that X is less than or equal to $\bar{k}$. However, this can be a problem for very large samples, owing to the lack of a simple formula for the cumulative distribution function $B_{n,p}(\bar{k})$.

An alternative is provided by the central limit theorem,[30] which says that the distribution of sample means is approximately normal for large samples. Thus, we can approximate $B_{n,p}(\bar{k})$ by an appropriate normal distribution $N_{\mu,\sigma}$. The relevant facts about the binomial distribution are given in Theorem 3.4.1: the mean is $\mu = np$ and the variance is $\sigma^2 = np(1-p)$. The key result is almost identical to Theorem 4.4.1, using $\mu = np$ and $\sigma_X = \sqrt{np(1-p)}$; however, a small correction is needed to convert the discrete data of the experiment to a continuous distribution. We state the result without proof.

Theorem 4.4.2 (Probability of Obtaining a Given Number of Successes from a Large Number of Bernoulli Trials). *Let $\bar{k}$ be the number of successful Bernoulli trials out of a total of n, each with success probability p. In the limit as $n \to \infty$, the probability of obtaining a success total at least as far from np as $\bar{k}$ is approximately $N(z)$, where*

$$z = \frac{-|\bar{k} - np| + \frac{1}{2}}{\sqrt{np(1-p)}}.$$ (4.4.2)

As with Theorem 4.4.1, the result in Theorem 4.4.2 is exactly correct only for the theoretical case where n is infinite and is approximate for finite n. The accuracy of the approximations depends on n being large enough, and also on p being neither too close to 0 nor too close to 1. In practice, statisticians use the rule of thumb that both np and $n(1-p)$ should be greater than 5.[31]

---

[30] Theorem 4.3.3.

[31] The real situation is far more complicated than this simplified narrative suggests and serves as a lesson on the dangers of using a result for a real case that is true only in some limit. The example that follows hints at the problem, which is explored in more detail in Problem 4.4.10.

Example 4.4.4. Suppose the incidence of heart attacks in the general population is about 1 in 1,000. A subgroup of 10,000 people has a total of 5 heart attacks. Is this significantly less than what would be expected from a random sample? Observe first that $p = 0.001$, which is quite small. However, the sample is large enough that $np = 10$, so it is reasonable to use a normal approximation. The appropriate normal distribution has mean $np = 10$ and variance $np(1 - p) = 9.99$. This data yields a z value of

$$z = \frac{-|\bar{k} - np| + \frac{1}{2}}{\sqrt{np(1-p)}} = \frac{-|5 - 10| + \frac{1}{2}}{\sqrt{9.99}} = -1.424 .$$

Thus,

$$P[X \le 5] \approx P[Z \le -1.424] = N(-1.424) = 0.0773 .$$

Of course, we obtained this answer using Theorem 4.4.2, which is only approximate. The correct answer is

$$P[X \le 5] = B_{10000, 0.001}(5) = 0.0670.$$

For a representative sample of size 10,000, there is almost a 7 % chance that there will be 5 or fewer heart attacks. While the actual incidence is suspiciously low, it is not so low that we can reach a scientific conclusion of significance. The approximation is off by about 13% even though the rule of thumb $np > 5$ holds. □

Note that Theorem 4.4.2 overestimates the probability in Example 4.4.4 by about 15 %. Had we not computed the exact answer, we would have drawn a weaker conclusion than is actually supported by the data. This is not troubling, as science should be conservative about drawing conclusions. However, we should be concerned about the possibility that the approximation in the theorem will lead us to draw conclusions that are stronger than the data can support. A clue for what happened lies in the relatively small value of $np = 10$. This means that there is a significant granularity to the distribution.[32] The next example is typical of cases where a much larger value of np indicates that granularity is minimal.

Example 4.4.5. Suppose you flip a coin 100 times and get heads 60 times. Is the coin fair?

Of course, we can't actually answer this question as posed. Instead we answer a related question: What is the probability of getting 60 or more heads on 100 flips of a fair coin? Theorem 4.4.2 provides an approximate answer to this latter question. The z value is -1.9, corresponding to a probability of 0.0287. Alternatively, we can calculate the exact probability of 60 or more heads with 100 tosses of a fair coin:

$$P[X \ge 60] = 1 - P[X \le 59] = 1 - B_{100, 0.5}(59) = 0.0284.$$

While this is a low probability, it means that if we had 10,000 fair coins and flipped each of them 100 times, we should expect about 284 of them (almost 3 %) to yield 60 or more heads. It seems likely that the coin is unfair, but not certain. More to the point of our mathematical study, the approximation is very accurate. □

Check Your Understanding 4.4.2:
Verify the calculations of Example 4.4.5.

---

[32] See Problems 4.3.8 and 4.4.10.

Problems

The data sets for Problems 4.4.2–4.4.5 are from [2] and can be found in both csv and xlsx form at http://www.math.unl.edu/~gledder1/MLS/, http://www.springer.com/978-1-4614-7275-9.

4.4.1. (a) A test group of 400 patients is given a new cholesterol-lowering medication for 2 months. The reduction of the patients' cholesterol level has a mean of 16 with a standard deviation of 60. Meanwhile, a control group is given a placebo. For this group, there is a reduction in cholesterol of 8 with a standard deviation of 40. Assuming that the control group is representative of the general population, determine the probability that the given improvement in the test subjects could have been measured in a random sample of 400 patients who did not get the medication.
(b) Repeat part (a) for a group of 10,000 patients given a different medication with resulting cholesterol reduction of mean 10 and standard deviation 40.
(c) Repeat part (a) for a group of 25 patients given a third medication with resulting cholesterol reduction of mean 24 and standard deviation 10.
(d) Which of the three medications would you say is most effective?
(e) In determining the effectiveness of a medication, what questions should be asked in addition to that about the probability of a significant difference between the test population and the general population?

4.4.2.* (Continued from Problems 3.1.3 and 4.2.1.)

(a) Are we justified in assuming the data in the file DOPAMINE.* to be normally distributed?
(b) If so, use it to investigate the question of whether dopamine concentrations for psychotic people are different from those for nonpsychotic people.

4.4.3. (Continued from Problems 3.1.4 and 4.2.2.)

(a) Are we justified in assuming the data in the file FRUITFLY.* to be normally distributed?
(b) If so, use it to investigate the question of whether selective breeding of fruit flies for DDT resistance decreases their fecundity.
(c) Repeat part (b) for selective breeding for susceptibility to DDT.
(d) Given the difference between the means of the control group and the resistant group, determine how large a sample is necessary to confirm that the resistant group is different at the 1 % significance level.

4.4.4. (Continued from Problems 3.1.5 and 4.2.3.) The file WEIGHT.* contains data on the weight gain of rats fed different diets. Assume that the standard diet is the low-protein cereal diet; hence, this column represents the general population.

(a) Determine the probability that the data for the low-protein beef diet represents a significant difference from the standard diet. What conclusion should we draw?
(b) Repeat part (a) for the high-protein beef diet.

4.4.5. (Continued from Problems 3.6.12, 3.6.13, and 4.2.4.)

(a) The file BUTTER.* contains measurements of butterfat content of the milk from several breeds of cows. Suppose we want to test the hypothesis that there is no significant difference between Jersey and Guernsey cows. It doesn't matter which of the sample means we take as μ and which as $\bar{x}$; however, we need to decide how to choose a value to represent the standard deviation of the general population. The most sensible choice is to use the average of the standard deviations of the two data sets. With this choice, determine the probability that the data from one of the two breeds could have come from cows of the other breed.

(b) There seems to be a contradiction between the result of part (a) and that of Problem 3.6.13. Discuss these results and draw appropriate conclusions about the use of the Cramer–von Mises test.

4.4.6. (Continued from Problem 3.1.6.)

The file BLOOD.* lists the systolic and diastolic blood pressures for 15 patients, both before and after treatment with the drug captopril.

(a) The data in the file is not sufficient for us to assess the effectiveness of the drug, because there is no data given for a control group on a placebo. For now, assume there is no change in the systolic blood pressure for patients on a placebo and use the standard deviation for the blood pressure change as an estimate of the standard deviation of the control group. With these assumptions, determine the probability that the systolic blood pressure data for the patients in the study could have been generated by a random sample of the general population.

(b) Repeat part (a) for the change in the diastolic blood pressure.

(c) Repeat part (a) for the average blood pressure change. Note that each patient now counts as two trials, one for each measurement.

(d) How much improvement might there be in the control group data if the conclusion of significance at the 1 % level is to be maintained? To answer this question, determine the z value corresponding to $P[Z \leq z] = 0.01$. Then use the standard deviation and mean from part (c) to compute the corresponding value of μ for the control group.

(e) Write a careful verbal statement of the methods and results of parts (a)–(d).

4.4.7.

(a) In Example 4.4.4, suppose the study found 70 heart attacks in a test group of 100,000 people. Assuming that Theorem 4.4.2 applies, is this result significant? Note that this larger group shows a smaller difference than the group in the example, yet the probability of this occurring by chance is smaller. Larger samples allow for stronger conclusions.

(b) Suppose a study of n patients has a heart attack rate of 0.0007. Determine how large n must be for the result to be significant at the 5 % level, according to Theorem 4.4.2.

(c) Repeat part (b) for the 1 % level.

4.4.8.* (Continued from Problem 3.4.5.)

Six of the 12 US presidents from Franklin Roosevelt to Barack Obama were left-handed. Is this significant?

(a) Use Theorem 4.4.2 to determine the probability that six or more of 12 presidents are left-handed, assuming that the overall rate of left-handedness is 15 %.

(b) Use the binomial distribution to answer the same question as part (a).

(c) Why do the answers to parts (b) and (c) differ by a factor of 3? Which one is correct?

(d) The question in parts (a)–(c) is not the correct one. In searching for evidence that left-handers are more likely to become president, we will naturally choose a sequence of presidents that begins and ends with a left-hander. Thus, the left-handedness of Roosevelt and Obama are built into the scenario, not part of the random data. Repeat parts (a) and (b) to determine the probability that four of ten consecutive presidents are left-handed simply by chance.[33]

4.4.9.

Turner's syndrome is a rare chromosomal disorder in which girls have only one X chromosome. About one of 2,000 girls in the United States has Turner's syndrome.

---

[33] Mark Twain said, "There are three kinds of lies: white lies, damn lies, and statistics." This appears to be an example of Twain's dictum. The apparent coincidence is not entirely random, because the person looking for a coincidence has deliberately chosen the consecutive string of presidents that makes the strongest statistical case. Add to that the general principle that left-handedness is merely one of many possible coincidences in the list of presidents, and it should not be surprising that some coincidence can be found.

(a) Suppose a group of 4,000 girls is tested, and four of these are found to have Turner's syndrome. Investigate the question of whether this group is unusual.

(b) Given a test group of size n with a frequency of 0.001 for the disease, how large must n be for significance at the 5 % level? What about the 1% level? Obtain approximate answers using Theorem 4.4.2.

(c) Consider a general setting in which the observed number of successes is twice the expected number. Assuming that p is very small (so that $1 - p \approx 1$) and n is large enough for Theorem 4.4.2 to apply, determine the value for the product np that is necessary for significance at the 1 % level.

4.4.10. Consider the scenario of Example 4.4.4, in which there are k successes in 10,000 trials. Assume that the general population has success probability 0.001 and that $2 \leq k \leq 8$.

(a) For each of the values of k, determine

  a. The actual probability from the binomial distribution.
  b. The approximate probability from Theorem 4.4.2.
  c. The approximate probability without the extra 1/2 in Theorem 4.4.2.

(b) Assuming the results from part (a) are typical of instances in which np is not much larger than 5, what can we say about the relationship between the approximate probability and the exact probability? When is the approximation an underestimate? When is it reasonably accurate? When is the approximate probability more accurate without the extra 1/2?

4.5 Estimating Parameters for Distributions

After studying this section, you should be able to:

- Find 95 % confidence intervals for normal distribution means.
- Define a likelihood function for parameter estimation.
- Use experimental data to determine the maximum likelihood estimate and 95 % confidence interval for success probability.
- Use mark-and-recapture data to determine the maximum likelihood estimate and 95 % confidence interval for population size.

In Chapter 3, we computed probabilities using probability distributions. Then in Section 4.4, we used probability distributions to make inferences about the relationship between a sample and a population. For both of these applications, we have to know the values of the parameters that characterize the probability distribution. This raises the question of how these parameter values are determined.

4.5.1 Confidence Intervals

Suppose we measure the mean of a sample we believe to be drawn from a normally distributed population. The sample mean $\bar{x}$ is the best estimate we can give for the population mean μ, but it is only an estimate—a different sample would produce a different result. Because we cannot be certain our measured value represents the population, it is important to quantify the degree of confidence we have in the estimate. The standard deviation makes this a simple matter.

In Section 3.6, we determined that 95 % of measurements made on a normally distributed population of mean μ and standard deviation σ lie within 1.96 standard deviations of the mean. The same calculation applies to the problem of inferring μ from sample data, with some important differences. In this latter setting, we have the sample mean $\bar{x}$ and standard deviation s. These parameters refer to the underlying population, not the distribution of sample means, but it is the latter that we are using to estimate μ. Hence, our best estimates for the population are $\mu \approx \bar{x}$ and $\sigma \approx s$, but the standard deviation we use to compute confidence intervals is that for the sample mean, which is $\sigma/\sqrt{n}$. We summarize the result here.

If the data for a sample of size n is normally distributed with mean $\bar{x}$ and standard deviation s, then the probability is 0.95 that the true population mean lies in the interval

$$\bar{x} - 1.960 \frac{s}{\sqrt{n}} < \mu < \bar{x} + 1.960 \frac{s}{\sqrt{n}}. \tag{4.5.1}$$

Similarly, the probability is 0.99 that the true population mean lies in the interval

$$\bar{x} - 2.576 \frac{s}{\sqrt{n}} < \mu < \bar{x} + 2.576 \frac{s}{\sqrt{n}}. \tag{4.5.2}$$

Example 4.5.1. In Section 3.1, we examined a data set for the number of hours of sleep gained by patients taking the drug L-hyoscine hydrobromide. A sample of size 10 had mean 2.33 and standard deviation 2.00. The 95 % confidence interval for the true mean number of hours of sleep gain is

$$2.33 - 1.96 \frac{2.00}{\sqrt{10}} < \mu < 2.33 + 1.96 \frac{2.00}{\sqrt{10}},$$

or

$$1.09 < \mu < 3.57.$$

With such a large 95 % confidence interval, we can have but little confidence in the measured mean of 2.33. The only cure for this problem is to use a much larger data set. With 1,000 patients instead of 10, the width of a confidence interval would drop by a factor of 10, from 2.48 to 0.25 for the 95 % case. It was the development of modern statistical methods that made scientists appreciate the need to collect more data.   □

Check Your Understanding 4.5.1:
Find the 99 % confidence interval for the number of hours of sleep gain induced by L-hyoscine hydrobromide using the data in Example 4.5.1.

4.5.2 Estimating Success Probability with a Likelihood Function

Suppose we flip a coin 100 times and get heads on 60 of the flips. We saw in Example 4.4.3 that the probability of getting as many as 60 heads out of 100 for a fair coin is 0.0287. This is such a low probability that we should abandon the assumption that the coin is fair. How do we determine the actual probability of heads? Our first guess is that we should use the experiment result: the probability of heads is most likely 0.6. This is the single best answer we can give, but we can also provide additional information.

Example 4.5.2.

Q) Suppose we consider two possibilities for the coin that yields 60 heads on 100 flips. One possibility is that the coin is fair, with a probability of $p = 0.5$ of getting heads. The other possibility is that the result of 60 out of 100 actually represents a success probability of $p = 0.6$. For each of these two assumptions, what is the probability that we actually get 60 heads out of 100?

A) This is a direct application of the binomial distribution. The probability of k successes in n Bernoulli trials with success probability p is[34]

$$b_{n,p}(k) = \binom{n}{k} p^k q^{n-k}, \qquad q = 1 - p, \qquad \binom{n}{k} = \frac{n!}{k!(n-k)!}. \qquad (4.5.3)$$

Given the known result $k = 60$ with $n = 100$, we have

$$b_{100,p}(60) = \binom{100}{60} p^{60}(1-p)^{40} \approx (1.375 \times 10^{28}) p^{60}(1-p)^{40}. \qquad (4.5.4)$$

In particular,

$$b_{100,\,0.6}(60) \approx 0.0812, \quad b_{100,\,0.5}(60) \approx 0.0108.$$

While there is a 1 % chance of getting 60 heads with a fair coin, there is an 8 % chance of getting 60 heads with a coin that hits heads 60 % of the time. □

Observe that the quantity $b_{100,p}(60)$ is a function of p. Clearly, the value of p that maximizes this function is the one that is most probable. However, the function does not tell us the probability that the maximizing value is actually correct. In the example, the specific value of 0.0812 for the probability of getting exactly 60 % heads, given $p = 0.6$, depends on the number of trials as well as the choice of the p value. Hence, the actual probability value is not important; what is important is the way the probability changes if we assume a different value of p. This idea is used in the definition of the likelihood function.

Likelihood function: *the function $L(p) = f_p(k)/f_{p^*}(k)$, where f_p is a probability distribution function with* unknown *parameter p, k is a* known *experiment result, and p^* is the value of p that gives the maximum of $f_p(k)$.*

---

[34] See Theorem 3.4.1.

Example 4.5.3. Equation (4.5.4) defines a likelihood function:

$$L(p) = \frac{b_{100,p}(60)}{b_{100,0.6}(60)} = \frac{\binom{100}{60} p^{60}(1-p)^{40}}{\binom{100}{60} 0.6^{60} 0.4^{40}} = \left(\frac{p}{0.6}\right)^{60} \left(\frac{1-p}{0.4}\right)^{40}.$$

This function appears in Figure 4.5.1.   □

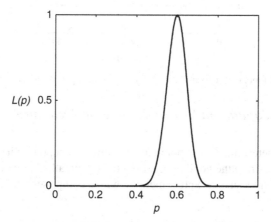

Fig. 4.5.1 The likelihood function for Example 4.5.3

In Example 4.5.3, we found $p^* = k/n$. This is a general result for binomial distributions.

We don't need a likelihood function for the purpose of finding the most likely parameter value, since that is simply the approximation determined by the experiment. However, likelihood functions can be used to identify confidence intervals.[35] We report the principal result here without proof[36].

> For distributions that are roughly normal, such as the binomial distribution with $np > 5$ and $n(1-p) > 5$, the 95 % confidence interval is approximately given by the range of values for which the likelihood is at least 0.15.

Example 4.5.4. Figure 4.5.2 shows the likelihood function for Example 4.5.3, along with lines marking likelihood values of 0.15 and the corresponding p values. The 95 % confidence interval is approximately $(0.503, 0.692)$. Note that this is not quite centered about the maximum likelihood estimator because the function is not symmetric around that point.   □

While we have focused on the success probability for the binomial distribution, we could define a likelihood function for any parameter in any probability distribution by similar methods.

---

[35] See Section 3.6.

[36] See [9], for example.

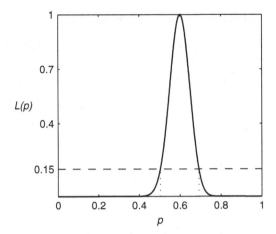

Fig. 4.5.2 The likelihood function for Examples 4.5.3 and 4.5.4, along with the 95 % confidence interval

4.5.3 Estimating Population Size by Mark and Recapture

A biologist wants to determine N, the population of turtles in a pond. She cannot capture all of the turtles, nor can she drain the pond in order to count them all. The standard approach to this problem is to use the mark-and-recapture method, which comprises the following procedure:

1. Capture and mark m individuals.
2. Return the marked individuals and wait for them to mix in with the rest of the population.
3. Capture n individuals and let k be the number of marked individuals in the collection.

To analyze a mark-and-recapture experiment, note that k/n is the fraction of marked individuals in the second capture group; the expected value for this fraction is m/N, where N is unknown. If the actual fraction matches the expected value, we have $k/n = m/N$; hence, the best estimate $N^\star$ is mn/k. The qualification on this statement is that mn/k is probably not an integer. If it isn't, then there are two possibilities. Let N_+ and N_- be the integers that satisfy

$$N_- < \frac{mn}{k} < N_+ , \qquad N_+ - N_- = 1 . \tag{4.5.5}$$

One of these values is the best estimate $N^\star$, but we need the likelihood function to determine which it is. We first need information about the correct probability distribution for mark-and-recapture experiments, which we give here without proof.

> The probability of observing k marked individuals in a group of n, given a total of m marked individuals in a population of N, is given by the **hypergeometric distribution**, which is defined by
>
> $$h_{n,N,m}(k) = \frac{\dbinom{m}{k}\dbinom{N-m}{n-k}}{\dbinom{N}{n}} , \qquad \dbinom{m}{k} \equiv \frac{m!}{k!(m-k)!} . \tag{4.5.6}$$

The quantities n, N, and m are parameters for the distribution, while k is the experiment outcome. However, in a likelihood context the outcome k is known and the parameter N is not. The maximum likelihood estimate is whichever of N_+ and N_- yields the larger value of h.

Example 4.5.5. Suppose we tag and release 25 animals and then get 8 tagged animals out of a subsequent sample of 30. Then $m = 25$, $k = 8$, and $n = 30$, so $mn/k = 93.75$; thus, $N_- = 93$ and $N_+ = 94$. From (4.5.6),

$$h_{30,93,25}(8) = \frac{\binom{25}{8}\binom{68}{22}}{\binom{93}{30}} \approx 0.1972\,,$$

and

$$h_{30,94,25}(8) = \frac{\binom{25}{8}\binom{69}{22}}{\binom{94}{30}} \approx 0.1971\,.$$

Getting 8 tagged animals is slightly more likely from a population of 93 than from a population of 94, so we take $N^\star = 93$. □

Once we know $N^\star$, we can define a likelihood function and use it to obtain confidence intervals. The likelihood function is

$$L(N) = \frac{h_{n,N,m}(k)}{h_{n,N^\star,m}(k)} = \frac{\binom{N-m}{n-k}\binom{N_0}{n}}{\binom{N^\star-m}{n-k}\binom{N}{n}}.$$

(4.5.7)

Example 4.5.6. In the mark-and-recapture experiment of Example 4.5.5, we have $m = 25$, $k = 8$, $n = 30$, and $N^\star = 93$. Thus, the likelihood function is

$$L(N) = \frac{\binom{N-25}{22}\binom{93}{30}}{\binom{68}{22}\binom{N}{30}} \approx 5.486 \times 10^6 \times \frac{\binom{N-25}{22}}{\binom{N}{30}}.$$

The 95 % confidence interval, which we take to be given by the inequality $L(N) > 0.15$, is $64 \leq N \leq 171$. The likelihood function is plotted in Figure 4.5.3, along with lines marking the 95 % confidence interval. □

4.5.4 A Final Observation

The 95 % confidence interval for Example 4.5.6 is much broader than that of Example 4.5.4. Our best estimate of the population size is 93, but we cannot rule out a population size almost twice as large with 95% confidence! This is typical in parameter estimation. Confidence intervals tend to be large. They can be reduced by significantly increasing the sample size, but this is not always practical. They also depend in a fundamental way on the nature of the parameter in question. Population size parameters tend to have a much higher degree of uncertainty than success probabilities because the probability of a given outcome generally depends much more on success probability than on population size. A given percentage error in population size gen-

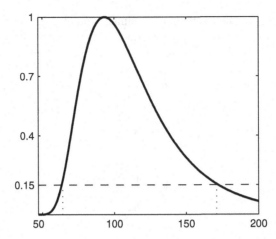

Fig. 4.5.3 The likelihood function for Example 4.5.5, along with the 95 % confidence interval

erally has less overall effect than the same error in success probability; this translates to a larger range of population size parameters that yield a probability within 15 % of the maximum.

Check Your Understanding Answers

1. $0.90 < \mu < 3.96$

Problems

The data sets for Problems 4.5.1–4.5.5 are from [2] and can be found in both csv and xlsx form at http://www.math.unl.edu/~gledder1/MLS/, http://www.springer.com/978-1-4614-7275-9.

4.5.1.* (Continued from Problem 4.2.1.)
Find the 95 % and 99 % confidence intervals for the mean dopamine level for nonpsychotic patients from the file DOPAMINE.*.

4.5.2. (Continued from Problem 4.2.2.)
Repeat Problem 4.5.1 for the mean fecundity of the control group of fruit flies from the file FRUITFLY.*.

4.5.3. (Continued from Problem 4.2.3.)
Repeat Problem 4.5.1 for the mean weight gain of rats fed a low-protein cereal diet from the file WEIGHT.*.

4.5.4. (Continued from Problem 4.2.4.)
Repeat Problem 4.5.1 for the mean butterfat content of milk from Guernsey cows in the file BUTTER.*.

4.5.5. (Continued from Problem 4.2.7.)
Repeat Problem 4.5.1 for the mean sepal lengths of *Iris virginica* from the file IRISES.*.

4.5.6.* In Example 4.4.3 we examined a famous data set consisting of frequencies of die rolls.

(a) Use the data to estimate the probability p of rolling a "5" or a "6."

(b) Define the likelihood function for the estimation of p from the data.

(c) Plot the likelihood function from part (b) and find the approximate 95 % confidence interval as in Example 4.5.4. [The large exponents involved in the calculation make this tricky, as you will get computational errors for p values that are not very close to $p^\star$.] You will have to experiment to find a reasonable range of p values for the calculations—the goal is to choose a range of p values large enough to see the part of the graph where $L > 0.05$.

(d) The connection between $L = 0.15$ and the 95 % confidence level was given in the text as a rule of thumb without justification. We can test the rule in this case, because the number of trials is so large that the binomial distribution can be approximated as a normal distribution. Find the 95 % confidence interval for the appropriate normal distribution and show that it is the same as that obtained in part (c).

(e) We can also find the value of L corresponding to the 95 % confidence interval theoretically by using the fact that the likelihood function for a normal distribution is simply the probability density function, suitably scaled. Use this idea to confirm the 15 % estimate. [You should find that 15 % is a little high.]

(f) Use the same argument as in part (e) to find the value of L that corresponds to the 99 % confidence interval for a normal distribution.

4.5.7. (a) Count the number of rolls of "5" or "6" in 25 rolls of a single six-sided die.

(b) Define the likelihood function for the estimation of p from the data.

(c) Plot the likelihood function from part (b) and find the approximate 95 % confidence interval as in Example 4.5.4.

(d) Repeat parts (a)–(c), but with 100 data points, using 25 rolls of 4 apparently identical dice. [In other words, count the number of successes independently of how they are distributed among the dice. This will take less time than rolling one die 100 times. The assumption that the dice are identical would need further testing.]

(e) Describe the effect of the larger sample size of part (d) on the likelihood function.

4.5.8. Choose an activity other than flipping coins or rolling dice that can be modeled with Bernoulli trials and doesn't take very long to perform, such as shooting free throws in basketball or flipping playing cards at a wastebasket. Repeat Problem 4.5.6 using this activity.

4.5.9. (Continued from Problem 3.7.6.)
The definition of a likelihood function requires an experiment that measures some quantitative outcome to be modeled by a known distribution with an unknown parameter value. In the case of a binomial distribution, the known result k is just np, where n is the number of trials and p is the unknown success rate. For other distributions, the measured mean can be used as the known result.

(a) Use the data from Table 3.7.2 to determine the sample mean $\bar{x}$ for the number of malaria parasites in erythrocytes. Use this result to estimate the parameter μ in the definition of the Poisson distribution.

(b) Determine the likelihood function for the estimation of the parameter μ, using the sample mean $\bar{x}$ as the known experiment result.

(c) Plot the likelihood function and find the values of μ corresponding to a likelihood of 0.15.

(d) Can we interpret the values where the likelihood is 0.15 as constituting a 95 % interval? Why or why not?

4.5.10. (Continued from Problem 3.7.7.)

(a) Repeat Problem 4.5.9 for the mean number of spiders found under boards in the experiment reported in Table 3.7.3.

4.5.11. Table 4.5.1 shows the data for a simulated mark-and-recapture experiment conducted in five stages [5]. The experiment began with ten individuals marked and returned to the population. In each of the four recapture rounds that followed, any previously unmarked individuals were returned to the population. Each of the rounds can be analyzed separately, and we should expect both the accuracy and precision of the population estimate to increase with each round.

(a) Determine the best estimate of the population based on the data for round 1.
(b) Repeat part (a) for the other rounds.
(c) Plot the likelihood function using the round 1 data. Use the range $\max(m,n) \leq N \leq 200$. Why is it important to choose $\max(m,n)$ for the minimum value of N?
(d) Add the likelihood function for the round 2 data to the plot.
(e) Repeat part (d) using the round 3 data.
(f) Repeat part (d) using the round 4 data.
(g) What effect do additional rounds have on the precision of the results? Note that we cannot know the accuracy without a count of the actual population.

Table 4.5.1 Mark-recapture data for Problem 4.5.11

Round #	m	n	k
1	10	13	1
2	22	18	5
3	35	16	8
4	43	20	14

4.5.12. (a) Repeat the investigation of Problem 4.5.11 using data that you collect in your own experiment using a 1-lb bag of dry white beans.

1. Separate out a small pile of beans that contains roughly 1/6 of the total. Do this visually, with 1/6 as a rough guideline rather than a strict requirement.
2. Count the smaller pile and mark those beans. Mix all of the beans in a paper bag. Make sure the marked beans are distributed throughout the population.
3. For each of four rounds:
   a. Remove a handful of beans that is roughly the same size as the pile that you marked in the first step. Do not count out a set number. Just grab a handful of roughly the right size.
   b. Count n and k for the new sample of beans.
   c. Mark any previously unmarked beans in the sample, return all of the beans to the bag, and mix the beans in preparation for the next round.

(b) Based on your experience conducting the experiment, what must be assumed about a population of animals if the mark-and-recapture method is to work?

4.6 Conditional Probability

After studying this section, you should be able to:

- Correctly identify a desired probability in terms of conditional probabilities.
- Calculate conditional probability from counts or probabilities of the component events.

- Explain the partition rule using concepts of probability.
- Use the partition rule to calculate probabilities.

So far, our study of probability has been restricted to cases of single random variables or collections of *independent random variables*.

A set of random variables is **independent** if the value each takes has no effect on the values of the others.

Experiments that can be thought of as sequences of independent random variables can be studied with the sequence rule.[37] Samples, which are collections of independent, identically distributed random variables, are studied with distributions of sample means.[38]

In this section and the next we turn to collections of random variables that are *not* independent. An example from the author's personal experience serves to illustrate a typical setting.

Example 4.6.1. Multiple sclerosis is a medical condition that is difficult to diagnose. Consider the following history of symptoms and medical tests.

1. A patient begins to experience strange symptoms that include loss of feeling in the legs, loss of balance, and difficulty in urination.
2. A neurologist examines the patient's medical history, performs a neurological exam, and orders MRI scans of the head and neck. The doctor makes a tentative diagnosis of multiple sclerosis (MS).
3. A second neurologist orders an MRI scan of the patient's middle back. These tests indicate a herniated thoracic disc.
4. A neurosurgeon repairs the herniated disc, and the patient's MS-like symptoms disappear.
5. Over the next 5 years, the patient has no recurrence of the MS symptoms.

Given that there is no definitive test, doctors can only collect information that helps them determine the likelihood that a patient has multiple sclerosis. Each bit of information changes that probability. The question "What is the probability that this patient has multiple sclerosis?" is not well defined without including a list of known information. At step 1, we can ask, "What is the probability of multiple sclerosis, given the patient's list of symptoms?" The existence of the symptoms significantly increases the probability as compared to the general population. The probability increases again after step 2, because the initial examination and tests rule out some other explanations. The information from step 3 significantly decreases the probability by suggesting an alternative explanation that can account for all of the facts. The probability remains slightly higher than for the general population, because we cannot rule out the possibility of two distinct conditions combining to create the symptoms. However, the additional knowledge in the final two steps reduces the probability to a final value near zero. □

Let's apply the language of probability theory to the thoracic MRI of step 3. Obviously the medical test counts as an experiment for these purposes. There are two possible outcomes: "has a herniated disc" and "does not have a herniated disc." If we are talking about the probability of MS, we must also imagine an experiment with two possible outcomes: "has MS" and "does not have MS." This experiment is imaginary, in that it cannot be conducted on a live patient, but it is a valid experiment for probability theory. We want to use the results of the medical test to improve our estimate of the probability that the patient has the condition of interest.

---

[37] See Section 3.2.

[38] See Sections 4.3 and 4.4.

4.6.1 The Concept of Conditional Probability

The concept of conditional probability is necessary to provide a setting in which to talk about the effects of additional information on the probability of an outcome or event.[39]

> **Conditional probability of B, given A, denoted $P[B|A]$:** *the probability of an event B in a population consisting of those individuals that comprise an event A from a prior experiment.*

Example 4.6.2. Let A be the outcome that an adult's height is 6 ft or greater, and let B be the outcome that a given person is female. Suppose 1 % of adult women are 6 ft tall or taller and 97 % of people as tall as 6 ft are men. Then

$$P[A|B] = 0.01, \qquad P[B|A] = 0.03 .$$

Note the distinction between these two conditional probabilities. Like $P[A]$, $P[A|B]$ is the probability of being tall, except that the latter is limited to the population of adult women. In contrast, $P[B|A]$ is the probability that a person chosen randomly from the population of 6-footers happens to be female.   □

> In the notation $B|A$, the event B is the one whose probability is of interest, and the event A restricts the population for which the probability of B is sought.

Example 4.6.3. In the situation of Example 4.6.1, let A be the event "has MS," let B be the event "loss of feelings in legs, loss of balance, and difficulty in urination," let C be the event "condition other than MS detected in initial screening," let D be the event "has a herniated thoracic disc," and let E be the event "symptoms disappear after surgery." At each stage in the diagnosis, we want to know the probability of A, but with different amounts of prior information. After step 1, the desired probability is $P[A|B]$; after step 2, it is $P[A|(B \cap C^c)]$,[40] and so on.   □

> **Check Your Understanding 4.6.1:**
> What is the desired conditional probability after step 3 in the situation of Example 4.6.1?

4.6.2 Formulas for Conditional Probability

Our definition gives meaning to the notation of conditional probability, but it does not tell us how to calculate it. An example will help us determine how to do this.

Example 4.6.4. Suppose we want to determine $P[A|B]$ in Example 4.6.3, assuming that we have a definitive procedure for diagnosing the disease. Starting with the full population, we proceed in two steps. Since we want $P[A|B]$ and not $P[A]$, we first dismiss all people who do not have the given list of symptoms (the group B^c). In the second step, we determine the fraction of symptomatic people (the group B) who have the disease; this fraction is $P[A|B]$.

---

[39] Because an event can have only one outcome, we can use the single term "event" to represent a single outcome as well as an event consisting of multiple outcomes.

[40] Recall that the event C^c is the set of outcomes not in C.

We can describe another procedure that yields the same result, again assuming that we have a definitive test. This time, we do the experiments of testing for the disease and testing for the symptoms simultaneously. While doing this, we count the number of individuals who are in both A and B and compare it with the total number of individuals who are in B. The ratio of these two counts is the probability that someone in B is also in A, which is the desired probability. Denoting the whole population as P, we have[41]

$$P[A|B] = \frac{|A \cap B|}{|B|} = \frac{|A \cap B|}{|P|} \frac{|P|}{|B|} = \frac{P[A \cap B]}{P[B]} .$$

□

If we want to determine conditional probabilities for infinite populations, we cannot use the procedure of Example 4.6.4, which relies on counting the size of sets. However, only a small modification is needed. To calculate $P[B|A]$, for example, we must consider only cases in which A occurs. Of these cases, successes require B to occur as well as A. Thus,

$$P[B|A] = \frac{P[A \cap B]}{P[A]} . \tag{4.6.1}$$

Other conditional probability formulas can be reasoned out in a similar manner.

Check Your Understanding 4.6.2:
Find a formula for $P[A^c|B]$.

We can also derive the formula (4.6.1) by considering the problem of calculating the probability of an intersection of events. Suppose we have a sequence of experiments, one that yields A or A^c followed by one that yields B or B^c. The probability of $A \cap B$ is the probability of getting A and then B in successive experiments. The probability of getting A in the first experiment is just $P[A]$, because no additional information that could affect the probability is available. The probability of getting B in the second experiment is the probability of getting B, given that the first experiment yielded A. This is $P[B|A]$, not $P[B]$. By the sequence rule,[42] the probability of the combined experiments is the product of the individual probabilities. Thus,

$$P[A \cap B] = P[A]P[B|A].$$

By doing the experiments in the reverse order, we obtain a similar result:

$$P[A \cap B] = P[B]P[A|B].$$

Taken together these formulas comprise the *multiplication rule*:

Multiplication Rule

$$P[A \cap B] = P[B]P[A|B] = P[A]P[B|A] . \tag{4.6.2}$$

Equation (4.6.2) generalizes (4.6.1).

---

[41] Recall that $|S|$ is the number of elements in the set S.

[42] See Section 3.2.

4.6.3 The Partition Rule

Conditional probabilities are related by another useful formula called the *partition rule*. This result is obtained by combining the multiplication rule with the extended complement rule,[43]

$$P[A \cap B] + P[A^c \cap B] = P[B] \ :$$

Partition Rule Given any events A and B,

$$P[B] = P[A] P[B|A] + P[A^c] P[B|A^c] \ . \tag{4.6.3}$$

Formally, we can derive the partition rule simply by substituting $P[A \cap B] = P[A] P[B|A]$ and $P[A^c \cap B] = P[A^c] P[B|A^c]$ into the extended complement rule. Conceptually, it is more useful to think of the partition rule as following from the fundamental principles of probability. Suppose two experiments are performed sequentially. The first experiment determines either the event A or the event A^c. The second experiment then determines B or B^c and yields probability $P[B|A]$ in the case of A and $P[B|A^c]$ in the case of A^c. Now consider all of those experiments in which B occurs. Some of these followed the event A, and the probability of first A and then B is $P[A] P[B|A]$. The others followed the event A^c, and the probability of this sequence is $P[A^c] P[B|A^c]$. Since these two sequences are disjoint and comprise all sequences that include the event B, the equality follows.

Example 4.6.5. Suppose it is known that 1/3 of men and 1 % of women are at least 6 ft tall and that the general population is evenly divided between men and women. What is the fraction of adults who are at least 6 ft tall?

Let A be the event that an adult's height is 6 ft or greater and let B be the event that a given person is female. The probability of being tall if male is $P[A|B^c] = 1/3$, the probability of being tall if female is $P[A|B] = 1/100$, and the probability of being female is $P[B] = 1/2$, from which we also have $P[B^c] = 1/2$. We can compute $P[A]$ from the partition rule (reversing the symbols A and B):

$$P[A] = P[B] P[A|B] + P[B^c] P[A|B^c] = \frac{1}{2}\frac{1}{3} + \frac{1}{2}\frac{1}{100} = \frac{103}{600} \ . \qquad \square$$

Check Your Understanding Answers

1. $P[A|(B \cap C^c \cap D)]$

2. $P[A^c|B] = \dfrac{P[A^c \cap B]}{P[B]}$

---

[43] See Equation (3.2.3).

Problems

4.6.1.* Suppose A, B, and C are the events of reaching ages 50, 55, and 60, respectively, with corresponding probabilities 0.91, 0.88, and 0.75. Find the conditional probabilities indicated below or explain why that probability does not make sense.

(a) $P[B|A]$.
(b) $P[C|B]$.
(c) $P[C|A]$.
(d) $P[B|C]$.

4.6.2. Table 4.6.1 shows data correlating heart disease with frequency of snoring [6].

(a) Find the conditional probabilities of heart disease for each of the three groups of snorers.
(b) What is the probability that a person in the study is a frequent snorer?
(c) Suppose a person has heart disease. From the data table, what is the probability that the person is a frequent snorer?

Table 4.6.1 Snoring and heart disease

Heart Disease	Non-snorers	Occasional	Frequent
Yes	24	35	51
No	1355	603	416

4.6.3.* Suppose 5 % of a population are diabetics and 1 % of diabetics are blind. Assume that the overall incidence of blindness in the population is 36 out of 10,000. Use conditional probability calculations to determine:

(a) The fraction of blind people who are diabetic.
(b) The fraction of non-diabetics who are blind.

4.6.4. For each of the following, indicate the correct notation using the events S, S^c, L, and L^c and use Table 4.6.2 to determine the given quantity.

(a) The probability of success, independent of size of stone.
(b) The probability that a randomly chosen patient had a large stone.
(c) The probability of success in treating a large stone.
(d) The probability that a successfully treated patient had a large stone.
(e) The probability that a randomly selected patient had a successful treatment of a large stone.
(f) The probability of failure in treating a small stone.
(g) The probability that a randomly selected patient had a successful treatment of a small stone.

Table 4.6.2 Outcomes of 700 kidney stone treatments grouped by size of stone

Size	Outcomes Success (S)	Failure	Total
small	315	42	357
large (L)	247	96	343
total	562	138	700

4.6.5. (a) Use the results of parts (a)–(e) of Problem 4.6.4 to verify the multiplication rule.
(b) Use the results of parts (e) and (g) of Problem 4.6.4 to determine the overall probability of success and compare the answer to Problem 4.6.4a.

4.6.6. Nine of 20 rats in a cage are infected with a disease. Suppose 7 infected rats and 4 uninfected rats are males. One rat is randomly selected from the cage.

(a) If the selected rat is infected, what is the probability that it is also male?
(b) If the selected rat is male, what is the probability that it is also infected?
(c) Are the events of being male and being infected independent?
(d) Suppose I of the 20 rats are infected and that A infected rats and B uninfected rats are males. What relationship must A, B, and I satisfy for the events of being male and being infected to be independent?

4.6.7. Suppose events A and B are mutually exclusive and B and C are independent. Suppose further that $P[A] = 0.3$, $P[B] = 0.2$, $P[C] = 0.6$, and $P[A \cap C] = 0.2$.

(a) Find $P[A \cap B \cap C]$ and $P[A \cap B \cap C^c]$.
(b) Find $P[A \cap B^c \cap C]$.
(c) Find $P[A \cap B^c \cap C^c]$.
(d) Find $P[A^c \cap B \cap C]$.
(e) Find $P[A^c \cap B \cap C^c]$.
(f) Find $P[A^c \cap B^c \cap C]$. [Hint: Start by calculating $P[A^c \cap B^c]$.]
(g) Find $P[A^c \cap B^c \cap C^c]$.
(h) Use the results from earlier parts to calculate $P[B|C]$. The result should be consistent with the assumptions.
(i) Repeat part (h) for $P[C|B]$.

4.7 Conditional Probability Applied to Diagnostic Tests

After studying this section, you should be able to:

- Explain the meaning of false positives and false negatives in medical diagnosis.
- Compute the probabilities of false positives and false negatives from medical test data and the estimated prevalence of the condition.

Medical tests are seldom definitive. Most commonly, the test involves the measurement of some quantity, with a positive or negative result being assigned according to its value. This is not an exact science. Because of demographic stochasticity, what is normal for one person might be abnormal for another. It is also possible for tests to be affected by minor factors; for example, you may test positive for opiates after eating a lot of poppy seeds. Thus, we usually have to consider having a condition and testing positive for a condition as the results of two different experiments. There is some probability of a false positive, where the test is positive but the condition is absent, or a false negative, which is the opposite. Imperfect medical tests can be useful, but it is important to understand their limitations. One of the most important limitations is a consequence of the basic mathematics of conditional probability and can be significant even for highly accurate tests. We explore this issue in this section, using a test for lead poisoning as a motivating example.

Example 4.7.1. The definitive test for lead in the blood requires blood from a vein. In 1992, a group of researchers from the Yale School of Medicine wanted to determine if this test could be replaced by a similar test using a single drop of blood from a fingerstick [8]. The researchers had doctors in four private practices[44] routinely test all of their patients aged 6 months to

---

[44] Nearly all the subjects were white and middle class, so it is not clear that the sample of subjects was representative of the population as a whole.

5 years using the fingerstick test. Those who tested positive were tested again using venous blood to determine if they actually had elevated lead in their systems.[45] The data allowed the researchers to estimate the probability of a positive fingerstick test, as well as the probability of a positive venous test given a positive fingerstick test. What we actually want to know is the probability of false positive and false negative results from the fingerstick test; these will have to be calculated. □

Let C be the event "positive for lead poisoning," which we assume is equivalent to a positive test on venous blood. Let T be the event "positive fingerstick test." Ideally, the events C and T are identical. When they differ, the fingerstick test is assumed to be the one that is wrong. These errors are important, because they lead to a misdiagnosis. In actual medical practice, screening tests such as the fingerstick test for lead are performed first, and we may or may not be able to detect the condition by some subsequent test. The physician must recommend a course of treatment or further testing based only on the screening test. Any patient who tests positive and does not actually have the condition will have endured unnecessary testing or potentially harmful treatment. Any patient who tests negative despite having the condition misses out on a potentially valuable treatment. In terms of the events C and T, we want to find:

1. $P[C^c|T]$, which is the probability of a false positive, also called a Type I error.
2. $P[C|T^c]$, which is the probability of a false negative, also called a Type II error.

These are the probabilities of interest to medical practitioners, who know the test result but not whether the condition is present. In analyzing experimental data from medical studies, it may be necessary to determine the probabilities of T and T^c, given knowledge of C.

Example 4.7.2. The study of Shonfeld et al. mentioned in Example 4.7.1 reported:

1. 35 positive fingerstick tests out of 1,085 subjects.
2. 30 % of subjects with positive fingerstick tests were confirmed to have lead poisoning.

Q) What are the probabilities that can be calculated directly from the given data, in terms of T and C? What are the probabilities of false positives and false negatives? What is $P[C^c \cap T]$ and what does it represent?

A) a. The first item is the test result, not segregated by the condition. The second item is the incidence of the condition among the group who tested positive. Thus,

$$P[T] = \frac{35}{1085} = 0.032, \qquad P[C|T] = 0.3 .$$

b. From the second item and the extended complement rule, we can calculate the probability of a false positive as $P[C^c|T] = 0.7$, or 70 %. We do not have enough information to calculate the probability of a false negative.

c. By the multiplication rule,

$$P[C^c \cap T] = P[T]\,P[C^c|T] = 0.022 .$$

This is the fraction of test subjects who tested positive that do not have the condition. □

The fingerstick test has a false positive rate of 70 %, which means that 70 % of *subjects who tested positive by fingerstick* were incorrectly diagnosed. However, these errors represent only a little more than 2 % of the subjects ($P[C^c \cap T]$) because only 3 % of all subjects test positive. It is easy to be confused about the terminology:

---

[45] Full analysis of the test would have required all of the subjects to be tested using venous blood. The value of the study is somewhat limited by this not having been done. On the other hand, it would have been difficult to get parents to agree to the additional test.

- $P[C^c|T]$ is the probability of a false positive *among those who test positive*.
- $P[C^c \cap T]$ is the probability of a false positive *among all subjects*.
- The official definition of false positive is $P[C^c|T]$.

Ideally, a test should make very few false positives or false negatives. In practice, tests can often be adjusted to decrease one of these errors, but only at the cost of increasing the other. In the lead poisoning case, the fingerstick test is taken to be positive if the lead content exceeds $15\,\mu g/dL$.[46] If we want to decrease the false negatives, we could lower the positive threshold to $13\,\mu g/dL$. Anyone who tests between 13 and 15 will then be judged as positive, rather than negative. Given that some of the subjects in this group actually have lead poisoning (C) and others do not (C^c), this change decreases the number of false negatives while increasing the number of false positives.

Which way we want to adjust the relative number of errors depends on which kind of error is more problematic. If the condition of interest is easily treated at low cost and with few side effects, then we are probably more worried about false negatives. If treatment is expensive and has significant side effects, or if the disease is not life-threatening or debilitating, we might be more worried about false positives. If T is the presence of the defendant's DNA in physical evidence from a crime scene, then we are certainly much more worried about false positives (assuming we accept the principle that a person should be considered innocent unless shown to be guilty beyond a reasonable doubt).

The medical test literature does not generally report the probability of false positives relative to the total of positives. Instead, it reports sensitivity and specificity, defined by

$$\text{Sensitivity} = P[T|C], \quad \text{Specificity} = P[T^c|C^c]. \tag{4.7.1}$$

The probabilities of false positives and false negatives must be calculated from these quantities along with some additional estimates.

4.7.1 The Standard Test Interpretation Problem

The predictive value of a test must be determined by a medical study. The study begins with patients who are known to have the condition and patients who are known not to have the condition. The patients in each group are tested, and the study then yields estimates of $P[T|C]$ and $P[T|C^c]$. (Note that these are different from the probabilities of false positives and false negatives.) The designer of the test measures its accuracy for patients whose condition is already known, while the diagnostician needs to assess the likelihood of having the disease for patients who test positive. Our goal is to use the probabilities determined in the medical study to calculate the probabilities of false positives and false negatives.

Data from a medical study is not sufficient for the task at hand. Information about how C affects $P[T]$ does not by itself give information about how T affects $P[C]$. The additional information needed to bridge the gap is the overall probability of C in the general population. In practice, we may have only a crude estimate for this probability; if it were easy to determine who has the condition and who doesn't, then we would not need to develop the medical test. Public health data can tell us the number of patients believed to have the condition, but this overestimates the correct answer for every incorrectly diagnosed patient and underestimates the number for every patient who is asymptomatic or undiagnosed. Nevertheless, in the model used to calculate the probabilities of false test results, we must assume that $P[C]$ is a known quantity.

---

[46] "dL" is a deciliter, equal to 100 cubic centimeters.

In general, we consider the following problem:

Given $P[T|C]$, $P[T|C^c]$, and $P[C]$, find $P[C^c|T]$ and $P[C|T^c]$.

In words, we know the overall prevalence of the condition and we know the probability of positive test results when the condition is present or absent, and our goals are to determine the probabilities of false positives and false negatives.

The solution of this test interpretation problem is provided by Bayes' formula, which can be found in most books on probability. However, it is more instructive to describe a general procedure for solving the problem.

There are eight different events associated with the combination of the condition experiment and the testing experiment.[47] These events are those for the individual experiments—C, C^c, T, T^c—and those for the various combinations of their results—$T \cap C$, $T \cap C^c$, $T^c \cap C$, and $T^c \cap C^c$. The four compound probabilities are of particular interest because we can determine all other probabilities easily in terms of them.[48] Figure 4.7.1 shows the four compound probabilities arranged in a 2×2 grid, with rows corresponding to T and T^c and columns corresponding to C and C^c. By the extended complement rule, we have

$$P[T \cap C] + P[T \cap C^c] = P[T] \, ,$$

and we write the latter to the right of the corresponding row. Each of the two rows and each of the two columns can be summed in similar fashion to obtain the four single-event probabilities. The total row and total column probabilities must both add up to 1. Once we have filled in the diagram, we can calculate any conditional probabilities using (4.6.1).

We will solve the standard test interpretation problem in two steps:

1. Use the given data to fill in the diagram.
2. Use the data shown in the diagram to determine the desired conditional probabilities.

Example 4.7.3 illustrates the procedure for filling in the diagram.

	C	C^c	
T	$P[T \cap C]$	$P[T \cap C^c]$	$P[T]$
T^c	$P[T^c \cap C]$	$P[T^c \cap C^c]$	$P[T^c]$
	$P[C]$	$P[C^c]$	

Fig. 4.7.1 A diagram indicating the compound events for the test and condition, with sums for each row and column

Example 4.7.3. Suppose a medical test has a sensitivity of 0.9 and a specificity of 0.95. Using the definitions (4.7.1), this means $P[T|C] = 0.90$ and $P[T^c|C^c] = 0.05$. Suppose the prevalence of the condition is 20 %; that is, $P[C] = 0.20$. These results give us the data needed to fill in the

---

[47] Remember to use the probabilistic definitions of these terms rather than the meanings associated with normal language use. An experiment is any observation of an outcome, so we have an experiment consisting of the observation of the condition and an experiment consisting of the observation of the test result. The "condition experiment" is a hypothetical experiment that results in knowledge of the presence or absence of the condition.

[48] Minimally, any three of them will suffice. The four events are mutually exclusive and all-encompassing; hence, their sum is 1 and any one of them is easily calculated from the other three.

four boxes in the diagram, after which the row and column sums can be added. The given data allows us to use the multiplication rule to calculate

$$P[T \cap C] = P[C] \times P[T|C] = 0.18 .$$

By the complement rule, we know $P[C^c] = 0.80$, from which the multiplication rule yields

$$P\left[T^\cap C^c\right] = P[C^c] \times P[T^c|C^c] = 0.76 .$$

Figure 4.7.2 shows the diagram at this point.

The remaining intersection probabilities can be determined by making the column sums match the totals for each column. For example, $P[T^c \cap C]$ must be 0.02 so that the first column correctly sums to $P[C]$. Once the probabilities of the combined events are known, the probabilities of T and T^c are computed by summing the rows. The result is shown in Figure 4.7.3.

□

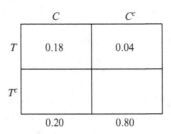

Fig. 4.7.2 The probability diagram for Example 4.7.3 after the initial calculations

	C	C^c	
T	0.18	0.04	0.22
T^c	0.02	0.76	0.78
	0.20	0.80	

Fig. 4.7.3 The full probability diagram for Example 4.7.3

Once the probability diagram is complete, any conditional probabilities can be calculated by the multiplication rule.

Example 4.7.4. From Figure 4.7.3,

$$P[C^c|T] = \frac{P[T \cap C^c]}{P[T]} = \frac{0.04}{0.22} \approx 0.18, \quad P[C|T^c] = \frac{P[T^c \cap C]}{P[T^c]} = \frac{0.02}{0.78} \approx 0.026 .$$

About 18 % of positive tests are false, and about 2.6 % of negative tests are false.   □

Example 4.7.4 shows a much higher rate of false positives than false negatives. Note that this was not because of the characteristics of the test itself. Because 5 % of those without the condition test positive as compared to 10 % of those with the condition who test negative, the test is, if anything, biased toward producing more false negatives. The reason for the higher rate of false positives is that the population consists mostly of people who do not have the condition; hence, both true positives and false negatives are rare. The number of false positives is based primarily on the accuracy of the test. For a very good test, this number can be quite small. However, the number of true positives depends on the prevalence of the condition. For an extremely rare condition, the number of true positives may well be small compared to the number of false positives.

Example 4.7.5. Tay-Sachs disease is a very rare chromosomal disorder that is initially asymptomatic. It is estimated that approximately one out of 10,000 individuals has the disease. Now consider a study to determine the effectiveness of a hypothetical diagnostic test. Suppose experiments show a sensitivity of 99.9 % and a specificity of 99 %. Calculations show that the probability of false positives is 0.9999.[49]   □

Example 4.7.5 points out the great difficulty associated with testing for rare conditions. Although the test in the example is 99 % accurate in correctly identifying people who do not have the disease, the overall population consists almost exclusively of this group, and 1 % of them test positive. The number of false positives is small because of the accuracy of the test, but the number of true positives is even smaller because of the rarity of the condition. Thus, a positive test result is usually nothing more than a failure to strongly indicate the absence of the disease.

The data used in the example is hypothetical, but even with considerably more accuracy, the probability of a false positive is large. In practice, the Tay-Sachs test is only used with people who are believed to be descended from a population (eastern European Jews) that has a high incidence of the disease. The risk of a false positive is much lower when the test is applied to a population in which the disease is not extremely rare.

Problems

4.7.1. Determine the probability of false positives for a Tay-Sachs test, assuming the data from Example 4.7.4.

4.7.2.* A school district has two high schools with 2,000 and 1,500 students. Assume that the obesity rates are 20 % at the smaller school and 30 % at the larger.

(a) Let A be the event of belonging to the smaller school and B be the event of being obese. Label each of the two given conditional probabilities in terms of A and B.
(b) Find $P[A]$ and $P[B]$.
(c) Determine the probability that a randomly selected obese student is from the larger school. Label this probability as a conditional probability in terms of A and B.

4.7.3.* In a given region, 20 % of the adult population are smokers, 1 % are smokers with emphysema, and 0.2 % are nonsmokers with emphysema.

(a) What is the probability that a randomly selected person has emphysema?
(b) If we select one smoker, what is the probability that this person has emphysema?
(c) If we select one nonsmoker, what is the probability that this person has emphysema?

4.7.4. One of the tests for multiple sclerosis consists of extracting cerebrospinal fluid from the spinal cord and looking for oligoclonal banding. This test has estimated sensitivity 0.914 and specificity 0.941 [4].

---

[49] See Problem 4.7.1.

(a) Determine the conditional probabilities of false positives and false negatives, given an estimated prevalence of 90 people per 100,000.
(b) Among patients who have one clinically diagnosed bout of optic neuritis with a negative MRI scan, the estimated prevalence of MS is 25 %. Repeat part (a) for this case.
(c) Suppose the prevalence of MS in a group with a particular clinical presentation is 10 %. Repeat part (a) for this case.
(d) Given the results of part (b), is the title of the article [4] fully warranted?
(e) In particular, what is likely to happen if liability concerns cause doctors to order medical tests in instances where there is a positive clinical finding that corresponds to a prevalence that is noticeably higher than the population at large but nevertheless fairly unlikely?

4.7.5. Suppose a medical test has sensitivity 95 % and specificity 90 %, which are typical values for a large number of tests.

(a) Determine the disease prevalence necessary for the conditional probability of false positives to be 50 %.
(b) Determine a general formula that expresses the conditional probability of false positives in terms of the prevalence.
(c) Plot the function obtained in part (b).
(d) Discuss the results.

4.7.6. A screening test is positive in 95 % of cases where a certain disease is present and in 10 % of cases where it is not. A second test can be performed, with results independent of the results of the first test. Assume that 2 % of a population have the disease.

(a) Determine the probabilities of false positives and false negatives if the test is applied once.
(b) Suppose the test is applied twice and both must be positive for a positive diagnosis. Determine the probabilities of false positives and false negatives for this case. [Hint: Because the test results are independent, think of the "test" as a sequence of two individual tests. Can you find the probabilities of a positive result for this double test?]

References

1. Durbin J, M Knott, and CC Taylor. Components of Cramer-von Mises statistics. *J. Royal Stat. Soc., B*, **37**: 216–237 (1975)
2. Hand DJ, F Daly, K McConway, D Lunn, and E Ostrowski. *Handbook of Small Data Sets*. CRC Press, Boca Raton, FL (1993)
3. Labby Z. Weldon's dice, automated. *Chance*, **22**: 6–13 (2009). Available online at http://galton.uchicago.edu/about/docs/labby09dice.pdf.
4. Masjuan J, JC Alvarez-Cemeño, N Garcia-Barragán, M Diaz-Sánchez, M Espiño, MC Sádaba, P González-Porqué, J Martinez San Millán, LM Villar. Clinically isolated syndromes: a new oligoclonal band test accurately predicts conversion to MS. *Neurology*, **66**: 576–578 (2006)
5. National Park Service. The Bean Counters: Mark-Recapture. Available on line at http://www.nps.gov/akso/parkwise/teachers/nature/wear_hoofinit/activities/MarkRecapture.htm.CitedJan2013.
6. Norton PG and EV Dunn.Snoring as a risk factor for disease: an epidemiological survey. *British Medical Journal*, **291**, 630–632 (1985)
7. Pearson K. On the criterion that a given system of deviations from the probable in the case of a correlated system of variables is such that it can be reasonably supposed to have arisen from random sampling. *Philosophical Magazine*, **5**, 157–175 (1900)
8. Shonfeld DJ, PM Rainey, MR Cullen, DR Showalter, and DV Cicchetti. Screening for lead poisoning by fingerstick in suburban pediatric practices. *Archives of Pediatrics and Adolescent Medicine*, **149**: 447–450 (1995)
9. Sprott DA. *Statistical Inference in Science* Springer (2000)

10. Stephens MA. Tests based on EDF statistics. In *Goodness-of-Fit Techniques*, eds. RB D'Agostino and MA Stephens. Marcel Dekker (1986)
11. Thode Jr HC. *Testing for Normality*. Marcel Dekker (2002)
12. Tietjen GL. The analysis and detection of outliers. In *Goodness-of-Fit Techniques*, eds. RB D'Agostino and MA Stephens. Marcel Dekker (1986)

Part III
Dynamical Systems

Chapters 5–7 provide a treatment of dynamical systems, beginning with one-variable discrete and continuous equations in Chapter 5, progressing to discrete linear systems in Chapter 6, and concluding with nonlinear continuous systems in Chapter 7. It is easy to see how to interpret discrete systems and use them for simulations. Their advantages end there; the remaining advantages lie with continuous models. Continuous systems have superior graphical methods and simpler mathematical properties. These advantages, which will become apparent in Chapter 5, more than offset the initial advantages of discrete models. In general, one should only use discrete models when synchronicity of events dictates discrete time.

The importance of good nondimensionalization of models is a recurring theme. This is one additional advantage of continuous models, which allow for nondimensionalization of time. At minimum, nondimensionalization reduces the number of parameters requiring estimated values for simulation or study in analysis. Beyond that, it can sometimes be used to reduce the number of essential components in a model. As will be seen in Chapter 7 in particular, analysis of models becomes more difficult as the number of components increases, and graphical methods are generally limited to one-component discrete models and two-component continuous models. Any reader who has skipped Section 2.6 so as to get to this point sooner is strongly urged to go back and study that section in detail before continuing.

The reader should have noticed that there is no mention of discrete nonlinear models in the description of the chapters in this part. I have not omitted these models entirely, but have relegated them to an appendix. This choice makes the material accessible to those who need it, while de-emphasizing it according to the author's professional judgment.

The accompanying sketch shows the interdependencies of the sections in Part III and connections to prerequisite topics.

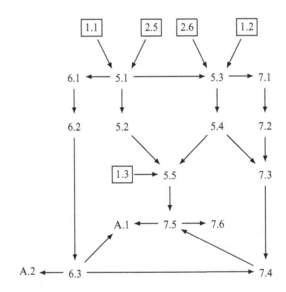

Chapter 5
Dynamics of Single Populations

A state wildlife management agency wants to create a sport fishery in a particular lake. After draining the lake and dredging the bottom, they will introduce desirable plants and create an ideal habitat for the chosen species of fish. They will then fill the lake with water and introduce a small population of fish. Fishing will be prohibited until the population has reached an acceptable level. Now suppose our job is to model the growth of the population of fish in this hypothetical lake. Our goal is to construct a mathematical model to predict how the fish population will change. The model will be useful in designing a strategy for managing the fishery.

Consider another example. An economic development program is planning to bring dairy cattle to small villages in developing countries. The cattle will be fed primarily on local grasses. Suppose our job is to decide how many cattle should be kept in a particular area so as to make full use of the local resources without overgrazing. As in the fishery scenario, we will need a mathematical model to predict the effect of the cattle on the grass resource.

In this chapter, we use the fishery and grazing scenarios as settings in which to develop and study a number of models for the change of populations with time. We restrict ourselves for now to models that require careful monitoring of only one population, the fish in the first case and the grass in the second. There are two main categories of population models, discrete and continuous, differing in the assumption made about how to mark time. *Discrete dynamic models* assume that time can be broken up into distinct uniform intervals. The length of the interval depends on the life history of the organism being modeled. Salmon have yearly spawning periods, so a time interval of 1 year is chosen for a discrete salmon model. Flour beetles have a complicated life cycle that can be approximated as a 2-week larval stage, a 2-week pupal stage, and an adult stage of variable length. Discrete models for flour beetles use a time interval of 2 weeks. *Continuous dynamic models* assume that time flows continuously from one moment to the next. The assumption of continuity in time is relative to the overall duration of the population. For example, it is common to ignore the diurnal-nocturnal variation of temperature and sunlight in a model that tracks a population of plants over a complete growing season.

A continuous dynamic model consists of a mathematical formula that describes the rate at which a population changes at any particular time and population level. While discrete models can also be based on rates of change, it is more common to prescribe the population at a given time in terms of the population at a previous time. This choice of form exaggerates the differences between the types of models. They are incorporated into a single chapter here to emphasize the similarities.

The primary goal of this chapter is to empower the reader with tools for the analysis of both discrete and continuous models. A secondary goal is to help the reader understand how to interpret discrete and continuous models and how to judge which kind of model is more suitable for a given biological setting. Generally, discrete models are preferred when similar

G. Ledder, *Mathematics for the Life Sciences: Calculus, Modeling, Probability, and Dynamical Systems*, 261
Springer Undergraduate Texts in Mathematics and Technology, DOI 10.1007/978-1-4614-7276-6_5,
© Springer Science+Business Media, LLC 2013

events happen simultaneously, such as in populations that have a specific season for births. Continuous models are generally preferable when similar events are spread over time and occur simultaneously with other events. In most disease situations, for example, some members of the population are just getting sick while others are recovering, and the sick individuals are at different stages of the disease. A continuous model is likely to be more appropriate than a discrete model in this case. Similarly, the flour beetle scenario described earlier should be modeled in continuous time because life history events are not synchronized.

Sections 5.1 and 5.3 serve as introductions to discrete models and continuous models, respectively. Each of these sections is followed by a section that presents the principal graphical method for that type: *cobweb analysis* for discrete models in Section 5.2 and *phase line analysis* for continuous models in Section 5.4. Section 5.5 presents *linearized stability analysis*, a particularly useful application of the derivative that works for both discrete and continuous models, although in different ways.

There are several sets of related problems:

Section	5.1	5.2	5.3	5.4	5.5
Beverton–Holt model	5.1.2	5.2.1			5.5.1
Ricker model	5.1.3	5.2.3			5.5.2
Hassell model	5.1.4	5.2.4			5.5.3
Pest control	5.1.5	5.2.5			5.5.4
Discrete logistic equation	5.1.6	5.2.2			5.5.5
Harvesting			5.3.2	5.4.2	5.5.7
Holling II resource model			5.3.10	5.4.3	5.5.8
Lake eutrophication			5.3.12	5.4.4	5.5.9
Allee effect				5.4.1	5.5.6

5.1 Discrete Population Models

After studying this section, you should be able to:

- Build a discrete population model from a set of assumptions about processes that contribute to each year's population.
- Run simulations with discrete population models.
- Find fixed points of discrete population models.

Suppose we want to model the growth of a fish population introduced in a newly renovated lake. How do we construct such a model? The model cannot be very complicated because we have only a limited knowledge of the relevant facts. Presumably we have estimates of the vital rates of the population—birth, survival, and growth—but only limited knowledge of details. We need a conceptual model that incorporates the essential biological processes, such as survival of adults from year to year, birth and survival to the following year, and perhaps also processes that change the population from outside, such as natural migration, stocking, accidental release, or fishing, but we also want to ignore relatively minor effects.

Population models are created by careful accounting. We need to posit a population census that occurs at a particular point in the yearly cycle. With this yearly cycle in mind, we need mathematical formulas to predict the contribution of each relevant biological process to the year $t + 1$ population. Even in the absence of a real yearly census, the model could have qualitative value; however, it will be most useful if the census can actually be conducted to check on the model predictions. Most likely, the census will need to rely on statistical methods, such as the mark-and-recapture method discussed in Section 4.5.

Consider a case in which there is no gain or loss from outside and suppose the census is taken early in the season, before the fish reproduce. Many of the fish counted in the year $t + 1$ census were already adults in the year t census. The number of such fish is the product of the total population N_t and the fraction S of adult fish that survive from 1 year to the next.[1]

Survival from the previous year cannot account for the entire fish population; since $S < 1$, the number of survivors from a given cohort decreases from each year to the next. Of course the year $t + 1$ census also includes fish that hatched from eggs laid after the year t census. This group includes only a small fraction of the newborn fish because of the high mortality rate.[2] To account for the size of the cohort of new adults, we need to look separately at the processes of reproduction and subsequent survival of the newborns. If the adults in year t lay an average of b eggs, then the total number of eggs is bN_t. The subsequent birth and survival of these eggs depends on the amount of competition among the newborns and with the adults. Hence, we assume that the probability that an egg laid in year t becomes an adult in year $t + 1$ is given by a function $p(N_t)$ to be specified later. Combining the number of eggs with the probability of birth and survival, we obtain the formula $bp(N_t)N_t$ for the number of new adults in year $t + 1$.

Having constructed models for each of the processes that contributes to the year $t + 1$ population, we can now construct a model that predicts the change in population from 1 year to the next. In words,

$$\text{next year's population} = \text{survivors from this year's population}$$
$$+ \text{recruitment from this year's eggs.}$$

Substituting the formulas for the two processes, we have the model

$$N_{t+1} = [bp(N_t) + S]N_t, \tag{5.1.1}$$

where

$$0 < S < 1, \quad b > 0, \quad 0 < p(N_t) < 1, \quad p' \le 0.$$

The unspecified function $p(N_t)$ represents a probability; hence, it must satisfy the range restriction $0 < p < 1$. The additional requirement $p' \le 0$ reflects the biological expectation that a larger population of adults decreases the survival probability for the newborn fish. This fishery model yields different variants depending on the choice of the function p.

5.1.1 Discrete Exponential Growth

For a first approximation, we assume that p is constant, with a value determined primarily by predation by other species of animals. In practice, competition for resources and cannibalism by adult fish make the survival probability a decreasing function of population size. Only when the overall population is low is it reasonable to assume a constant p; hence, we should not expect the model results to remain meaningful as the population grows.

For reasons that will only become clear in hindsight, it is often better to think of the model as a formula for the change in the population rather than the population itself. The model in this form is

---

[1] This is why it is important to know where the census comes in the yearly cycle. Only adult fish are counted in the year t census because the census occurs before reproduction. We could have posited a post-reproduction census instead; the resulting model would have been equivalent, but different.

[2] The combination of birth and survival to the next census is called *recruitment*.

$$N_{t+1} - N_t = (bp + S - 1)N_t.$$

For convenience, let $R = bp + S - 1$. We then arrive at the simplest mathematical model for population growth,

$$N_{t+1} = (1 + R)N_t, \qquad R \geq -1. \tag{5.1.2}$$

The reader should be wondering what makes the introduction of the parameter R "convenient." We could instead have defined a parameter $k = bP + S$, and obtained the model $N_{t+1} = kN_t$. This is a matter of taste. The argument for (5.1.2) is that the condition for the population to be growing is then $R > 0$. Thus, the parameter R is the net relative growth rate of the population.

Some additional insight into the model follows from rewriting the condition $R > 0$ as $bp > 1 - S$. In this inequality, bp is the number of recruits per the previous year's adults, and $1 - S$ is the fraction of the previous year's adults that die. The condition $R > 0$ means that the number of recruits exceeds the number of deaths in the adult population.

If the initial population N_0 and the net relative growth rate R are known, we can use (5.1.2) to determine each year's population from the previous year's. This equation is simple enough that we can also obtain an explicit solution formula for the population in year t. Given N_0, we have $N_1 = (1 + R)N_0$, $N_2 = (1 + R)N_1 = (1 + R)^2 N_0$, $N_3 = \cdots = (1 + R)^3 N_0$, and so on. The pattern suggests the general formula

$$N_t = (1 + R)^t N_0, \tag{5.1.3}$$

a conjecture that is easily confirmed. Since the explicit solution is an exponential function of the independent variable t, the model is called the *discrete exponential growth model*.

We now have two equations for the discrete exponential growth model. The original model (5.1.2) defines the sequence recursively, meaning that N_{t+1} is given in terms of N_t rather than in terms of t directly. Mathematicians use the term *difference equation* for this type of definition to emphasize that it is a problem to be solved rather than a solution formula. The sequence of populations is defined explicitly by the solution (Equation (5.1.3)) of the difference equation. The ready determination of an explicit formula for a discrete model is a luxury reserved only for a few of the simplest models. Most of the models worth studying are ones for which we have only the original recursive definition, so we will need methods for studying such equations without a solution formula.

The solution formula reveals an unrealistic property of the discrete exponential growth model, which follows from the unrealistic assumption that the survival probability of newborns is independent of population size. Observe that

$$\lim_{t \to \infty} N_t = \infty.$$

Eventually, there will be so many fish that the model lake will have no room for water, but the model population will still be growing rapidly. Real populations do not behave this way, and we certainly prefer models that don't do so either. The discrete exponential model can be used successfully for a short time only—during the period of growth when space and resources are plentiful. Eventually, the growth of the population leads to crowding and limitation of resources. A better model should include a mechanism for slowing the net relative growth rate for larger populations.

Before we turn to more realistic models, we note one other insight that can be gained from manipulating the appearance of the model. If we rearrange the original model to isolate R, we obtain the form

$$\frac{N_{t+1} - N_t}{N_t} = R. \tag{5.1.4}$$

The quantity $N_{t+1} - N_t$ is the change in population from time t to time $t + 1$. Since one time unit has elapsed, we can also think of $N_{t+1} - N_t$ as the rate of change of population per unit time. Dividing by N_t yields the *relative rate of change* of the population, which is the constant R. It is often helpful to think of a discrete population model as a statement about the relative rate of change of the population rather than as a formula for computing the new population from the current population. For the discrete exponential growth model, the resulting equation (5.1.4) provides a clear interpretation of the model, which is that the relative rate of change of the population is constant. Given $R > 0$, the model predicts that the population growth rate will be unabated as the population grows. While this conclusion is confirmed by the solution formula (5.1.3), the point is that we could have obtained the same conclusion without actually knowing the formula.

5.1.2 The Discrete Logistic Model

Mathematical models in physics and chemistry are almost always derived from first principles. However, it is seldom possible to derive ecological models from first principles, leading us to settle for heuristic models instead. A *heuristic* model is one in which the specific details of the formulas that appear in the model are chosen because they have the right qualitative behavior. We can develop a heuristic fishery model that improves on the discrete exponential growth model by modifying (5.1.4). We want the relative rate of growth to fall to 0 at some large value that represents the capacity of the environment. The easiest way to do that is to make the relative rate of growth linear with a negative slope. Using the parameters specified in Figure 5.1.1, we obtain the *discrete logistic model*,

$$\frac{N_{t+1} - N_t}{N_t} = R\left(1 - \frac{N_t}{K}\right).$$
(5.1.5)

The relative rate of change in the logistic model is the same as in the exponential model

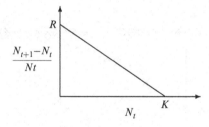

Fig. 5.1.1 The relative rate of change for the discrete logistic model

for $N_t \approx 0$, but it decreases as the population grows. Eventually, the relative rate of change becomes 0 when $N_t = K$. The discrete logistic model is the simplest population model that can account for the effect of a finite amount of space and resources. More realistic models include the Beverton–Holt model and the Ricker model, which appear in the problems. Both of these models can be derived from first principles.

For the purpose of simulations, we can rewrite the discrete logistic model as a formula for N_{t+1} in terms of N_t:

$$N_{t+1} = N_t + RN_t\left(1 - \frac{N_t}{K}\right).$$
(5.1.6)

5.1.3 Simulations

If the parameters in a model are assigned values, we can run simulations by choosing different starting populations and calculating the results.

Example 5.1.1. Consider the model

$$N_{t+1} = N_t + 0.2N_t \left(1 - \frac{N_t}{1,000}\right), \qquad N_0 = 100.$$

This is the discrete logistic model with $R = 0.2$ and $K = 1,000$ and an initial population of 100, which might be realistic for our fishery scenario. From (5.1.5), we find the initial relative growth rate to be $0.2(1 - 100/1,000) = 0.18$, which is only slightly different from the initial growth rate in the discrete exponential model. As the population increases, this relative growth rate decreases, so it must be recalculated at each step. Nevertheless, it is a simple matter to use the model to project the population to any future time. For example,

$$N_1 = 100 + 20(1 - 100/1,000) = 118,$$

$$N_2 = 118 + 23.6(1 - 118/1,000) = 138.8152.$$

The solution is illustrated in Figure 5.1.2 along with the populations obtained with initial values of 0, 10, 500, and 1,000. Note that the model generally yields fractional values. This may seem to be a problem, but it really isn't. At best, the results of a mathematical model only approximate reality. Most models include errors that are far more significant than that of allowing fractional fish.   □

Check Your Understanding 5.1.1:
Find N_1, N_2, and N_3 for the model of Example 5.1.1, given $N_0 = 500$.

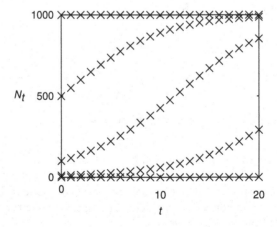

Fig. 5.1.2 Population results for Example 5.1.1

5.1.4 Fixed Points

Example 5.1.2. Consider the model

$$N_{t+1} = N_t + 0.2N_t \left(1 - \frac{N_t}{1{,}000}\right), \quad N_0 = 1{,}000.$$

Using the model to determine population values for times greater than 0, we begin with $N_1 = 1{,}000 + 200(1 - 1{,}000/1{,}000) = 1{,}000$. Thus, $N_1 = N_0$. Similarly, we have $N_{t+1} = N_t$ for all t; that is, the population in the model will be 1,000 in all subsequent years. This is just what we should have expected, since we set the initial population N_0 to be the same as the environmental capacity K. We also have $N_1 = N_0$ and $N_{t+1} = N_t$ for $N_0 = 0$. Since there is no migration process, there is no mechanism to start population growth if there is no initial population. □

A value of N for which $N_{t+1} = N_t$ is called a *fixed point*. Fixed points play a special role in the study of discrete dynamic models; hence, determination of fixed points is a key part of the analysis of any discrete dynamic model. In general, fixed points can be determined by simple algebra. Suppose $N_t = N$, where N is a fixed point. Then by definition, $N_{t+1} = N$ also. The substitutions $N_{t+1} = N_t = N$ into the discrete logistic model (5.1.6) yield the equation

$$N = N + RN \left(1 - \frac{N}{K}\right).$$

This equation defines fixed points in terms of the model parameters R and K. After simplifying, we obtain

$$N \left(1 - \frac{N}{K}\right) = 0.$$

There are two fixed points, $N = K$ and $N = 0$. You can quickly check that these are fixed points by computing $N_1 = 0$ from $N_0 = 0$ and $N_1 = K$ from $N_0 = K$.

Example 5.1.3. Consider the model

$$N_{t+1} = \frac{RN_t^2}{1 + N_t^2},$$

where R is a positive constant. To find the fixed points of this model, we set $N_{t+1} = N$ and $N_t = N$ and obtain the equation

$$N = \frac{RN^2}{1 + N^2}.$$

As with most population models, $N = 0$ is a fixed point.[3] If $N \neq 0$, the equation becomes

$$1 = \frac{RN}{1 + N^2},$$

which we can rewrite as

$$1 + N^2 = RN.$$

---

[3] This is always true if there is no migration from outside.

This is a quadratic equation for N. Applying the quadratic formula to the form $N^2 - RN + 1 = 0$, we obtain

$$N = \frac{R \pm \sqrt{R^2 - 4}}{2}.$$

These fixed points exist only if $R > 2$. It can be shown[4] that $R < 2$ means that the population at time $t + 1$ is always less than the population at time t; hence, $R > 2$ is a requirement for viability. Assuming this requirement is met, the model then has three fixed points[5]:

$$N = 0, \quad N = \frac{R - \sqrt{R^2 - 4}}{2}, \quad N = \frac{R + \sqrt{R^2 - 4}}{2}.$$

$\square$

Fixed points are of two different types, both of which are illustrated in Figure 5.1.2. The fixed point $N = 0$ is said to be *unstable* because sequences that start near it move away from it. The fixed point $N = K$ in the logistic growth model is said to be *asymptotically stable* because sequences that start near it move toward it.[6] When parameters in a discrete dynamic model are left unspecified, it is a simple matter to find the fixed points, but it is more difficult to determine if a fixed point is stable or unstable. This issue will be addressed in Sections 5.2 and 5.5. Discrete models do not necessarily converge to a fixed value, as will be seen in the problems. Nevertheless, those models that do so must converge to a stable fixed point; hence, the determination of stable fixed points is always a key component of model analysis.

Check Your Understanding Answers

1. $N_1 = 550, N_2 = 599.5, N_3 = 647.52$.

Problems

5.1.1.* For the model

$$N_{t+1} = \frac{RN_t^2}{1 + N_t^2}$$

of Example 5.1.3:

(a) Find the formulas for the fixed points if $R > 2$.
(b) Show that the population is necessarily decreasing if $R < 2$.

5.1.2.* One specific fishery model is the Beverton–Holt model, which can be written in dimensionless form as

$$N_{t+1} = \left(S + \frac{A}{1 + N_t} \right) N_t, \quad A > 0; \quad 0 \leq S < 1.$$

---

[4] See Problem 5.1.1a.

[5] Only nonnegative fixed points have biological meaning. Given $R > 2$, all three fixed points are nonnegative, as is shown in Problem 5.1.1b.

[6] Graphs of N versus t have horizontal asymptotes at asymptotically stable fixed points.

(a) Find the relative rate of change for the model.

(b) What is the biological significance of the assumption $S = 0$?

(c) Find a general formula for the fixed points of the model. Note that some fixed points may only exist for certain ranges of parameter values.

(d) Plot (on one set of axes) graphs of the fixed points in the SN plane, given $A = 0.5$ and $A = 2$.

(e) Run a simulation with the model using $A = 0.8$, $S = 0$, and $N(0) = 0.2$. Discuss the simulation results, emphasizing the connection to the results of part (c).

(f) Repeat part (e), but with $A = 2$.

(g) Repeat part (e), but with $S = 0.5$.

(This problem is continued in Problem 5.2.1.)

5.1.3. Another fishery model is the Ricker model, which can be written in dimensionless form as

$$N_{t+1} = [S + Ae^{-N_t}]N_t.$$

(a) Find the relative rate of change for the model.

(b) What is the biological significance of the assumption $S = 0$?

(c) Find a general formula for the fixed points of the model. Note that some fixed points may only exist for certain ranges of parameter values.

(d) Plot (on one set of axes) graphs of the fixed points in the SN plane, given $A = 0.5$ and $A = 2$.

(e) Run a simulation with the model using $S = 0$, $A = 2$, and $N(0) = 0.2$. Discuss the simulation results, emphasizing the connection to the results of part (c).

(f) Repeat part (e), but with $A = 8$.

(g) Repeat part (e), but with $A = 14$.

(This problem is continued in Problem 5.2.3.)

5.1.4. The Hassell model, which is sometimes used for population dynamics of insect pests, can be written in dimensionless form as

$$N_{n+1} = \left(\frac{A}{1 + N_n}\right)^b N_n, \qquad A, b > 0.$$

(a) Find a general formula for the fixed points of the model. Note that some fixed points may only exist for certain ranges of parameter values.

(b) Run a simulation with the model using $b = 0.5$, $A = 10$, typical values for field populations of many insect species. Discuss the simulation results, emphasizing the connection to the results of part (a).

(c) Repeat part (b) using $b = 2$, $A = 6$, which roughly matches a laboratory population of a species of weevils.

(d) Repeat part (b) using $b = 3$, $A = 4$, which roughly matches a field population of Colorado potato beetles.[7]

(This problem is continued in Problem 5.2.4.)

5.1.5. One method for controlling insect pests is to release a population of sterile males into a field population. The method results in a reduced population, because females that mate with sterile males do not contribute to the next generation.

(a) Suppose a field population of insects consists of N_n females, all normal, and $N_n + S$ males, S of which are sterile. [Note: We are interpreting the quantity N in the Hassell model to

---

[7] Parts (b)–(d) use parameter values adapted from [1].

include only females and assuming that the population of males is the same as that of females.] Obtain a simple population model that incorporates the sterile population into the Hassell model for the specific case $b = 1$. [Hint: Assume that the new population of females predicted by the Hassell model must be multiplied by the probability that a female's randomly chosen mate is not sterile.]

(b) The model of part (a) has two parameters, A and S. Plot a graph showing how the fixed points of the model vary with S when $A = 4$. [Hint: The equation for the fixed points is more easily solved for S than for N. You can still graph the result in the SN plane.]

(c) Run a simulation of the model with $A = 4$ and $S = 2/3$. For N_0, use the fixed point corresponding to $A = 4$ and $S = 0$. Discuss the simulation results, emphasizing the connection to the results of part (b).

(d) Repeat part (c) with $S = 1.2$ and $N_0 = 1$.

(This problem is continued in Problem 5.2.5.)

5.1.6. The discrete logistic model has some interesting mathematical properties. Among these is the possibility of periodic solutions. To explore this possibility algebraically, we consider a dimensionless version of the model, given as

$$x_{t+1} = x_t + x_t(R - x_t).$$

(a) Use the model to derive the formula

$$x_{t+2} = x_t + x_t(R - x_t) + x_t(1 + R - x_t)(R - x_t)(1 - x_t).$$

(b) Suppose X is a number for which $x_t = X$ implies $x_{t+2} = X$, but $x_{t+1} \neq X$. Thus, X represents points that repeat in alternate time steps, but are *not* true fixed points. Use the formula from part (a) and the fixed points of the model to derive the equation

$$0 = 1 + (1 - X)(1 + R - X).$$

Rearrange this equation to obtain the simpler form

$$X^2 - BX + B = 0, \qquad B = R + 2.$$

(c) Determine approximate values of X for the particular cases $R = 2.2$ and $R = 2.5$.

(d) Run a simulation with $R = 2.2$ and $x_0 = 1$.

(e) Run a simulation with $R = 2.2$ and $x_0 = 1.8$.

(f) Run a simulation with $R = 2.5$ and $x_0 = 1.8$.

(g) Run a simulation with $R = 2.7$ and $x_0 = 1.8$.

(h) Discuss the simulation results, emphasizing the connection to the theoretical results.

(This problem is continued in Problem 5.2.2.)

5.2 Cobweb Analysis

After studying this section, you should be able to:

- Construct cobweb diagrams.
- Use cobweb diagrams to run simulations.
- Use cobweb diagrams to find fixed points and determine their stability.

Cobweb diagrams are graphical representations of discrete dynamic models. These diagrams are used to draw conclusions regarding the long-term behavior of a one-component discrete model. The model must meet two requirements:

1. The formula that determines the population in the next time step from that of the previous time step must not refer directly to the time.[8] This requirement rules out models that account for yearly changes in the environment.[9]
2. The formula that prescribes the model dynamics must be of *first order*. This means that the formula for the population at time $t + 1$ can only refer to the population at time t and not to earlier population data. This requirement rules out models for organisms with a life history that involves stages of comparable duration, as is common among insect species. Suppose it takes two time units for an individual to become an adult. Individuals that survive the first period are juveniles for the second period. In this scenario, the adult population at time $t + 1$ depends on the juvenile population at time t, which in turn depends on the adult population at time $t - 1$. This is an example of a *second-order* model. We defer consideration of higher order models to Chapter 6.

Fortunately, there are a lot of useful models that are both autonomous and first order. In particular, the Beverton–Holt and Ricker models introduced in the Section 5.1 problems are autonomous first-order models.

5.2.1 Cobweb Diagrams

Figure 5.2.1 shows a cobweb diagram for the discrete exponential growth model

$$N_{t+1} = 1.5N_t.$$

At first glance, cobweb diagrams seem to be a jumble of lines going in different directions. Interpreting them is a skill that takes practice. We focus first on the construction of the diagram and then fill in details to help with the interpretation.

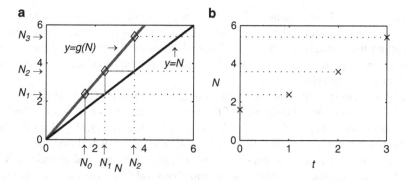

Fig. 5.2.1 (a) Cobweb diagram and (b) time history for $N_{t+1} = 1.5N_t$

---

[8] Processes that depend only on the state of a system, and not directly on the time, are called *autonomous*. Much of the analysis we can do with population models only works for autonomous models.

[9] This does not mean that we cannot study the effects of global climate change. We can use analysis to study its effects on fixed points and stability. If we want to study the short-term effects of climate change, however, we need to run simulations.

Any autonomous first-order discrete model can be written in the form

$$N_{t+1} = g(N_t), \tag{5.2.1}$$

where the function g specifies the particular model. For example, the diagram in Figure 5.2.1 is for the model defined by $g(N) = 1.5N$. The basis of a cobweb diagram is a graph that displays $y = g(N)$ and $y = N$ on the same set of axes.[10]

To minimize confusion, I label the horizontal axis as N, and in Figure 5.2.1a, the heavy line is $y = g(N) = 1.5N$ and the medium line is $y = N$. The light lines with a stair-step shape are a graphical rendition of a simulation. The procedure for creating these lines on a plot that already contains $y = g(N)$ and $y = N$ is specified in Algorithm 5.2.1.

Algorithm 5.2.1 *Cobweb diagram for $N_{t+1} = g(N_t)$*

1. *Mark the point on the horizontal axis corresponding to a chosen initial population N_0. (The diagram in Figure 5.2.1a was constructed using $N_0 = 1.6$.)*
2. *Draw a vertical line segment from the point marked in step 1 to the curve $y = g(N)$. Mark the intersection point.*
3. *Draw a horizontal line segment from the point marked in step 2 to the line $y = N$. Note that this line segment could go either to the left or to the right, depending on the relative locations of the curve $y = g(N)$ and the line $y = N$. Mark the intersection point.*
4. *Continue extending the diagram by alternating vertical and horizontal line segments, always moving up or down to $y = g(N)$ and left or right to $y = N$.*
5. *Emphasize the intersection points on the curve $y = g(N)$ that end the vertical line segments. (These are the points marked with diamonds in Figure 5.2.1a.)*
6. *For a time history plot, as in Figure 5.2.1b, extend the horizontal lines from the cobweb diagram to another set of axes. Mark a point on each of these horizontal lines by choosing consecutive time values, starting with $t = 1$ for the first horizontal line. Also mark the initial condition at time $t = 0$. (These are the points marked with "x" in Figure 5.2.1b.)*

Just as in a simulation by calculation, all of the parameters must be chosen before constructing the cobweb diagram, including the initial value N_0. Once the parameters are chosen, the process of constructing the cobweb diagram is mechanical. Just plot the two functions and draw the line segments in the prescribed manner. Think of the construction and interpretation as separate steps, and the construction is easy.

We begin the interpretation of the cobweb diagram with the points marked by diamond symbols. These points are the endpoints of the vertical line segments. The coordinates of the three marked points in Figure 5.2.1a are (N_0, N_1), (N_1, N_2), and (N_2, N_3). Because the heavy curve is the equation $y = g(N)$ and the model is $N_{t+1} = g(N_t)$, it follows that any point on the curve that has horizontal coordinate N_t for some t must have vertical coordinate N_{t+1}. The purpose of the vertical line segments is to compute new values of the population. The dotted horizontal lines connecting Figure 5.2.1a to b are used to transfer the y coordinates of the points on the $y = g(N)$ curve to the corresponding points (t, N_t) on the time history plot.

At this point, the reader should be asking a key question: "How do we know that the horizontal coordinates of the marked points are N_0, N_1, and N_2?" We chose the value N_0 before

---

[10] It is not clear how the vertical axis should be labeled, since $g(N)$ and N represent different things. Eventually we will identify values of N on this axis; however, for the purpose of constructing the cobweb diagram, it seems best to think of the vertical axis as a generic "y" coordinate.

we began the process, but the other points must be determined by the process itself. This is where the horizontal line segments come in. The first marked point has coordinates (N_0, N_1), so the first horizontal line segment marks the line $y = N_1$. The segment ends on the line $y = N$. The intersection of the lines $y = N_1$ and $y = N$ is at $N = N_1$, and this is precisely the N value that we need to use to be able to draw a vertical line segment to the point (N_1, N_2). Thus, we can see the sequence of N values by looking at *either* the vertical or the horizontal coordinates of the marked points. The vertical coordinates are N_1, N_2, ..., and the horizontal coordinates are N_0, N_1, ... You should now pause to review what has been said. Once you understand the reason for the alternating vertical and horizontal lines, the cobweb diagram loses its mystery and becomes a simple tool.

5.2.2 Stability Analysis

Up to this point, we have focused on explaining the process of the cobweb diagram and its interpretation as a simulation. Now we turn to the use of the diagram for analysis. What information does it contain that is not found in the corresponding time history?

In a time history, we can only see the current point and past points. In a cobweb diagram, we can see the future as well; that is, we can predict the trend of future populations from the relative positions of the graphs of $y = g(N)$ and $y = N$. For the model of Figure 5.2.1, the future is that the sequence will continue to increase as the diagram is extended to larger values of N and y. Every horizontal line segment will go to the right, so each value of N will be larger than the preceding value.

Figure 5.2.2 shows a cobweb diagram for

$$N_{t+1} = \frac{2.5 N_t^2}{1 + N_t^2}.$$

This diagram tells a different story. The graphs of $y = g(N)$ and $y = N$ cross at the fixed point $N = 2$, and the cobweb diagram indicates that the marked points will simply get closer to the fixed point. Thus, the cobweb diagram shows that the equilibrium solution $N = 2$ is asymptotically stable.

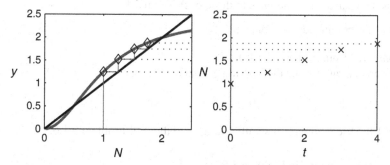

Fig. 5.2.2 Cobweb diagram and time history for $N_{t+1} = \frac{2.5 N_t^2}{1 + N_t^2}$ with $0.5 < N_0 < 2$

Note that not all starting values give the same long-term behavior. Starting values greater than 0.5 result in a cobweb diagram with the same general features as Figure 5.2.2, where the corresponding sequence of population values approaches 2 as $t \to \infty$. However, initial values less than 0.5 lead to a qualitatively different picture, as shown in Figure 5.2.3. This time, the horizontal line segments all move to the left, so the marked points approach the fixed point $N = 0$. Thus, $N = 0$ is a second asymptotically stable fixed point. The unstable fixed point $N = 0.5$ serves as a dividing point between solutions that go to 0 and solutions that go to 2. This is typical behavior for dynamic models with one constituent; the stable fixed points serve as sequence limits for ranges of initial values delimited by the unstable fixed points.

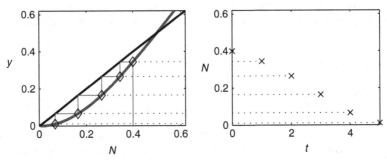

Fig. 5.2.3 Cobweb diagram and time history for $N_{t+1} = \frac{2.5N_t^2}{1+N_t^2}$ with $0 < N_0 < 0.5$

Problems

5.2.1. (Continued from Problem 5.1.2.)
Plot cobweb diagrams for the Beverton–Holt model

$$N_{t+1} = \left(S + \frac{A}{1+N_t}\right)N_t, \quad A > 0, \quad 0 \le S < 1,$$

with $N_0 = 0.2$, using

(a) $A = 0.8$, $S = 0$.
(b) $A = 2$, $S = 0$.
(c)* $A = 0.8$, $S = 0.5$.
(d) Discuss the results with reference to those of Problem 5.1.2.

(This problem is continued in Problem 5.5.1.)

5.2.2. (Continued from Problem 5.1.6.)
Plot a cobweb diagram for the discrete logistic model

$$x_{t+1} = x_t + x_t(2.2 - x_t)$$

using $x_0 = 1.8$. Discuss the results with reference to those of Problem 5.1.6.

(This problem is continued in Problem 5.5.5.)

5.2.3. (Continued from Problem 5.1.3.)
Plot cobweb diagrams for the Ricker model

$$N_{t+1} = AN_t e^{-N_t}$$

with $N_0 = 0.2$, using

(a) $A = 2$.

(b) $A = 8$.

(c) $A = 14$.

(d) Discuss the results with reference to those of Problem 5.1.3.

(This problem is continued in Problem 5.5.2.)

5.2.4. (Continued from Problem 5.1.4.)
Plot cobweb diagrams for the Hassell model

$$N_{n+1} = \left(\frac{A}{1+N_n}\right)^b N_n, \qquad A, b > 0,$$

with $N_0 = 0.2$, using

(a) $b = 0.5, A = 10$.

(b) $b = 2, A = 6$.

(c) $b = 3, A = 4$.

(d) Discuss the results with reference to those of Problem 5.1.4.

(This problem is continued in Problem 5.5.3.)

5.2.5. (Continued from Problem 5.1.5.)
Plot cobweb diagrams for the pest control model

$$N_{n+1} = \frac{4N_n^2}{(1+N_n)(N_n+S)}, \qquad S > 0,$$

using

(a) $S = 2/3, N_0 = 0.5$.

(b) $S = 2/3, N_0 = 0.2$.

(c) $S = 1.2, N_0 = 1$.

(d) Discuss the results with reference to those of Problem 5.1.5.

(This problem is continued in Problem 5.5.4.)

5.3 Continuous Population Models

After studying this section, you should be able to:

- Build continuous dynamic models from assumptions about rates of change.
- Use computer simulations to study continuous dynamic models.
- Find equilibria of continuous dynamic models.

Suppose we want to build a model for the growth of a plant population. We could use the number of individuals as a measure of the population, but there are some obvious difficulties. Two small plants make the same contribution to the overall size as a single plant that is twice as large. In many settings, it makes more sense to measure a plant population in terms of its biomass rather than the number of individuals. Biomass changes on a continuous basis, so it makes sense to use a model that is continuous in time rather than the discrete models we have been studying. In this section and the next, we examine some basic techniques for the development and analysis of continuous models.

5.3.1 Exponential Growth

The simplest discrete model is the exponential growth model, which is characterized by a constant relative growth rate. The continuous case also has an exponential growth model, characterized by a constant relative growth rate. To derive this model, we note that the derivative dN/dt of a function $N(t)$ of continuous time is the *absolute rate of change*. The *relative rate of change* is the ratio of the absolute rate of change to the magnitude $|N|$; since $N > 0$, a constant relative rate of change corresponds to the equation

$$\frac{1}{N}\frac{dN}{dt} = r. \qquad (5.3.1)$$

We can rewrite the equation in the standard form as

$$\frac{dN}{dt} = rN. \qquad (5.3.2)$$

Some insight into the relationship between the discrete and continuous exponential models comes from comparing the different forms in which the models can be written. The relative growth form (5.3.1) corresponds nicely with the relative growth form of the discrete model $(N_{t+1} - N_t)/N_t = R$. The more common form (5.3.2) also appears to compare closely to the common discrete form $N_{t+1} = kN_t$; however, this apparent similarity is misleading. While both calculate some quantity as being proportional to the current population, the continuous quantity rN is a rate of change while the discrete quantity kN_t is an updated population value. I have specifically chosen the constants r and R to represent the two relative rates of change to facilitate the appropriate comparison between the models, without giving the impression that the relative growth rate parameter should have the same value for the two cases. Instead, there is a mathematical relationship we can use to compute either value from the other.

Although solution methods for differential equations are beyond the scope of this book, (5.3.2) can be solved by inspection. We are looking for a function whose derivative is a constant times the function. A quick review of a table of derivatives[11] reveals that the exponential function has this property; specifically, the function $N(t) = N_0 e^{rt}$ has the required property and also satisfies the initial condition $N(0) = N_0$. As in the discrete case, the exponential model is only useful while resources are abundant. The unbounded solution of the equation makes the model unsuitable for any situations in which growth should be restricted.

> **Check Your Understanding 5.3.1:**
> Determine an appropriate mathematical relationship between the discrete and continuous relative growth rates R and r by comparing the solution of the continuous model (5.3.2) with that of the corresponding discrete model $N_{t+1} - N_t = RN_t$.

5.3.2 Logistic Growth

A more realistic model is the *logistic growth model*, which we can define in a manner analogous to the discrete logistic model of Section 5.1. We again assume a relative rate of population

---

[11] See Table 1.3.1.

change that is given by a linear relationship, as previously depicted in Figure 5.1.1. Using the continuous definition of the relative rate of change yields the equation

$$\frac{1}{N}\frac{dN}{dt} = r\left(1 - \frac{N}{K}\right),$$

which we can rewrite as
$$\frac{dN}{dt} = rN\left(1 - \frac{N}{K}\right). \tag{5.3.3}$$

Unlike the discrete logistic equation, the continuous logistic equation can be solved explicitly. However, it is not particularly advantageous to do so. All important analytical results can be obtained directly from the differential equation, as we shall see in the remainder of this chapter. Simulations can be conducted with any degree of accuracy by numerical methods, which work by computing the solutions to discrete approximations of the differential equation.

Example 5.3.1. Analogous to Example 5.1.1, consider the model

$$\frac{dN}{dt} = 0.2N\left(1 - \frac{N}{1,000}\right), \quad N(0) = 100.$$

We can choose a sequence of time values t_j for which we seek corresponding approximate population values N_j. With $t_0 = 0$, we obtain a starting value of $N_0 = 100$. Numerical methods work by using a discrete approximation to the differential equation to calculate each successive function value. The simplest of these methods is Euler's method, which is based on the forward difference approximation

$$\frac{dN}{dt}(t_j) \approx \frac{N_{j+1} - N_j}{t_{j+1} - t_j}. \tag{5.3.4}$$

Substituting this approximation into the differential equation yields the discrete dynamic model

$$N_{j+1} = N_j + 0.2hN_j\left(1 - \frac{N_j}{1,000}\right), \quad N_0 = 100,$$

where we have assumed time intervals of equal length $t_{j+1} - t_j = h$. It can be shown that any desired degree of accuracy can be achieved by choosing small enough time steps, although the amount of calculation required increases as well. Euler's method is only first order, which means that the error is cut approximately by a factor of 2 when the time step is halved. Much more sophisticated methods are available; the most popular method is fourth order,[12] which means that the error is cut approximately by a factor of $2^4 = 16$ when the time step is halved. Sophisticated methods of this type can easily be programmed by hand, and are also built into computer algebra systems such as Maple and mathematical software such as Matlab. While they are not part of R's core package, numerical methods for R are readily available as add-on packages. With a sufficiently accurate method, graphs of the simulation result, as shown in Figure 5.3.1, are indistinguishable from the exact solution. As we might expect, the behavior of the continuous logistic model is similar to that of the corresponding discrete model seen in Figure 5.1.2.   □

---

[12] See Appendix C.

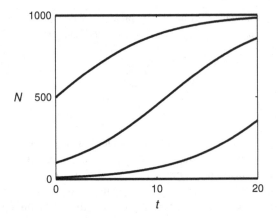

Fig. 5.3.1 Population results for Example 5.3.1

5.3.3 Equilibrium Solutions

Analogous to the fixed points of discrete models are *equilibrium solutions* of continuous models, which are values of N for which $dN/dt = 0$. Finding equilibrium solutions is a simple algebraic problem.

Example 5.3.2. Setting $dN/dt = 0$ in the logistic model (5.3.3) yields the algebraic equation

$$rN\left(1 - \frac{N}{K}\right) = 0\,;$$

the solutions $N = 0$ and $N = K$ are the equilibrium solutions for the logistic equation, the same as the fixed points for the discrete logistic model.   □

5.3.4 A Renewable Resource Model

So far, we have considered general models for growth in a continuous setting. With some modifications to the logistic equation, we can create a renewable resource model capable of reproducing behavior seen in populations of endangered species and helping us predict the outcomes of conservation efforts. The key feature that must be added to turn an ordinary continuous growth model into a renewable resource model is the consumption of the resource by a fixed population. In general, the simplest models follow from the assumptions that the number of consumers is constant and that the amount of resource consumption by each individual does not depend on the total number of consumers. These assumptions yield the generic model

$$\frac{dV}{dT} = RV\left(1 - \frac{V}{K}\right) - Cq(V), \tag{5.3.5}$$

where $V(T)$ is the biomass of the resource,[13] C is the number of consumers, K is the carrying capacity, and $q(V)$ is the rate of consumption per consumer. One specific scenario for which

---

[13] The choice of symbols is seldom significant; I have used V because vegetation is the resource in the motivating scenario. Of course, the model can be used for animal populations as well. Also, the usual lowercase t has been replaced by the uppercase T to allow for systematic use of upper and lower cases in nondimensionalization.

the assumptions are well justified is a grassland used to graze cattle. The model is appropriate for this scenario because the rancher can directly control the size of the cattle population. If the consumer population size depends on the availability of the resource, then (5.3.5) needs to be augmented by a differential equation that models those dynamics. Such systems are considered in Chapter 7.

A variety of choices are possible for the consumption function $q(V)$. The simplest is the linear model $q = QV$. Another possibility is the Holling type II model from Section 2.5. For cattle grazing on grass, a more realistic model is

$$q(V) = \frac{QV^2}{A^2 + V^2}, \qquad Q, A > 0, \tag{5.3.6}$$

which is known as the Holling type III model.[14] Figure 5.3.2 shows a plot of the model in dimensionless form. The graph of the Holling type III model is similar but not identical to that of the Holling type II model (Figure 2.5.1a). The most important difference is the behavior for small resource levels. In the Holling type III model, the consumption curve at low resource levels has a small slope, which is appropriate in cases where the consumer has an alternate food supply. As a simple example, consider the behavior of bears. Bears eat a variety of foods found in different places. If there are very few fish available, they will spend more of their time looking for other sources of food; consequently, an increase in the supply of fish will not make as much difference as it would if the bears had no alternatives. In the case of cattle on a grassland, there are often relatively low quality plants that the cattle can eat if their primary food is in short supply.

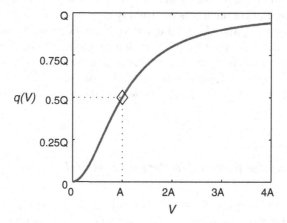

Fig. 5.3.2 The consumption rate function $q = QV^2/(A^2 + V^2)$

With the Holling type III model for q, we have the renewable resource model

$$\frac{dV}{dT} = RV\left(1 - \frac{V}{K}\right) - C\frac{QV^2}{A^2 + V^2}. \tag{5.3.7}$$

This form of the model has five parameters, which is too many for a systematic analysis. The number can be reduced significantly by nondimensionalization. There are always choices in nondimensionalization, although they are not always obvious. Here, there are no obvious alternatives to $1/R$ for defining a dimensionless time. However, the dependent variable V could

---

[14] The properties of this model are explored in Problem 5.3.1.

be made dimensionless by dividing by either A or K. Which we choose depends on what we think is the primary process that limits the resource population. If there is only a small amount of consumption, then the population of the resource will be only slightly less than the carrying capacity, so we should choose K as the representative value for nondimensionalization. Alternatively, we should choose A if we expect consumption to be the primary mechanism that limits the amount of the resource. This seems more likely for a model whose purpose is to understand the effect of consumption on the resource population. Using these choices, we define the dimensionless quantities

$$v = \frac{V}{A}, \qquad t = RT$$

and systematically replace all of the variables in the model using the substitutions

$$V \to Av, \qquad \frac{d}{dT} \to \frac{dt}{dT}\frac{d}{dt} = R\frac{d}{dt}.$$

After simplification,[15] we obtain the dimensionless form

$$v' = v\left(1 - \frac{v}{k}\right) - \frac{cv^2}{1+v^2}v, \qquad k = \frac{K}{A}, \qquad c = \frac{CQ}{RA}. \tag{5.3.8}$$

Once dimensionless variables are substituted into a dimensional model, the remaining parameters sort themselves into dimensionless combinations. Usually it is best to think of these combinations as scaled versions of key original parameters, and it is good modeling practice to label them accordingly. The parameter k is the dimensionless carrying capacity relative to the reference population value A. The parameter c is a dimensionless measure of the number of consumers. This choice of parameters is particularly convenient if we are studying the model for the purpose of exploring the effects of conservation policies on the resource. Of the two dimensionless parameters, k is strictly a property of the biological system, whereas c is, at least in theory, completely subject to control. The natural question to ask is, "Given some objective criterion and a specific value of k, what is the optimal choice of c?" This is a much simpler and simultaneously more general question than any we could address using the dimensional form of the model.[16]

5.3.5 Equilibria of the Renewable Resource Model

Analysis of the renewable resource model (5.3.8) will be a recurring theme of this chapter. For now, we consider the problem of finding the equilibrium solutions. Setting $v' = 0$, we obtain the equation

$$0 = v\left(1 - \frac{v}{k}\right) - \frac{cv^2}{1+v^2} = v\left(1 - \frac{v}{k} - \frac{cv}{1+v^2}\right).$$

Clearly, $v = 0$ is an equilibrium value regardless of k and c. This makes biological sense, as our model has no mechanism for starting a new population. Other equilibrium values of v are solutions of the equation

$$1 - \frac{v}{k} - \frac{cv}{1+v^2} = 0, \tag{5.3.9}$$

and these solutions will depend on the parameter values.

---

[15] See Problem 5.3.3.

[16] See Problem 5.3.9.

We have several choices for finding the solutions of (5.3.9). With specific values of k and c, we can use a numerical method to obtain approximate equilibria; however, we get more information from the model if we leave the parameter values unspecified. Alternatively, we could use a computer algebra system to solve the equation. While this sounds like the logical strategy, it is actually not the best plan for a number of reasons. The resulting formula, if one is given, will be complicated and difficult to interpret. The solution formula generated by Maple 9.5 takes up most of a page. Complicated solution formulas can lead to subsequent errors, such as documenting "results" that do not exist for any set of parameter values.[17]

When working with messy algebraic equations, it is helpful to recognize that the equation can take a number of different forms, and some of these may suggest strategies that do not work for the others. One possibility for (5.3.9) is to multiply by all of the denominators to obtain a polynomial equation:

$$v^3 - kv^2 + (1 + ck)v - k = 0. \tag{5.3.10}$$

This equation is an improvement on the original, but it is still not very manageable. There is a general formula for the solution of an arbitrary cubic polynomial equation, but that gives the same unwieldy result as the original equation, which is why you did not learn that formula along with the quadratic formula in school. One conclusion we can draw from a cubic polynomial is that there can be no more than three solutions for any given pair of k and c values.

5.3.6 A Graphical Method

Solutions of algebraic equations can be found by graphing. If we have specific values for k and c, we can graph the cubic polynomial of (5.3.10) and look for places where the graph crosses the v axis; however, this is problematic without given values for the parameters. Alternatively suppose we put the complicated third term in (5.3.9) by itself on one side of the equation and divide by c. These manipulations change the equation to the form

$$\frac{1}{c}\left(1 - \frac{v}{k}\right) = \frac{v}{1 + v^2}. \tag{5.3.11}$$

What is so great about this version? Note first that any equation can be solved by graphing the functions on both sides of the equation together on one graph and looking for the intersections. In (5.3.11), we have one function that is complicated but parameter-free, and another function that has both parameters but is especially simple. We only need to graph the complicated function once because it has no parameters. The simple function is a straight line, which means that it is easy to see how the parameters affect the graph. To clarify this notion, let $f(v)$ represent the function

$$f(v) = \frac{1}{c}\left(1 - \frac{v}{k}\right).$$

Setting $f(v) = 0$ yields the result $v = k$, which means that k is the v intercept of the straight line. Similarly, setting $v = 0$ shows that $1/c$ is the vertical intercept. Given any pair of parameter values, we can mark k on the horizontal axis and $1/c$ on the vertical axis and connect those points with a straight line. Where that line intersects the graph of $v/(1 + v^2)$, we have an equilibrium value of v.

Figure 5.3.3 shows the curve $v/(1 + v^2)$ and the straight lines with $k = 10$ and $c = 1, 2$, and 3. These values of c show the different possibilities for equilibria. Note that there will be at least one positive solution no matter what points $(k, 0)$ and $(0, 1/c)$ are used to define the line. For the $k = 10$, $c = 2$ line, there are three positive solutions, while there is only one positive

---

[17] I've seen a published research paper in which this occurred.

solution for each of the other values of c with $k = 10$. In general, there will usually be either one or three positive solutions. Exactly two positive solutions can occur, but this requires that the line be tangent to the curve at one point. As for the magnitude of the solutions, the most striking feature of the graph is the great difference between the very small solution we get when c is large and the relatively large solution we get when c is small. The implications of this feature will be discussed in subsequent analysis.

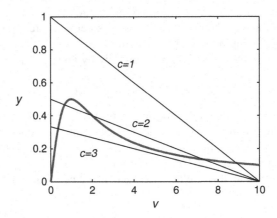

Fig. 5.3.3 Graphs of $v/(1+v^2)$ and $f(v)$, with $k = 10$

Check Your Understanding Answers

1. $1 + R = e^r$

Problems

5.3.1. Consider the Holling type III predation/ consumption model,

$$q(V) = \frac{QV^2}{A^2 + V^2}.$$

(a) Explain the graphical significance of the parameter Q.
(b) Determine the dimension of A, given that V is biomass per unit area.
(c) Explain the graphical significance of the parameter A. [Hint: Use the answer to part (b).]
(d) Compute the derivative $q'(V)$.
(e) Explain the significance of the result for $q'(0)$ in terms of the feeding habits of the consumer.

5.3.2. The Schaefer equation,

$$\frac{dX}{dT} = RX\left(1 - \frac{X}{K}\right) - QX, \qquad R, K, Q > 0,$$

is often used to model the impact of harvesting on a fish population, where X is either the number of fish or the biomass of fish.

(a) Compare the Schaefer model with the resource consumption model (5.3.7). Describe the key difference between the models in biological terms. In other words, explain the differences between the conceptual models underlying the two mathematical models. [Hint: Ignore superficial differences, such as the identities of the resources and the consumers. Look at the assumptions about the biological processes.]

(b) Derive the dimensionless form of the Schaefer model,

$$x' = x(1-x) - Ex,$$

by using $x = X/K$ and $t = RT$ for the dimensionless population and time and using an appropriate definition for E. The symbol E is commonly used in this context to suggest that it represents the harvesting "effort."

(c) Determine the critical points for the dimensionless model. In particular, what restriction is there for E if there is to be a positive equilibrium population level?

(d) The term Ex in the differential equation represents the yield rate of the fishery. If the fishery is managed, the goal might be to maximize the sustained yield $y = Ex^*(E)$, where $x^*(E)$ is the equilibrium population for any particular level of effort. Use the differential equation of part (b) to obtain a formula for y that depends only on x^*.

(e) Find the equilibrium value x^* that maximizes the yield; then find the effort E^* that achieves the maximum yield.

(This problem is continued in Problem 5.4.2.)

5.3.3. Obtain the dimensionless renewable resource model (5.3.8) from the original model (5.3.7) by using the substitutions $V = Av$ and $T = t/R$. Note that the chain rule of differentiation implies

$$\frac{dy}{dT} = \frac{dy}{dt}\frac{dt}{dT}$$

for any quantity y that is a function of time.

Problems 5.3.4–5.3.9 refer to the dimensionless resource consumption model

$$v' = v\left(1 - \frac{v}{k}\right) - \frac{cv^2}{1+v^2}, \qquad k, c > 0.$$

5.3.4.* Find the equilibrium solutions for the following cases:

(a) $k = 10, c = 1$
(b) $k = 10, c = 2$
(c) $k = 10, c = 3$

5.3.5. Run computer simulations for the following cases:

(a) $k = 10, c = 1, v(0) = 0.9$
(b) $k = 10, c = 2, v(0) = 0.9$
(c) $k = 10, c = 3, v(0) = 0.9$
(d) $k = 10, c = 2, v(0) = 0.5$
(e) $k = 10, c = 2, v(0) = 0.1$
(f) Describe the behaviors that can be seen in the model, using these simulations as examples.

5.3.6. Suppose $k = 20$. Determine (approximately) all equilibrium solutions for the cases $c = 1.5$, $c = 3.5$, and $c = 5.5$.

5.3.7.(a) Let $k = 4$. Equilibrium values of v can be anything in the interval $(0,4)$, with the actual value(s) depending on c. Rearrange the equilibrium equation so that it takes the form $c = g(v)$. Note that g calculates the value of c necessary for any particular equilibrium value. Plot the function g and observe that it is always decreasing. Explain why this property means that there is always one unique positive equilibrium when $k = 4$.

(b) Repeat the calculations of part (a) with $k = 10$, drawing the appropriate conclusion for this case.

5.3.8. Problem 5.3.7 indicates a way to determine the range of k values for which multiple positive equilibria are possible. Use calculus to determine the largest value of k for which there is always a unique positive equilibrium.

5.3.9.* A reasonable goal in managing the renewable grass resource is to maximize the grass consumption.

(a) Explain why increasing the number of cows does not necessarily increase the grass consumption.
(b) What part of the (dimensionless) model represents the rate of grass consumption by the cows?
(c) Assume that the vegetation biomass reaches a stable equilibrium solution. Use the equation for the equilibria to determine a function of v that represents the total rate of grass consumption at equilibrium, but does not include the parameter c.
(d) Determine the level of vegetation that maximizes the grass consumption rate.
(e) Explain why the result seems reasonable.
(f) Determine (in terms of k) the value of c that is needed to achieve the maximum consumption rate.
(g) Use the definition of c in (5.3.8) to obtain a formula for the optimum number of cows.

5.3.10.(a) Derive the dimensionless Holling type II resource consumption model

$$v' = v\left(1 - \frac{v}{k}\right) - \frac{cv}{1+v}, \qquad k, c > 0,$$

by writing down the appropriate dimensional model and nondimensionalizing as in Problem 5.3.8.
(b) Sketch a graph that shows the equilibrium solutions v^* as a function of c for the case $k = 2$. For what range of c values are there two positive solutions? For what range are there no positive solutions? [Hint: Solving the equation for v^* is possible, unlike in the Holling type III case, but it is still very messy. It is easier to solve the equilibrium equation for c, graph c versus v^*, and then switch the axes.]
(c) For arbitrary k, find the level of vegetation v^* and the corresponding consumer population c that maximizes the renewable yield $y = cv^*/(1+v^*)$. [Hint: See Problem 5.3.9.]

(This problem is continued in Problem 5.4.3.)

5.3.11. Problems 5.3.2, 5.3.9, and 5.3.10 are concerned with harvesting in which the goal is to produce the maximum sustainable yield. In some cases, it is more realistic to use a bioeconomic approach. Suppose there is a cost $C(E)$ associated with the harvesting effort that appears in the Schaefer model of Problem 5.3.2. It is reasonable to assume that the optimal effort will be the one that maximizes the revenue, which we can think of as the sustained yield Ex^* minus the cost. It is possible to obtain some general results for realistic cost functions (in particular, $C' > 0$ and $C'' < 0$); however, it is more useful for our purposes to consider a specific example, $C(E) = A\sqrt{E}$.

(a) The goal is to maximize revenue $R(E) = Ex^*(E) - C(E)$, where $x^*(E)$ is the equilibrium solution of the Schaefer equation. Use this information to eliminate x^* from the formula for R.

(b) Obtain an equation for the critical points of $R(E)$. Ideally, we would like to solve this equation for the optimal effort as a function of the parameter A. Instead, solve the equation for A and plot the result in the EA plane. You may use a computer to assist in the graphing, but you should use hand computations to identify the E intercepts and the coordinates $(\bar{E}, \bar{A})$ of the local maximum.

(c) Use the formula for R' to determine the optimal effort for the case $A > \bar{A}$.

(d) If $A < \bar{A}$, then there are two critical points, and it becomes necessary to identify which of them yields maximum R. The easiest approach in this case is to show that one is a local maximum and the other is a local minimum. Do this by plotting the curve $R'' = 0$ on the graph from part (b). [Note: We are talking about local extrema of the function $R(E)$, not the local extremum of the graph in the EA plane. The graph of $R'' = 0$ passes through the point $(\bar{E}, \bar{A})$. This means that one of the two E values on the graph of $R' = 0$ is a local maximum and the other is not.]

(e) Summarize the results by plotting the optimal E versus A. [The graph will be discontinuous because of the different results of parts (c) and (d).]

5.3.12. If you visit a lot of lakes of different sizes in different parts of the world, you will see that there are two common physical states. Some lakes are relatively clean and fresh-smelling— these are called *oligotrophic*. In contrast, *eutrophic* lakes are overgrown with algae and have a rank smell. Eutrophic lakes have a phosphorus content that is many times higher than that of oligotrophic lakes. Biologists have long known that there does not appear to be a gradation of intermediate states; indeed, rapid eutrophication of formerly oligotrophic lakes is an environmental problem that can be caused by runoff of fertilizer from farms. Why there are no intermediate states, how eutrophication occurs, and the possibility of restoring a eutrophic lake are all questions that can be explored with a simple mathematical model due to a 1999 paper by Carpenter et al. [2]:

$$\frac{dP}{dT} = B - SP + R\frac{P^q}{M^q + P^q}, \quad B, S, R, M > 0, \quad q \geq 2.$$

This model incorporates three mechanisms for change in phosphorus content in a lake. The first term represents the influx of phosphorus from the environment; this can include artificial sources such as farm runoff as well as natural sources such as decomposition of plants.[18] The second term represents the combined processes of sedimentation, outflow, and absorption by plants, all of which remove phosphorus from the water. The last term represents the recycling of phosphorus from sediments. This term plays a significant role in the physical system, because the large values typical of q (from 2 for a cold deep lake to as much as 20 for a warm shallow lake) mean that the recycling rate is roughly R when $P > M$ and very small when $P < M$. Thus, eutrophic lakes, which are high in phosphorus, have large recycling rates that keep the phosphorus concentration high; oligotrophic lakes, in contrast, have very little recycling of sedimentary phosphorus.

(a) Obtain a dimensionless form of the model in terms of the variables $p = P/M$ and $t = ST$ and the parameters $b = B/SM$ and $r = R/SM$ in addition to q. We will generally consider q, r, and a minimum value $b = b_0$ to be fixed by environmental factors. Nondimensionalization reduces the number of parameters by two, making it easier to analyze the model and reducing the number of uncertain parameters that must be estimated for simulations.

---

[18] The standard use of phosphates in laundry and dishwashing detergents was linked to lake eutrophication in the late 1960s and spawned one of the early conflicts between the environmental movement and industry. Phosphates are still used in some detergents, but smaller amounts and better treatment of wastewater have significantly reduced their contribution to eutrophication.

(b) Arrange the equilibrium equation so that it takes the form

$$\frac{p^q}{1+p^q} = f(p),$$

where the function on the right is a linear function with a slope that depends on r and a p intercept that depends on b.

(c) Plot the nonlinear function in part (b) using $q = 8$ and the linear function using $r = 5$ along with the b values 0.25 and 0.75. Use the graph to determine approximate values for the equilibrium points. [These parameter values illustrate typical model behavior and are plausible for a variety of lakes.][19]

(d) Another approach to understanding the equilibria of the model is to fix values of q and r and then plot the equilibria as a function of b. Do this for $q = 8$ and $r = 5$, for the interval $0 \le b \le 1$. [Hint: The desired plot should be in the bp plane, but points on the plot can still be found by calculating b values from selected p values. The plot from part (c) is helpful for determining a suitable range of p to try. Restrict the plot to realistic values of b.]

(e) Discuss the pattern of equilibrium solutions. We will determine the stability of these solutions in Section 5.4; for now, simply note the presence or absence of large, small, and intermediate values of p at equilibrium for the given values of b.

(f) The results of part (e) raise an additional question: For what combinations of r and b values is there only one (large p) equilibrium? These combinations of parameters can only result in a eutrophic lake. To address the question, consider that we are interested in specific cases where the two plots in part (c) are tangent. The requirement for equilibrium and the requirement for tangency comprise two algebraic equations relating p^*, r, and b. Eliminate r from these equations to obtain the formula

$$b = p - \frac{p(1+p^q)}{q}.$$

Given $q = 8$, use calculus to find the maximum value of b that can be obtained from this relationship. A larger value of b means that equilibria with small p are impossible.

(This problem is continued in Problem 5.4.4.)

5.4 Phase Line Analysis

After studying this section, you should be able to:

- Plot a phase line for a continuous dynamic model and use it to determine stability of equilibria.

The *phase line* is a very simple graphical representation that permits easy determination of the long-term behavior of a one-component continuous dynamic model. As with cobweb diagrams for discrete models, the phase line method has two basic requirements:

1. The formula that prescribes the rate of change of the population must not refer directly to the time. This requirement rules out models that account for seasonal changes in the environment.

---

[19] Carpenter, Ludwig, and Brock estimate $q = 7.8$ and $r = 8$ for Lake Mendota, which is adjacent to the campus of the University of Wisconsin at Madison and is probably the most thoroughly studied lake in the world.

2. The differential equation must be of *first order*, meaning that it contains a first derivative, but no higher order derivatives.

The logistic model and renewable resource models of Section 5.3 are autonomous first-order continuous models; hence, the phase line can be used for both.

5.4.1 The Phase Line

Figure 5.4.1 shows a graph of the function $f(N) = N(1-N)$ along with the phase line representation of the logistic equation

$$\frac{dN}{dt} = N(1-N).$$

The phase line representation consists of a number line with some points and arrows that show the direction of change in terms of the current state. A simple procedure is all that is needed for this powerful tool.

Algorithm 5.4.1 *Phase line representation for $dN/dt = f(N)$*

1. Find the equilibrium solutions ($f(N) = 0$) and mark these points on the phase line.
2. The equilibrium solutions partition the interval $[0,\infty)$ into regions.[20] Each of these regions needs an arrow. The arrowhead points to the right for regions in which $f(N) > 0$ and to the left in regions where $f(N) < 0$.

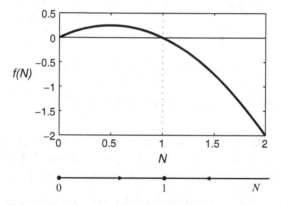

Fig. 5.4.1 The function $f(N) = N(1-N)$ and the phase line for $N' = N(1-N)$

In spite of its simplicity, the phase line is a powerful tool. The arrows in each segment indicate the direction of change of the dynamic variable; for example, the phase line representation of Figure 5.4.1 indicates that N increases when less than one and decreases when greater than one. These arrows point toward $N = 1$ from both sides, marking $N = 1$ as stable. Similarly, $N = 0$ is seen to be unstable. The phase line representation has a generality that the cobweb diagram of

---

[20] This assumes that the dependent variable cannot be negative in the model, as is the case when the dependent variable represents a population. If the model makes sense for negative values of the dependent variable, then use the interval $(-\infty, \infty)$.

Section 5.2 lacks. Initial conditions need not be specified, as the phase line shows the behavior for all possible cases. Parameters can often be left unspecified as well, since the method requires only a rough sketch of the graph of f.

Example 5.4.1. Figure 5.4.2 shows the phase line for the general logistic model

$$\frac{dN}{dt} = rN\left(1 - \frac{N}{K}\right).$$

The model predicts that K is a stable equilibrium value and that all populations that are initially positive approach K in time.   □

5.4.2 The Phase Line for the Holling Type III Renewable Resource Model

We return now to the renewable resource model of Section 5.3, given in dimensionless form as

$$v' = v\left(1 - \frac{v}{k} - \frac{cv}{1+v^2}\right). \tag{5.4.1}$$

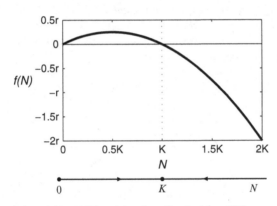

Fig. 5.4.2 The function $f(N) = rN(1 - N/K)$ and the phase line for $N' = f(N)$

We found several cases differing in the number and position of the equilibria. The phase line is just the tool needed to completely determine stability for each case. While we could use (5.4.1) to determine arrow directions directly, it is easier to first rewrite the equation in a different form.

Example 5.4.2. In Section 5.3, we manipulated the equation for equilibria of the renewable resource model into the form

$$\frac{1}{c}\left(1 - \frac{v}{k}\right) = \frac{v}{1+v^2}. \tag{5.4.2}$$

To best determine the phase line arrows for (5.4.1), we first rearrange it to reveal a direct comparison of the quantities in (5.4.2). This requires that we remove a factor of c, yielding the equation

$$v' = cv\left[\frac{1}{c}\left(1 - \frac{v}{k}\right) - \frac{v}{1+v^2}\right]. \tag{5.4.3}$$

The point of this manipulation is to give us a direct connection between the relative positions of the curve and the line in the equilibrium plot of Figure 5.3.3 and the direction of arrows in the phase line. Specifically, (5.4.3) says that v is increasing whenever the value of the linear function $f(v) = (1/c)(1 - v/k)$ exceeds the value of the nonlinear function $v/(1 + v^2)$. The phase line arrows point to the right for just those regions where the line is above the curve and to the left when the curve is above the line.

The equilibrium plots for three sets of parameter values are illustrated in Figure 5.4.3 along with the corresponding phase line plots. A comparison of the three cases shows the range of possible behaviors that can be exhibited by the Holling type III resource utilization model. There is one large equilibrium solution for the small c case and one small equilibrium solution for the large c case; both of these equilibria are stable and represent the only possible long-term behavior. In contrast, there are three equilibria for medium values of c. The largest and smallest are stable and match the corresponding equilibria for small and large values of c. The middle equilibrium is unstable. This scenario has two possible long-term behaviors: the resource level evolves to one or the other of the two stable states, as determined by the initial amount. $\square$

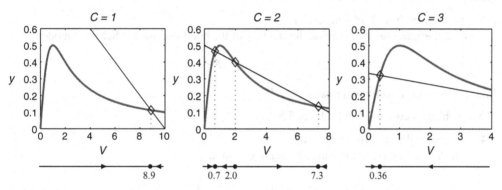

Fig. 5.4.3 Phase line representations for the renewable resource model (Equation (5.4.1)), with $k = 10$ and several values of c

5.4.3 Comparison of Graphical Methods

In this section and in Section 5.2, we have seen the principal graphical methods for analysis of single-component dynamic models: the cobweb diagram for discrete models and the phase line for continuous models. A review of the similarities and differences of these methods reveals some important general differences between discrete and continuous models.

Cobweb diagrams display a simulation graphically, allowing us to determine the stability of fixed points and explore the long-term behaviors for all possible starting values. They do not give any information about the general case, but they do give complete information about the model for any fixed set of parameter values. In contrast, phase line representations focus directly on stability and long-term behavior without reference to a simulation. Hence, they do not require specification of parameter values, and the results they provide are completely general. We will see in the following chapters that the phase line can be extended to two dimensions for the analysis of systems consisting of two dynamic variables, while the cobweb diagram cannot.

These comparisons between corresponding discrete and continuous models and methods are typical. Discrete models seem simpler than continuous ones because they can be defined without calculus and are easily used for simulations; however, the mathematical properties of continuous models are generally simpler than those of the corresponding discrete models. The take-home message is that one should avoid using discrete models for continuous biological processes, as doing so can introduce mathematical complications that do not appear in the corresponding biological systems.

Problems

5.4.1.* Use phase line analysis to determine the stability of the equilibria for the model

$$\frac{dN}{dt} = -N\left(1 - \frac{N}{T}\right)\left(1 - \frac{N}{K}\right), \qquad 0 < T < K.$$

The parameter K has the same biological meaning that it has in the logistic model. Assuming that N is a population, explain the biological meaning of T.

(This problem is continued in Problem 5.5.6.)

5.4.2. (Continued from Problem 5.3.2.)
Use phase line analysis to determine the stability of the equilibria for the Schaefer model

$$x' = x(1 - x) - Ex.$$

You will need to consider two cases.

(This problem is continued in Problem 5.5.7.)

5.4.3. (Continued from Problem 5.3.10.)
Consider the Holling type II resource consumption model

$$v' = v\left(1 - \frac{v}{k} - \frac{c}{1+v}\right).$$

(a) Plot the phase line for the case $k = 2$, $c = 2$.
(b) Given $k = 2$, determine the stability of the equilibria for all $c > 0$. Note: These equilibria were plotted in Problem 5.3.10b. There are three possible phase line sketches, each valid for some range of c.
(c) Suppose there is initially a small number of cattle and a large amount of vegetation. Explain what the model predicts will happen during the following sequence of events (assume that the vegetation does not fully disappear):

1. The number of cattle is increased to the intermediate of the three ranges found in part (b).
2. The number of cattle is further increased to the larger of the three ranges.
3. The number of cattle is decreased back to the intermediate range.

(d) Use the results of part (c) to discuss the inherent difficulty in restoring resource stocks that have been depleted by overconsumption.

(This problem is continued in Problem 5.5.8.)

5.4.4. (Continued from Problem 5.3.12.)

(a) Use phase line analysis to determine the stability of the critical points for the lake eutrophication model

$$p' = b - p + \frac{5p^8}{1+p^8},$$

where p is the phosphorus content of the water, with $b = 0.75$.

(b) Repeat part (a) with $b = 0.25$.
(c) Explain the model predictions for a lake that experiences the following sequence of events, assuming that the lake is initially at equilibrium with $b = 0.25$ and a small amount of phosphorus.

   1. Runoff of fertilizer from farms raises b to 0.75, where it remains for a long period of time.
   2. Environmental restrictions reduce b back down to 0.25, where it remains for a long period of time.

(This problem is continued in Problem 5.5.9.)

5.5 Linearized Stability Analysis

After studying this section, you should be able to:

- Use the derivative to determine the stability of fixed points of discrete dynamic models and equilibria of continuous dynamic models.

In Sections 5.2 and 5.4, we learned how to analyze first-order autonomous population models graphically, with cobweb diagrams for discrete models $N_{t+1} = g(N_t)$ and phase line analysis for continuous models $N'(t) = f(N)$. Now we consider a method that is based on calculations. Asymptotic stability is a *local* property, which means that it depends only on the properties of the model very close to the point of interest. Local properties can be analyzed using calculus.

5.5.1 Stability Analysis for Discrete Models: A Motivating Example

We begin with an example.

Example 5.5.1. Note that $N = 2$ is a fixed point for the model

$$N_{t+1} = \frac{2.5N_t^2}{1 + N_t^2}.$$

To analyze the stability of this fixed point, we need to know the local behavior of the function

$$g(N) = \frac{2.5N^2}{1 + N^2}$$

at the point $N = 2$. We have

$$g'(N) = \frac{(5N)(1 + N^2) - (2.5N^2)(2N)}{(1 + N^2)^2} = \frac{5N}{(1 + N^2)^2}.$$

Specifically, $g'(2) = 0.4$, so the linear approximation[21] to g at $N = 2$ is

$$g(N) \approx g(2) + g'(2)(N - 2) = 2 + 0.4(N - 2).$$

In the vicinity of $N = 2$, the original model is therefore approximated by the linear model

$$N_{t+1} = 2 + 0.4(N_t - 2).$$

We can simplify this model by defining the population perturbation $x = N - 2$. Systematically replacing N by $2 + x$ yields the *linearized* model

$$x_{t+1} = 0.4x_t.$$

This is the exponential model,[22] whose solution we can write as

$$x_t = 0.4^t x_0.$$

From this result, we can determine the long-time behavior of the model:

$$\lim_{t \to \infty} x_t = 0.$$

Since x_t is the approximate difference between N_t and 2, we also have

$$\lim_{t \to \infty} N_t = 2,$$

which marks the fixed point $N = 2$ as asymptotically stable.   □

To see why the method of Example 5.5.1 works, consider the stability question using a cobweb diagram. Figure 5.5.1 shows the basic elements of the cobweb diagram, including both the actual curve $y = g(N)$ and the linear approximation $y = 2 + g'(2)(N - 2)$. As we zoom in on the fixed point, the curve and the line come closer together, so the simulation lines in the cobweb diagrams for the linear and nonlinear models will be indistinguishable, provided we choose an initial value close enough to the fixed point. The advantage of using the linearized model is that we don't need to analyze it with the cobweb diagram, because it can be solved explicitly. In general, this is how analytical stability methods work. Instead of using a graphical method for the original nonlinear model, we replace the model with a linear approximation that can be analyzed with simple calculations.

Linearized stability analysis is much quicker than graphical analysis and works without the need to specify parameter values. The only drawback is that the conclusions are a little weaker. With cobweb analysis, we can determine the entire range of initial conditions that converge on a particular fixed point in the long term. With linearized analysis, we can only conclude that sequences converge to the stable fixed point if they start "close enough" to it. The mathematics of linearization eliminates the possibility of a definite interval of initial conditions. This is not surprising, as we should not expect to be able to answer global questions by using local analysis.

---

[21] See Section 1.4.

[22] See Section 5.1.

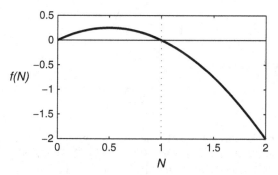

Fig. 5.5.1 $y = g(N)$ (*heavy*), $y = N$ (*medium*), and $y = 2 + 0.4(N - 2)$ (*dashed*) for $N_{t+1} = \frac{2.5N_t^2}{1 + N_t^2}$

5.5.2 Stability Analysis for Discrete Models: The General Case

The power of mathematics is in its ability to obtain general results from motivating examples. In Example 5.5.1, the conclusion of asymptotic stability for $N = 2$ derived from the solution formula $x_t = 0.4^t x_0$, which has a limit of 0 because $|0.4| < 1$. The number 0.4 originally came from the calculation of $g'(2)$, which is the slope of the linear approximation of $g(N)$ at 2. If we change the problem, the details will be different, but the general result will be the same. Given a fixed point $N^\star$ for the equation $N_{t+1} = g(N_t)$, the *linearized model* at that fixed point is

$$x_{t+1} = g'(N^\star)x_t, \qquad x = N - N^\star.$$

Thus, the difference between N and $N^\star$ is approximately

$$x_t = [g'(N^\star)]^t x_0.$$

This quantity vanishes as $t \to \infty$ if $|g'(N^\star)| < 1$, which serves as a sufficient condition for asymptotic stability. Every example will work the same way, with only superficial differences owing to different g functions. Instead of repeating the full calculation every time, we can summarize the result in a theorem.

Theorem 5.5.1 (Stability of Discrete Dynamic Variables). *Let $N^\star$ be a fixed point for the sequence defined by $N_{t+1} = g(N_t)$, where g is a differentiable function. Let $x_t = N_t - N^\star$ be a small perturbation from the fixed point. Then the linearized model near $N = N^\star$ is*

$$x_{t+1} = g'(N^\star)x_t ;$$

furthermore,

- *The fixed point $N^\star$ is asymptotically stable if $|g'(N^\star)| < 1$ and unstable if $|g'(N^\star)| > 1$;*
- *x_t alternates in sign whenever $g'(N^\star) < 0$ and retains a uniform sign whenever $g'(N^\star) > 0$.*

Theorem 5.5.1 allows for a very efficient determination of stability, as long as $g'(N^\star) \neq \pm 1$. With a combination of linearized stability analysis and cobweb diagrams, we can often obtain a complete picture of long-term behavior for a discrete model with an arbitrary initial point.

Example 5.5.2. Consider again the model

$$N_{t+1} = \frac{2.5N_t^2}{1+N_t^2}.$$

The fixed points[23] are the solutions of

$$N = \frac{2.5N^2}{1+N^2} \; ;$$

hence, either $N = 0$ or $1 = 2.5N/(1+N^2)$. The latter simplifies to the quadratic equation $N^2 - 2.5N + 1 = 0$, which yields the solutions $N = 0.5$ and $N = 2$. The fixed points are thus $N^\star = 0$, $N^\star = 0.5$, and $N^\star = 2$. To determine the stability of each fixed point, we first need to determine g'. From

$$g(N) = \frac{2.5N^2}{1+N^2},$$

we have

$$g'(N) = \frac{(5N)(1+N^2) - (2.5N^2)(2N)}{(1+N^2)^2} = \frac{5N}{(1+N^2)^2}.$$

Specifically, $g'(0) = 0$, $g'(0.5) = 1.6$, and $g'(2) = 0.4$. By Theorem 5.5.1, 0 and 2 are asymptotically stable and 0.5 is unstable. Combined with the cobweb diagram of Figure 5.2.2, we can conclude that the population approaches 2 if $N_0 > 0.5$ and approaches 0 if $N_0 < 0.5$. This is an example of a model in which extinction can occur when the population is below some threshold value, a phenomenon sometimes called the *Allee effect*. □

So far, we have used the linearization technique only for a model with fixed parameters. It can also be applied without fixing the values of parameters.

Example 5.5.3. Consider the discrete logistic model

$$N_{t+1} = N_t + RN_t\left(1 - \frac{N_t}{K}\right), \qquad R > -1.$$

We found the fixed points to be $N = 0$ and $N = K$ in Section 5.1. For the stability analysis, we have

$$g(N) = N + RN - \frac{RN^2}{K} \; ;$$

thus,

$$g'(N) = 1 + R - \frac{2RN}{K}.$$

At the fixed points, we have $g'(0) = 1 + R$ and $g'(K) = 1 - R$. The stability requirement for 0 is $-1 < 1 + R < 1$, from which we have that 0 is asymptotically stable if R satisfies both inequalities $R > -2$ and $R < 0$. The first of these is satisfied automatically for biologically meaningful values of R. The stability requirement for K is $-1 < 1 - R < 1$, which reduces to the inequalities $R < 2$ and $R > 0$. Our conclusions are as follows:

- For $-1 < R < 0$, 0 is asymptotically stable and K is unstable. The population will approach 0 from any starting value.

---

[23] See also Example 5.1.3.

- For $0 < R < 2$, 0 is unstable and K is asymptotically stable. The population will approach K from any positive starting value. More specifically, we can determine whether the population will oscillate by checking the sign of the derivative $g'(K) = 1 - R$. It will oscillate as it approaches K if $1 < R < 2$, but not if $0 < R < 1$.
- For $R > 2$, both 0 and K are asymptotically unstable. The long-term behavior is not clear from this analysis. There are a number of possibilities, including periodic behavior and chaotic behavior.[24] □

5.5.3 Stability Analysis for Continuous Models

Stability analysis for continuous models is also based on examination of a linearized model. The idea is exactly the same as for discrete models, but the result is different.

Let $N^\star$ be an equilibrium solution for the problem $dN/dt = f(N)$. As with the discrete case, we can replace f near $N = N^\star$ by a linear approximation $f(N) \approx f(N^\star) + f'(N^\star)(N - N^\star) = f'(N^\star)(N - N^\star)$, where we have used the fact that $f(N^\star) = 0$ because $N^\star$ is an equilibrium solution. Thus, we have

$$\frac{dN}{dt} \approx f'(N^\star)(N - N^\star).$$

With $x(t) = N(t) - N^\star$, we have the linearized model

$$\frac{dx}{dt} = f'(N^\star)x.$$

This is the exponential growth equation, and it has the solution

$$x = x(0)e^{f'(N^\star)t}.$$

Thus, $x \to 0$ whenever $f'(N^\star) < 0$. This gives us a theorem analogous to Theorem 5.5.1.

Theorem 5.5.2 (Stability of Continuous Dynamic Variables). *Let $N^\star$ be an equilibrium solution for the differential equation $N' = f(N)$, where f is a differentiable function. Let $x = N - N^\star$ be a small perturbation from the fixed point. Then the linearized system near $N = N^\star$ is*

$$x' = f'(N^\star)x.$$

Furthermore:

- *The equilibrium solution $N^\star$ is asymptotically stable if $f'(N^\star) < 0$ and unstable if $f'(N^\star) > 0$.*

Example 5.5.4. The model

$$N' = f(N) = rN\left(1 - \frac{N}{K}\right)$$

---

[24] See Problem 5.5.5.

has equilibrium solutions $N^\star = 0$ and $N^\star = K$. We have

$$f' = r\left(1 - \frac{N}{K}\right) + rN\left(-\frac{1}{K}\right) = r\left(1 - \frac{2N}{K}\right).$$

Specifically, $f'(0) = 1 > 0$ and $f'(K) = -1 < 0$. By Theorem 5.5.2, 0 is unstable and K is asymptotically stable.   ☐

5.5.4 Similarities and Differences

The standard forms for discrete and continuous models are $N_{t+1} = g(N_t)$ and $N' = f(N)$. The functions g and f do not represent corresponding quantities; one is the new population, while the other is the rate of change. This means that discrete results stated in terms of g and continuous results stated in terms of f look considerably different. If we want to understand the similarities and differences of discrete and continuous models, we must first rewrite the standard form for discrete models to focus on the rate of change. As discussed earlier,[25] the rate of change in a discrete model is the same as the difference between consecutive values; this corresponds to a general form

$$N_{t+1} - N_t = F(N_t), \tag{5.5.4}$$

for some function F. This is equivalent to the standard form version

$$N_{t+1} = N_t + F(N_t).$$

Thus, the function F that represents the relative rate of change in a discrete model is related to the function g that represents the new population value by the equation

$$g(N) = N + F(N).$$

This identification allows us to recast Theorem 5.5.1 in terms of the rate of change. (The derivation of this result is left as Problem 5.5.11.)

> **Theorem 5.5.3 (Stability of Discrete Dynamic Variables).** *The fixed point $N^\star$ for the discrete model $N_{t+1} - N_t = F(N_t)$, with F continuous, is asymptotically stable if and only if $F'(N^\star) < 0$ and $F'(N^\star) > -2$.*

Compare Theorems 5.5.2 for the continuous case and 5.5.3 for the discrete case. The requirements $f'(N^\star) < 0$ and $F'(N^\star) < 0$ both say that the derivative of the rate of change must be negative at the point $N^\star$. This is enough in the continuous case, but there is an extra requirement in the discrete case. While $N^\star$ is stable in the continuous case no matter how negative $f'(N^\star)$ is, stability is lost in the discrete case if $F'(N^\star)$ is too negative. The additional stability requirement for the discrete case has an important biological consequence: populations whose dynamics are governed by synchronous processes, such as fish that reproduce in one short period out of the year, are more likely to be unstable than those whose dynamics are governed by asynchronous processes.

---

[25] See Section 5.1.

The additional stability requirement also has an important mathematical consequence. Numerical simulations for continuous models are actually done on a discretized version of the model.[26] The extra stability requirement makes it possible for the discretized version used for simulation to be unstable even though the original continuous model is stable. Thus, simulations for continuous models can show instability that is an artifact of the simulation method rather than an actual property of the system. Fortunately, this problem can be dealt with effectively, provided one is aware of it. Numerical analysts use the term *stiff* to describe a differential equation or a system of differential equations that is prone to loss of stability when discretized. If you suspect that the problem you are working on is stiff, you can use special numerical methods designed to prevent the loss of stability. In practice, stiffness is seldom encountered in one-component models, but it is rather common in multicomponent models in biology and chemistry.

Problems

5.5.1.* (Continued from Problems 5.1.2 and 5.2.1.)

(a) Determine the stability of the fixed points for the Beverton–Holt model

$$N_{t+1} = \left(S + \frac{A}{1+N_t}\right) N_t, \quad A > 0; \ 0 \leq S < 1.$$

(b) Discuss the results with reference to Problems 5.1.2 and 5.2.1.

5.5.2. (Continued from Problems 5.1.3 and 5.2.3.)

(a) Determine the stability of the fixed points for the Ricker model

$$N_{t+1} = \left[S + Ae^{-N_t}\right] N_t.$$

(b) Discuss the results with reference to Problems 5.1.3 and 5.2.3.

5.5.3. (Continued from Problems 5.1.4 and 5.2.4.)

(a) Determine the stability of the fixed points for the Hassell model

$$N_{n+1} = \left(\frac{A}{1+N_n}\right)^b N_n, \quad A, b > 0,$$

for the cases $b \leq 1$, $b = 2$, and $b = 3$.
(b) Determine whether the stable solutions found in part (a) oscillate.
(c) Discuss the results with reference to Problems 5.1.4 and 5.2.4.

5.5.4. (Continued from Problems 5.1.5 and 5.2.5.)

(a) Determine the stability of the fixed points for the pest control model

$$N_{n+1} = \frac{4N_n^2}{(1+N_n)(N_n+S)}, \quad S > 0.$$

---

[26] As in Example 5.3.1.

Three cases must be considered; these correspond to the two portions of the parabola and the straight line in the plot of Problem 5.1.5b. Assume that you can use the three fixed points for $S = 2/3$ to indicate stability on their respective curves in the SN plane.[27]

(b) Discuss the results with reference to Problems 5.1.5 and 5.2.5.

5.5.5.* (Continued from Problems 5.1.6 and 5.2.2.)

Two steps of the discrete logistic model $x_{t+1} = x_t + x_t(R - x_t)$ yield the formula

$$x_{t+2} = x_t + x_t(R - x_t)(x_t^2 - Bx_t + B), \quad B = R + 2.$$

While this formula is far more complicated than the original, it makes it possible for us to study solutions in which fixed x values repeat over cycles of two time steps.

(a) Use the function

$$f(x) = x + (Rx - x^2)(x^2 - Bx + B)$$

to obtain a formula for $f'(X)$ at a cycle point X (that is, a point that satisfies $x^2 - Bx + B = 0$). [Hint: If you apply the product rule to the second term of f, one of the resulting terms is immediately 0.]

(b) Find the cycle points X and the derivatives $f'(X)$ for $R = 2.2$. Use the results to determine the stability of the 2-cycle.

(c) Repeat part (b) for $R = 2.5$ and $R = 2.7$.

(d) Show that the critical value for stability of the 2-cycle is $R = \sqrt{6}$.

(e) Discuss the results with reference to Problems 5.1.6 and 5.2.2.

5.5.6. *(Continued from Problem 5.4.1.)
Use Theorem 5.5.2 to determine the stability of the equilibria for the model

$$\frac{dN}{dt} = -N\left(1 - \frac{N}{T}\right)\left(1 - \frac{N}{K}\right), \quad 0 < T < K.$$

Discuss the results with reference to Problem 5.4.1.

5.5.7. (Continued from Problems 5.3.2 and 5.4.2.)
Determine the stability of all equilibria for the Schaefer model

$$x' = x(1 - x) - Ex, \quad E > 0.$$

Discuss the results with reference to Problems 5.3.2 and 5.4.2.

5.5.8. (Continued from Problems 5.3.10 and 5.4.3.)
Determine the stability of all equilibria for the Holling type II resource consumption model

$$v' = v\left(1 - \frac{v}{k} - \frac{c}{1+v}\right), \quad k = 2, \quad c > 0.$$

[Hint: For the positive equilibria, use the equilibrium relation $v' = 0$ to eliminate c from the formula for $f'(v^*)$. Then solve $f'(v^*) = 0$ for v^*. Argue that $f'(v^*) < 0$ when v^* is above this critical value. This will mark one portion of the parabola in the cv plane as stable and the other as unstable.] Discuss the results with reference to Problems 5.3.10 and 5.4.3.

---

[27] Technically, we should have to determine stability with arbitrary S, but the algebra in this case is rather messy.

5.5.9. (Continued from Problems 5.3.12 and 5.4.4.)
Use stability analysis to confirm the results of Problem 5.4.4, parts (a) and (b).

5.5.10. (a) Use Theorem 5.5.2 to show that a positive equilibrium of the renewable resource
model (5.3.7) is stable if

$$\frac{v}{k} + cv\frac{1-v^2}{(1+v^2)^2} > 0.$$

(b) Use the equation for equilibria to eliminate c from the condition of part (a) and rearrange
the result to get the form

$$1 + v^2\left(\frac{2v}{k} - 1\right) > 0.$$

(c) Use the result from part (c) to show that any equilibrium that satisfies $v \geq 0.5k$ must be
asymptotically stable.[28]

5.5.11. Derive the result of Theorem 5.5.3 by defining the appropriate function g for the model
$N_{t+1} - N_t = F(N_t)$ and applying Theorem 5.5.1.

5.5.12. Consider a model for a fish population that runs over a 2-year cycle:

$$Y_{t+1} = g(A_t), \qquad A_{t+1} = f(Y_t),$$

where

$$f(0) = g(0) = 0, \qquad f', g' > 0, \qquad f'', g'' < 0, \qquad f(Y) < Y, \qquad g(A) < g_m.$$

(a) Sketch graphs of f and g.
(b) Derive a model that tracks adults only with a 2-year census interval.
(c) Determine the restriction necessary on the functions f and g for the population to persist
(in other words, for the fixed point $A^* = 0$ to be unstable).

References

1. Britton NF. *Essential Mathematical Biology*. Springer, Berlin and New York (2004)
2. Carpenter SR, D Ludwig, and WA Brock. Management of eutrophication for lakes subject to potentially
   reversible change. *Ecological Applications*, **9**: 751–771 (1999)

---

[28] This is an example of a useful algebra calculation that could not reasonably be done with a computer algebra
system. Of course, one could get a CAS to do the calculation by giving it the sequence of steps in the calculation,
but not by expecting the CAS to do algebra with human ingenuity.

Chapter 6
Discrete Dynamical Systems

In Chapter 5, we considered the dynamics of single quantities changing in either discrete or continuous time. Here we consider the dynamics of systems of several related quantities changing in discrete time. These arise in a variety of settings and can exhibit quite complicated behavior. This chapter deals exclusively with linear systems, which are used to represent dynamics of structured populations divided into classes by age, size, or stage. Discrete linear systems are a major tool in conservation biology modeling, where the primary goal is to determine the effects of parameters on the growth rate of a population. Discrete nonlinear systems appear in Appendix A.1 and require some background from Chapter 7.

We begin in Section 6.1 with a simulation-driven introduction to the dynamics of structured populations. The models we obtain are analogous to exponential growth models for single quantities, except that the possibility of various distributions of population among the classes makes the exponential growth rate difficult to determine. However, we can demonstrate for any example that the model does eventually tend toward exponential growth; for problems with a limited number of classes, we can prescribe an intuitive method for determining both the eventual growth rate and the stable distribution of the population.

It is often true that problems that can be solved by intuition are more easily solved by a formal mathematical procedure based on prior conceptual development. The analysis of discrete linear dynamical systems is an outstanding example of this phenomenon. In Section 6.2, we develop some of the basic mathematical theory of matrix algebra, and then we apply this theory in Section 6.3 to the problem of determining the eventual growth rate and stable population distribution for structured models.

The matrix algebra theory of Sections 6.2 and 6.3 is essential background for the analysis of continuous systems in Chapter 7. Indeed, this material is essential for much of the mathematical analysis performed in biology, and the reader is advised to aim for mastery.

6.1 Discrete Linear Systems

After studying this section, you should be able to:

- Construct a structured discrete linear population model from a narrative description.
- Identify a narrative description from a structured discrete linear population model.
- Describe the general behavior of two-component discrete linear population models in terms of the growth rate and stable population ratios.
- Determine the long-term growth rate and stable population ratios for two-component discrete linear population models.

G. Ledder, *Mathematics for the Life Sciences: Calculus, Modeling, Probability, and Dynamical Systems*,   301
Springer Undergraduate Texts in Mathematics and Technology, DOI 10.1007/978-1-4614-7276-6_6,
© Springer Science+Business Media, LLC 2013

In this section, we extend the basic ideas of the single-component linear model to analogous models of populations with structure. We begin with a brief summary of the earlier model, adapted from material in Chapter 5. The most basic form of this model is

$$N_{t+1} = \lambda N_t , \tag{6.1.1}$$

which can be rearranged into the form

$$\lambda = \frac{N_{t+1}}{N_t} . \tag{6.1.2}$$

This version reveals that λ is the constant factor by which the population is augmented at each time step.

The model (6.1.1) has the explicit solution

$$N_t = \lambda^t N_0 , \tag{6.1.3}$$

which indicates that the population grows without bound if $\lambda > 1$, stays constant if $\lambda = 1$, and shrinks toward 0 if $\lambda < 1$. Thus, the unique fixed point $N = 0$ is asymptotically stable if $\lambda < 1$, neutrally stable if $\lambda = 1$, and unstable if $\lambda > 1$.

It is helpful to keep these basic facts about single-component models in mind, as multicomponent models exhibit behaviors that are similar, albeit more complicated.

6.1.1 Simple Structured Models

Simple models of single (unstructured) populations track only the total population size. For reasons that will become clear through the examples, such models can sometimes fail to capture important aspects of population behavior. Better models may be obtained by adding structure to a population model.

Structured population model: *a population model in which individuals are categorized according to some discrete or continuous property.*

We restrict consideration to models in which individuals are divided into discrete classes.

Example 6.1.1. A population consists of juveniles and adults and changes from year to year through survival and reproduction:

- Ten percent of the juveniles survive to become adults.
- All adults die after 1 year.
- Adults produce an average of 20 juveniles in their single season of life, through reproduction.
- Juveniles also reproduce, with an average of one juvenile offspring each.

For simplicity, we assume the population has only one sex. Our model requires equations that compute J_{t+1} and A_{t+1} in terms of the populations of both at time t. All juveniles at time $t + 1$ result from reproduction, but we must include both those produced by adult parents and those produced by juvenile parents. All adults at time $t + 1$ result from survival of the previous year's juveniles. Thus, we have the model

$$J_{t+1} = J_t + 20A_t , \qquad A_{t+1} = 0.1J_t . \tag{6.1.4}$$

□

As with one-component dynamic models, we can study the model (6.1.4) with simulations, provided we prescribe initial population values at time 0.

Example 6.1.2. Suppose we have an initial population of 200 juveniles and 10 adults that changes in accordance with (6.1.4). The prescribed initial populations are

$$J_0 = 200, \qquad A_0 = 10.$$

The populations at time 1 are then found using $t = 0$:

$$J_1 = J_0 + 20A_0 = 400, \qquad A_1 = 0.1J_0 = 20.$$

Similarly, we find

$$J_2 = J_1 + 20A_1 = 800, \qquad A_2 = 0.1J_1 = 40.$$

In this simulation, the populations of both classes double at each time step. In other words,

$$\frac{J_{t+1}}{J_t} = 2, \qquad \frac{A_{t+1}}{A_t} = 2.$$

Alternatively, we may write an exact solution as

$$J_t = 2^t J_0, \qquad A_t = 2^t A_0.$$

□

In Example 6.1.2, the structured model (6.1.4) displays exactly the same behavior as the unstructured model (6.1.1). However, we should be careful not to draw too strong a conclusion from this. The simulation results may be more complicated if we start with different initial populations.

Example 6.1.3. Consider the model (6.1.4) with an initial population of 100 juveniles and 10 adults. Table 6.1.1 shows the population growth over several time steps. The population does *not* double at each time step, which would require $J_{t+1}/J_t = 2$ and $A_{t+1}/A_t = 2$ for each t. However, the ratios of successive populations do approach 2 gradually as time increases. □

Table 6.1.1 Populations for (6.1.4) with $J_0 = 100$, $A_0 = 10$

Time	0	1	2	3	4	5
J	100	300	500	1,100	2,100	4,300
A	10	10	30	50	110	210
J_{t+1}/J_t	3	1.67	2.2	1.91	2.05	
A_{t+1}/A_t	1	3	1.67	2.2	1.91	
J/A	10	30	16.7	22	19.1	20.5

Check Your Understanding 6.1.1:
Verify the data in Table 6.1.1.

The comparison between Examples 6.1.2 and 6.1.3 is instructive. In both cases, the model predicts a growth rate of 2; however, the simulation produces a growth rate of exactly 2 in the first instance and only approaches 2 gradually in the second. We can gain crucial insight into the behavior of structured models by examining the ratio of J to A. In Example 6.1.2, this ratio is initially 20:1; because both population classes double at each time step, the ratio of 20:1 is maintained. However, in Example 6.1.3, the ratio of J to A is initially 10:1. Over the course of the simulation, the ratio changes, because the growth rates of the two component populations are not exactly the same. However, the ratio appears to approach 20:1 as the growth rate approaches 2.

Examples 6.1.2 and 6.1.3 illustrate the basic behavior of discrete linear structured models. There is a special ratio of initial populations, 20 to 1 in this case, for which the model shows growth at a constant rate $\lambda > 0$.[1] If a simulation starts with a different ratio of initial populations, then the model populations will only gradually settle into a pattern with the characteristic growth rate and proportions.

6.1.2 Finding the Growth Rate and Stable Distribution

The model of Examples 6.1.2 and 6.1.3 seems to have a characteristic growth rate and population ratio that solutions approach regardless of the initial population values. Is there some method that can be used to find these characteristic features without having to resort to simulations? The answer is "yes," but some theory of matrix algebra is needed before the method can be fully developed; this is addressed in the next two sections. Meanwhile, it is instructive to solve the problem for small models, such as that of the example, using basic principles rather than a sophisticated mathematical technique.

Example 6.1.4. To find the characteristic growth rate and population ratios for the model (6.1.4), we begin by assuming that the initial populations are $J_0 = j$ and $A_0 = a$, and that their ratio produces a growth rate of λ that is correct at each time step. We may freely take $a = 1$, which will leave us with two unknowns, λ and j. We know from our simulations that the results are $\lambda = 2$ and $j = 20$, but our current goal is to determine these values without recourse to simulations. The key idea is that we can calculate J_1 and A_1 in two different ways by utilizing all of the information that is given.

- The model equations allow us to compute J_1 and A_1 for any initial values, including $J_0 = j$ and $A_0 = 1$. Thus,

$$J_1 = J_0 + 20A_0 = j + 20, \qquad A_1 = 0.1J_0 = 0.1j.$$

- If we choose j correctly, then both component populations will change by a factor of λ in each time step; hence,

$$J_1 = \lambda J_0 = \lambda j, \qquad A_1 = \lambda A_0 = \lambda.$$

These two calculations must yield the same answers, which gives us two equations, one for each of the component populations J_1 and A_1:

$$j + 20 = \lambda j, \qquad 0.1j = \lambda.$$

---

[1] Of course there could be "growth" at a rate less than 1 for some models.

For reasons that will become clearer later, the best way to proceed is to solve one of the equations for j and substitute into the other to obtain an equation for λ. The second equation gives us $j = 10\lambda$, and then the first becomes

$$10\lambda + 20 = 10\lambda^2 \,,$$

or

$$0 = 10\lambda^2 - 10\lambda - 20 = 10(\lambda - 2)(\lambda + 1) \,.$$

Curiously, we get more than one answer for λ. We know which one is correct from our simulation: $\lambda = 2$ with its corresponding value $j = 10\lambda = 20$.   □

6.1.3 General Properties of Discrete Linear Structured Models

Example 6.1.4 suggests two important mathematical questions.

1. In general, what sort of equation do we get for λ and how many solutions does it have? If there is more than one solution for λ, how do we know which one is correct?

These issues will be explored in more depth in the next two sections; in short, here is what we will discover:

Discrete linear structured population models with n components always have a long-term growth rate λ, which is the largest positive root of a polynomial of degree n.

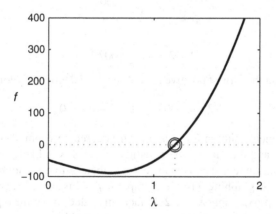

Fig. 6.1.1 The polynomial $f(\lambda)$ for Example 6.1.5

Example 6.1.5. A population consists of larvae, young adults, and older adults. One percent of larvae grow into young adults each year, and 30% of young adults survive to become older adults. Young adults have an average of 104 offspring each year, while older adults have an average of 160 offspring. Find the long-term growth rate and population ratio.

From the description, we obtain the model

$$L_{t+1} = 104Y_t + 160A_t, \tag{6.1.5}$$
$$Y_{t+1} = 0.01L_t, \tag{6.1.6}$$
$$A_{t+1} = 0.3Y_t . \tag{6.1.7}$$

Now suppose the initial populations are l, y, and 1, and the long-term growth rate is λ. Using the model, we obtain

$$L_1 = 104y + 160, \qquad Y_1 = 0.01l, \qquad A_1 = 0.3y .$$

The assumption that l:y:1 is the ratio of populations corresponding to growth at rate λ yields

$$L_1 = \lambda l, \qquad Y_1 = \lambda y, \qquad A_1 = \lambda .$$

Combining these two sets of equations yields three equations for λ, l, and y:

$$104y + 160 = \lambda l,$$
$$0.01l = \lambda y,$$
$$0.3y = \lambda .$$

We have

$$y = \frac{1}{0.3}\lambda$$

from the third equation and

$$l = 100\lambda y = \frac{100}{0.3}\lambda^2$$

from the second. The first equation then becomes

$$\frac{104}{0.3}\lambda + 160 = \frac{100}{0.3}\lambda^3 ,$$

or

$$104\lambda + 48 = 100\lambda^3 .$$

Hence, we need to find the largest positive root of the third degree polynomial equation

$$f(\lambda) = 100\lambda^3 - 104\lambda - 48 = 0 . \tag{6.1.8}$$

There is no convenient solution formula for third degree polynomial equations. Sometimes one can guess a root and then factor the equation, but this is unlikely unless the parameters have been chosen for arithmetical convenience. Nevertheless, there is no difficulty in getting an approximate solution by graphing. From the graph of f in Figure 6.1.1, we see that the largest positive solution is approximately $\lambda = 1.2$. In fact, this value solves the equation exactly.

Once λ is known, we can quickly recover the other variables:

$$l = \frac{100}{0.3}\lambda^2 = 480, \qquad y = \frac{1}{0.3}\lambda = 4 .$$

The stable age distribution has 480 larvae and 4 young adults for every adult. We can also compute the fractions of the adults in the population as $1/485 \approx 0.002$ older adults and $4/485 \approx 0.008$ younger adults. Figure 6.1.2 shows the populations and the relative growth rates for a simulation beginning with 1,000 larvae, 50 young adults, and 5 older adults. Figure 6.1.2a shows the actual populations and Figure 6.1.2b shows the ratios of successive populations. □

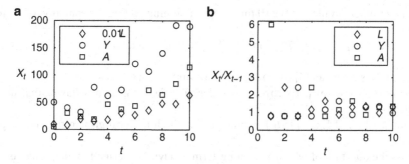

Fig. 6.1.2 (a) Populations and (b) population ratios for Example 6.1.5

Note that plots of populations show the trend of accelerating growth, while plots of population ratios give more detailed information about the manner of growth.

Problems

6.1.1.* Consider a population with the same life history as that of Example 6.1.1, but with only 5% survival of juveniles and average reproduction rates of 11 per year for adults and 0.6 per year for juveniles.

(a) Run a computer simulation for 10 time steps, assuming an initial population of 10 adults.
(b) Plot the ratios J_t/J_{t-1}, and A_t/A_{t-1} together on a common set of axes, starting at $t = 1$.
(c) Plot the ratio J_t/A_t.
(d) Based on the plots of parts (b) and (c), describe the long-term behavior that the model predicts.
(e) Use the method of Example 6.1.5 to determine the long-term growth rate and population ratios.
(f) Discuss the question of whether the long-term behavior described by the model is biologically realistic. Could it be realistic in the short term?

6.1.2. Repeat Problem 6.1.1, but with adult reproduction at 5.4 rather than 11.

6.1.3. Consider a population of juveniles and adults for which 5% of juveniles survive to become adults, 60% of adults survive in any given year, and adults have an average of 11 offspring per year.

(a) Write down the equations for the model based on the given assumptions.
(b) Run a computer simulation for 10 time steps, assuming an initial population of 10 adults.
(c) Plot the ratios J_t/J_{t-1} and A_t/A_{t-1} together on a common set of axes, starting at $t = 1$.
(d) Plot the ratio J_t/A_t.
(e) Based on the plots of parts (c) and (d), describe the long-term behavior that the model predicts.
(f) Use the method of Example 6.1.5 to determine the long-term growth rate and population ratios.
(g) Discuss the question of whether the long-term behavior described by the model is biologically realistic. Could it be realistic in the short term?

6.1.4. Consider a population with the same life history as that of Example 6.1.5, but with 4% survival of larvae, 60% survival of young adults, and average reproduction rates of 11 per year for young adults and 50 per year for older adults.

(a) Run a computer simulation for 10 time steps, assuming an initial population of 10 for each group.
(b) Plot the ratios L_t/L_{t-1}, Y_t/Y_{t-1}, and A_t/A_{t-1} together on a common set of axes, starting at $t = 1$.
(c) Plot the ratios L_t/A_t and Y_t/A_t together on a common set of axes.
(d) Based on the plots of parts (b) and (c), describe the long-term behavior that the model predicts.
(e) Use the method of Example 6.1.5 to determine the long-term growth rate and population ratios.
(f) Discuss the question of whether the long-term behavior described by the model is biologically realistic. Could it be realistic in the short term?

6.1.5.* Consider a population with the same life history as that of Example 6.1.1, but with r for the survival probability of juveniles and average reproduction rates of f per year for adults and s per year for juveniles. Determine the relationship that the parameters must satisfy to achieve a long-term growth rate of exactly 1.

6.1.6. Consider a population with the same life history as that of Problem 6.1.3, but with r for the survival probability of juveniles, b for the survival probability of adults, and an average reproduction rate of f per year for adults.

(a) Determine the relationship that the parameters must satisfy to achieve a long-term growth rate greater than 1.
(b) Suggest a biological interpretation of the requirement in part (a). [Hint: Write the requirement in the form $g(b, r, f) \geq 1$ and determine the biological meanings of the terms in the function g.]

(This problem is continued in Problem 6.2.5.)

6.1.7. Consider a population with the same life history as that of Problem 6.1.4, but with r for the survival probability of juveniles, p for the survival probability of young adults, and reproduction rates of f_1 per year for younger adults and f_2 per year for older adults.

(a) Determine the relationship that the parameters must satisfy to achieve a long-term growth rate greater than 1.
(b) Suggest a biological interpretation of the requirement in part (a). [Hint: Write the requirement in the form $g(p, r, f_1, f_2) \geq 1$ and determine the biological meanings of the terms in the function g.]

6.1.8.* Let R_t be the number of red blood cells in circulation on day t and let M_t be the number produced by the bone marrow on day t. Assume that a fraction f of red blood cells are removed from the circulation by the spleen each day and that all red blood cells produced by the marrow become part of the circulation on the following day. Also assume that the number produced by the marrow on a given day is γ times the number that were removed from circulation on the previous day [5].

(a) Write down the equations for the model based on the given assumptions.
(b) Suppose the number of red blood cells is approximately constant over long periods of time. Determine the value of γ necessary for this result.

(This problem is continued in Problem 6.2.6.)

6.1.9. Peregrine falcons (*Falco peregrinus anatum*) were placed on the endangered species list in 1970 due to a combination of poisoning from the pesticide DDT, habitat loss, and hunting. The population recovered because of bans on DDT and hunting along with fostering of baby falcons to improve survival rates. By 1999, there were more than 2,000 breeding pairs in the United States, which was deemed sufficient to remove peregrine falcons from the endangered list. In 2001, the U.S. Fish and Wildlife Service established a regulation permitting the harvesting of up to 5% of newborn peregrine falcons for use by falconers.[2] Several environmental groups objected to the harvesting permits and unsuccessfully appealed the regulation in 2005. Shortly thereafter, the issue was taken up by a group of undergraduate mathematics students and their mentors at the University of Nebraska-Lincoln, who constructed a mathematical model to assess the viability of the falcon population with and without harvesting [4].[3] The falcon population model considers the population to consist of three classes: fledglings (B), juveniles (J), and adults (A), with only females considered. These populations are governed by the equations

$$B_{t+1} = s_2 f A_t ,$$
$$J_{t+1} = s_0 B_t ,$$
$$A_{t+1} = s_1 J_t + s_2 A_t .$$

Parameter value estimates (without harvesting) from a population in Colorado are $f = 0.830$, $s_0 = 0.544$, $s_1 = 0.670$, and $s_2 = 0.800$.

(a) Describe the life history assumptions of the model. In particular, explain what each of the parameters represents and how populations of the different classes are connected to populations in the previous year. Note that the incorporation of s_2 into the fledgling equation says something about the timing of the census.

(b) Run a computer simulation for 15 time steps, assuming an initial population consisting of 2,000 adults, 500 fledglings, and 300 juveniles. The real starting populations of fledglings and juveniles in 2001 would have been higher than this, so (assuming the parameters are reliable) this simulation should underestimate falcon populations in 2016 in the absence of harvesting. Describe the model's prediction for population growth.

(c) Repeat part (b), but with f reduced by 5% to account for harvesting at the maximum allowable rate.

(d) Plot the ratios B_{t+1}/B_t, J_{t+1}/J_t, and A_{t+1}/A_t for the data from (c) together on a common set of axes.

(e) Plot the ratios B_t/A_t and J_t/A_t for the data from (c) together on a common set of axes.

(f) Based on the plots of parts (d) and (e), describe the long-term behavior that the model predicts.

(g) Use the method of Example 6.1.5 to determine the long-term growth rate and population ratios for the case without harvesting.

(h) Repeat part (g) for the case with harvesting. What percentage change in the population growth rate results from 5% harvesting?

(i) Determine the relationship that the parameters must satisfy to achieve a long-term growth rate greater than 1. It is convenient at this point to describe a single recruitment parameter $r = s_0 s_1 f$. The viability condition is then a simple inequality involving s_2 and r.

(j) Assuming base values for all parameters, what percentage of harvesting would produce a steady population size?

---

[2] In fact, much of the fostering and general population increase was due to the efforts of falconers, who then felt that their investment of time and money should be rewarded by being permitted to harvest baby falcons for their sport. This case serves as an example of the difficult issues involved in trying to create conservation policies that satisfy diverse interests.

[3] The work presented in the paper focuses on the reliability of population projections when parameter values are uncertain, which is beyond the scope of what we can do here. Nevertheless, our methods allow us to obtain population growth rate predictions for a variety of circumstances and to determine critical parameter values for population viability.

(k) Parameter values are only estimates, and the adult survival probability s_2 is particularly important. Assuming 5% harvesting, how large must s_2 be to correspond to a steady population?

(This problem is continued in Problem 6.2.7.)

6.1.10. BUGBOX-population [6] is a virtual biology laboratory that gives students experience in observing biological systems, formulating mathematical models, and collecting data to determine model parameters. The BUGBOX is populated by organisms called "boxbugs," which have a life history that makes them more suited to population dynamics experiments than any real insects.[4] All boxbugs are female. The three life stages—larvae, pupae, and adults—are distinct in appearance. All stages are immobile; the fact that they neither move nor even rotate makes it easy to identify an individual as it moves through its life. There are four species, each with a more complicated life history than the previous one. Each can be modeled by a system of equations that predict the populations of the three stages at time $t + 1$ in terms of the populations at time t. The details of these equations must be determined by observation.

(a) Run experiments to determine the correct model for Species 1. The equations are very simple, with only one parameter. Use the letter f to represent this parameter.
(b) Determine the correct models for Species 2, 3, and 4, using r, b, and s to represent the new parameters in turn.
(c) Complete the model for Species 4 by estimating the values of the parameters.
(d) Run a computer simulation for Species 4 for 20 time steps, assuming an initial population of 10 adults.
(e) Plot the ratios L_{t+1}/L_t, P_{t+1}/P_t, and A_{t+1}/A_t together on a common set of axes, starting at $t = 2$.
(f) Plot the ratios L_t/A_t and P_t/A_t together on a common set of axes.
(g) Based on the plots of parts (e) and (f), describe the long-term behavior that the model predicts.
(h) Use the method of Example 6.1.5 to determine the long-term growth rate and population ratios.
(i) Run several virtual experiments for 20 time steps. Is the model reasonably accurate for small numbers of time steps? How about large numbers?
(j) What biological feature of the "real" boxbug population is not built into the model?

(This problem is continued in Problem 6.2.8.)

6.2 A Matrix Algebra Primer

After studying this section, you should be able to:

- Multiply matrices and vectors.
- Write discrete linear systems in matrix-vector form.
- Compute determinants of 2×2 and 3×3 matrices.
- Determine whether an equation $\mathbf{Ax} = \mathbf{0}$ has nonzero solutions for a given nonzero matrix $\mathbf{A}$.
- Compute nonzero solutions of an equation $\mathbf{Ax} = \mathbf{0}$ when they exist.

---

[4] This software was written when the author was co-teaching an interdisciplinary research course. The research involved population dynamics of aphids and coccinellids (ladybird beetles). Boxbug biology combines the biology of these two real insect types.

A working knowledge of stage-structured models in biology requires some understanding of topics in matrix algebra. We begin with the mathematical definitions needed so that discrete linear structured models can be written in matrix-vector form. We then study what for us is the central problem of matrix algebra: given a nonzero matrix $\mathbf{A}$, find (if possible) a nonzero vector $\mathbf{x}$ such that $\mathbf{Ax} = \mathbf{0}$.

6.2.1 Matrices and Vectors

In section 6.1, we considered models that had two or more related dynamic quantities, each with a separate symbol and each computed by its own formula. These quantities can be advantageously combined into a single multicomponent quantity called a *vector*, in which the individual (or *scalar*) quantities are arranged in a column.[5]

Example 6.2.1. In Example 6.1.3, we considered a model of a population with classes J and A and initial populations $J_0 = 100$ and $A_0 = 10$. We can define a two-dimensional vector $\mathbf{x}$ whose components are J and A. Thus,

$$\mathbf{x} = \begin{pmatrix} J \\ A \end{pmatrix}, \qquad \mathbf{x}_0 = \begin{pmatrix} J_0 \\ A_0 \end{pmatrix} = \begin{pmatrix} 100 \\ 10 \end{pmatrix}. \qquad \square$$

A system of linear equations involving the components of a vector contains several coefficients in addition to the variables. These are grouped together into *matrices*, which are two-dimensional arrays of scalar quantities, with one row for each equation and one column for each variable. Matrices have a dimension $m \times n$, where m is the number of rows and n the number of columns. In a population biology context, we always have equal numbers of equations and variables; $n \times n$ matrices are said to be *square*.

Example 6.2.2. The model of Example 6.1.3 consists of two scalar equations, which we can write as

$$J_{t+1} = 1J_t + 20A_t,$$
$$A_{t+1} = 0.1J_t + 0A_t$$

to emphasize that there is a coefficient for each possible term in the linear functions on the right sides of the equations. The four coefficients can be arranged as a 2×2 matrix:

$$\mathbf{M} = \begin{pmatrix} 1 & 20 \\ 0.1 & 0 \end{pmatrix}. \qquad \square$$

In Example 6.2.2, we constructed the matrix $\mathbf{M}$ by systematically putting the coefficient of variable j from equation i into the matrix in row i and column j. This allows us to use matrices to represent the corresponding systems of equations. Before we can do so, we need a few more definitions. In general, we use the notation m_{ij} to refer to the entry in row i and column j of the

---

[5] The advantage of the formalism of vectors will only become apparent after we have defined arithmetic operations that allow for vector calculations to faithfully reproduce the corresponding scalar calculations.

matrix **M**. We use the term *main diagonal* to refer to the set of matrix entries whose row and column numbers are the same; that is, entries of the form m_{kk}. This allows us to define the $n \times n$ *identity matrix* to be the matrix of appropriate size in which all entries on the main diagonal are 1 and all other entries are 0; for example, the 3×3 identity matrix is

$$\mathbf{I} = \begin{pmatrix} 1 & 0 & 0 \\ 0 & 1 & 0 \\ 0 & 0 & 1 \end{pmatrix}$$

The identity matrix is particularly helpful when we want to subtract a common value λ from each entry on the main diagonal of a matrix **M**.[6]

Example 6.2.3. Let **M** be the matrix in Example 6.2.2. Let λ be an unknown number. Define a matrix **A** by $\mathbf{A} = \mathbf{M} - \lambda \mathbf{I}$. Then

$$\mathbf{A} = \begin{pmatrix} 1 & 20 \\ 0.1 & 0 \end{pmatrix} - \lambda \begin{pmatrix} 1 & 0 \\ 0 & 1 \end{pmatrix} = \begin{pmatrix} 1 & 20 \\ 0.1 & 0 \end{pmatrix} - \begin{pmatrix} \lambda & 0 \\ 0 & \lambda \end{pmatrix} = \begin{pmatrix} 1 - \lambda & 20 \\ 0.1 & -\lambda \end{pmatrix}.$$

Notice that **A** is actually a family of matrices, with λ as the parameter.   □

So far, our matrices and vectors have merely served to combine quantities into a data structure. The power of these structures becomes apparent only after we have endowed them with mathematical operations. The addition operation is defined for pairs of vectors or matrices of the same size. We compute a vector $\mathbf{u} + \mathbf{v}$ by adding the corresponding components of the individual vectors, and matrix addition works the same way. Multiplication is somewhat more complicated. For the moment, we consider only multiplication of a vector of size n by a matrix of size $n \times n$, with the matrix on the left.

Matrix product of an $n \times n$ square matrix A and an n-vector x: *the vector*

$$\mathbf{Ax} = \begin{pmatrix} a_{11} & a_{12} & \cdots & a_{1n} \\ a_{21} & a_{22} & \cdots & a_{2n} \\ \vdots & \vdots & \vdots & \vdots \\ a_{n1} & a_{n2} & \cdots & a_{nn} \end{pmatrix} \begin{pmatrix} x_1 \\ x_2 \\ \vdots \\ x_n \end{pmatrix} = \begin{pmatrix} a_{11}x_1 + a_{12}x_2 + \cdots + a_{1n}x_n \\ a_{21}x_1 + a_{22}x_2 + \cdots + a_{2n}x_n \\ \vdots \\ a_{n1}x_1 + a_{n2}x_2 + \cdots + a_{nn}x_n \end{pmatrix}. \qquad (6.2.1)$$

Example 6.2.4. Let **M** be the matrix of Examples 6.2.2 and 6.2.3 and let $\mathbf{x}_0$ be as in Example 6.2.1. We can multiply $\mathbf{x}_0$ on the left by **M**:

$$\mathbf{Mx}_0 = \begin{pmatrix} 1 & 20 \\ 0.1 & 0 \end{pmatrix} \begin{pmatrix} 100 \\ 10 \end{pmatrix} = \begin{pmatrix} (1)(100) + (20)(10) \\ (0.1)(100) + (0)(10) \end{pmatrix} = \begin{pmatrix} 300 \\ 10 \end{pmatrix}.$$

                                                                                                                        □

Keep in mind that the product $\mathbf{Ax}$ of a square matrix **A** and a vector **x** is defined only when the matrix is on the left and both matrix and vector have the same size n. The product is also a vector of size n.

---

[6] The reason for wanting to do this will become clear in Section 6.3.

6.2.2 Population Models in Matrix Notation

The alert reader may have noticed the correspondence between the preceding examples and Example 6.1.3. Using the matrix $\mathbf{M}$ and vector $\mathbf{x}$, we observe the relationship

$$\mathbf{M}\mathbf{x}_t = \begin{pmatrix} 1 & 20 \\ 0.1 & 0 \end{pmatrix} \begin{pmatrix} J_t \\ A_t \end{pmatrix} = \begin{pmatrix} J_t + 20A_t \\ 0.1J_t \end{pmatrix} = \begin{pmatrix} J_{t+1} \\ A_{t+1} \end{pmatrix} = \mathbf{x}_{t+1} \ .$$

Hence, the structured population model can be written in a very simple form using matrix-vector notation.

> All discrete linear population models can be written in the form $\mathbf{x}_{t+1} = \mathbf{M}\mathbf{x}_t$. The matrix M is called a **population projection matrix**.

Example 6.2.5. In Example 6.1.5, we encountered the model

$$L_{t+1} = 104Y_t + 160A_t,$$
$$Y_{t+1} = 0.01L_t,$$
$$A_{t+1} = 0.3Y_t \ .$$

This model takes the form $\mathbf{x}_{t+1} = \mathbf{M}\mathbf{x}_t$, with

$$\mathbf{x} = \begin{pmatrix} L \\ Y \\ A \end{pmatrix}, \qquad \mathbf{M} = \begin{pmatrix} 0 & 104 & 160 \\ 0.01 & 0 & 0 \\ 0 & 0.3 & 0 \end{pmatrix}.$$

If we start with 1,000 larvae, 50 young adults, and 5 old adults, we can compute the populations at time 1 by matrix multiplication:

$$\begin{pmatrix} L_1 \\ Y_1 \\ A_1 \end{pmatrix} = \begin{pmatrix} 0 & 104 & 160 \\ 0.01 & 0 & 0 \\ 0 & 0.3 & 0 \end{pmatrix} \begin{pmatrix} 1,000 \\ 50 \\ 5 \end{pmatrix} = \begin{pmatrix} 6,000 \\ 10 \\ 15 \end{pmatrix}.$$

$\square$

6.2.3 The Central Problem of Matrix Algebra

Let $\mathbf{A}$ be an $n \times n$ matrix with at least one nonzero entry and let $\mathbf{0}$ be the n-vector whose entries are all 0. For any given matrix $\mathbf{A}$, we are interested[7] in finding nonzero solutions $\mathbf{x}$ for the equation

$$\mathbf{A}\mathbf{x} = \mathbf{0} \ .$$

It is instructive to examine the corresponding scalar problem. If the equation is $ax = 0$ and $a \neq 0$, then of course the only solution is $x = 0$. This is not necessarily the case for the matrix equation.

---

[7] The reason for this interest will become clear in Section 6.3.

Example 6.2.6. Let $\mathbf{A}$ and $\mathbf{x}$ be given by

$$\mathbf{A} = \begin{pmatrix} 0 & 1 \\ 0 & 0 \end{pmatrix}, \qquad \mathbf{x} = \begin{pmatrix} 1 \\ 0 \end{pmatrix}.$$

Then

$$\mathbf{Ax} = \mathbf{0},$$

which demonstrates that there are nonzero solutions for this particular matrix. $\square$

Example 6.2.7. Are there any nonzero solutions to $\mathbf{Ix} = \mathbf{0}$, where $\mathbf{I}$ is the identity matrix?
Suppose

$$\mathbf{x} = \begin{pmatrix} x_1 \\ x_2 \end{pmatrix},$$

with x_1 and x_2 to be determined. Then

$$\mathbf{Ix} = \begin{pmatrix} 1 & 0 \\ 0 & 1 \end{pmatrix} \begin{pmatrix} x_1 \\ x_2 \end{pmatrix} = \begin{pmatrix} x_1 \\ x_2 \end{pmatrix}.$$

The equation $\mathbf{Ix} = \mathbf{0}$ corresponds to the equation pair $x_1 = 0$, $x_2 = 0$. Clearly, these equations have no nonzero solutions. $\square$

Examples 6.2.6 and 6.2.7 demonstrate that nonzero solutions exist for some matrices, but not for others. Is there some way to predict whether a given matrix has nonzero solutions without trying to find them first? Before we can answer this crucial question, we need to develop the computational tool called the *determinant*.

6.2.4 The Determinant

The determinant of a matrix is a number that is calculated using the value of its entries using a complicated formula. Here we consider only the determinants of 2×2 and 3×3 matrices, which are relatively simple.

Determinant of a 2×2 matrix A: *the quantity*

$$\det(\mathbf{A}) = ad - bc,$$

given

$$\mathbf{A} = \begin{pmatrix} a & b \\ c & d \end{pmatrix}.$$

The determinant of a 3×3 matrix A: *the quantity*

$$\det(\mathbf{A}) = (aei + bfg + cdh) - (ceg + bdi + afh),$$

given

$$\mathbf{A} = \begin{pmatrix} a & b & c \\ d & e & f \\ g & h & i \end{pmatrix}.$$

These formulas may seem almost random at first, but there is a pattern. Each of the positive terms is a product of elements aligned diagonally from top left to bottom right,[8] and each of the negative terms is a product of elements aligned diagonally from top right to bottom left.[9]

Example 6.2.8. Let

$$\mathbf{A} = \begin{pmatrix} -1.2 & 104 & 160 \\ 0.01 & -1.2 & 0 \\ 0 & 0.3 & -1.2 \end{pmatrix}.$$

Then

$$\det(\mathbf{A}) = (-1.2)(-1.2)(-1.2) + (104)(0)(0) + (160)(0.01)(0.3) - (160)(-1.2)(0)$$
$$-(104)(0.01)(-1.2) - (-1.2)(0)(0.3) = -1.728 + 0.48 + 1.248 = 0.$$

□

6.2.5 The Equation Ax = 0

The determinant is an efficient way to identify matrices for which $\mathbf{Ax} = \mathbf{0}$ has nonzero solutions. We state the principal result without proof.

> **Theorem 6.2.1 (Singular Matrices).** *The equation $\mathbf{Ax} = \mathbf{0}$ has nonzero solutions for $\mathbf{x}$ if and only if $\det(\mathbf{A}) = 0$. Such a matrix is said to be **singular**.*

In some settings, we merely want to know if a matrix is singular; in other settings we want to be able to calculate the solutions. We illustrate one procedure for solving $\mathbf{Ax} = \mathbf{0}$ when $\det(\mathbf{A}) = 0$ with an example.

Example 6.2.9. Let $\mathbf{A}$ be the matrix from Example 6.2.8. This matrix is singular, so we know the equation $\mathbf{Ax} = \mathbf{0}$ has nonzero solutions. Suppose x_1, x_2, and x_3 are the components of a solution. Then $\mathbf{Ax} = \mathbf{0}$ is equivalent to the scalar equations

$$-1.2x_1 + 104x_2 + 160x_3 = 0, \quad 0.01x_1 - 1.2x_2 = 0, \quad 0.3x_2 - 1.2x_3 = 0.$$

In general, it is difficult to solve systems of equations with lots of variables. However, we can simplify the task by looking specifically for a solution with $x_3 = 1$. This seems like a wild guess, but in this case it works. With $x_3 = 1$, the third equation yields $x_2 = 4$. The second equation then yields $x_1 = 480$. We have apparently found a solution

$$\mathbf{x} = \begin{pmatrix} 480 \\ 4 \\ 1 \end{pmatrix}.$$

Note that we never used the first of the three equations. This equation is available as a check. Substituting the results into the first equation yields

---

[8] Lightly copying the first two columns of the 3×3 determinant to the right of the matrix, giving the appearance of a 3×5 matrix, will help you to see this.

[9] These patterns do NOT hold in higher dimensional determinants. The reader who wants to work with higher dimensional matrices should consult a linear algebra book for a complete definition of the determinant.

$$(-1.2)(480) + (104)(4) + (160)(1) = 0 ,$$

which confirms that our solution is indeed correct.[10]   □

Was the choice of $x_3 = 1$ special? No, it was convenient, but not necessary. We could have chosen any nonzero value for any of the three variables. We would have obtained a different answer, but all the solutions of $\mathbf{Ax} = \mathbf{0}$ are simple multiples of each other. For example, had we started with $x_2 = 1$, we would have found $x_3 = 0.25$ and $x_1 = 120$. The ratio $x_1 : x_2 : x_3$ is $480 : 4 : 1$ in all cases.

In general, the procedure of Example 6.2.9 could have failed only if the correct solutions required $x_3 = 0$. Had that been the case, we would simply have tried again with $x_3 = 0$ rather than $x_3 = 1$.

A Warning

By setting one of the components of the solution vector in Example 6.2.9 to an arbitrary value, we obtained a solution that was otherwise unique. This does not happen with all matrices, but it is guaranteed for matrices that arise in population models. A full development of matrix algebra is beyond the scope of our treatment.

Problems

In Problems 6.2.1–6.2.4, compute the determinant of the indicated matrix.

6.2.1.* $\begin{pmatrix} 1 & 2 & 3 \\ 0 & 1 & 2 \\ 3 & 0 & 1 \end{pmatrix}$

6.2.2. $\begin{pmatrix} 2 & 3 & -1 \\ 0 & 5 & 3 \\ -4 & -6 & 2 \end{pmatrix}$

6.2.3. $\begin{pmatrix} a & b & 0 \\ 0 & a & b \\ a & 0 & b \end{pmatrix}$

6.2.4. $\begin{pmatrix} a & 0 & 2a \\ 0 & b & 3b \\ 3c & c & 2c \end{pmatrix}$

6.2.5.* (Continued from Problem 6.1.6.)
Write the model of Problem 6.1.6 in matrix-vector form. Then find the determinant of the matrix.
(This problem is continued in Problem 6.3.5.)

6.2.6. (Continued from Problem 6.1.8.)
Write the model of Problem 6.1.8 in matrix-vector form. Then find the determinant of the matrix.
(This problem is continued in Problem 6.3.6.)

---

[10] Note that what we have done is essentially equivalent to the part of Example 6.1.5 that followed the calculation of λ. The whole procedure will be made systematic in Section 6.3.

6.2.7. (Continued from Problem 6.1.9.)
Write the model of Problem 6.1.9 in matrix-vector form. Then find the determinant of the matrix.

(This problem is continued in Problem 6.3.7.)

6.2.8. (Continued from Problem 6.1.10.)
Write the species 4 boxbug model (see Problem 6.1.10) in matrix-vector form.

(This problem is continued in Problem 6.3.8.)

6.2.9.* Let $\mathbf{A} = \begin{pmatrix} -\lambda & 1 \\ 3 & 2-\lambda \end{pmatrix}$.

(a) Find all values of λ for which the equation $\mathbf{A}\mathbf{x} = \mathbf{0}$ has nonzero solutions.
(b) Find one nonzero solution for each λ in part (a).

6.2.10. Let $\mathbf{A} = \begin{pmatrix} 3-\lambda & 2 \\ 1 & 2-\lambda \end{pmatrix}$.

(a) Find all values of λ for which the equation $\mathbf{A}\mathbf{x} = \mathbf{0}$ has nonzero solutions.
(b) Find one nonzero solution for each λ in part (a).

6.2.11. Let $\mathbf{A} = \begin{pmatrix} 2 & 0 & c \\ 0 & 1 & 2 \\ 2 & 0 & 1 \end{pmatrix}$.

(a) Find all values of c for which the equation $\mathbf{A}\mathbf{x} = \mathbf{0}$ has nonzero solutions.
(b) Find one nonzero solution for each c in part (a).

6.2.12. Let $\mathbf{A} = \begin{pmatrix} 0 & 1 & 2 \\ c & 1 & 0 \\ 1 & 3 & 2 \end{pmatrix}$.

(a) Find all values of c for which the equation $\mathbf{A}\mathbf{x} = \mathbf{0}$ has nonzero solutions.
(b) Find one nonzero solution for each c in part (a).

6.3 Long-Term Behavior of Linear Models

After studying this section, you should be able to:

- Explain the biological significance of eigenvalues and eigenvectors.
- Compute (real-valued) eigenvalues and their associated eigenvectors.
- Determine the dominant eigenvalue of a matrix.
- Describe the long-term behavior of discrete linear models.

Population models were introduced in Section 6.1 with scalar notation and ad hoc methods. Both of our examples exhibited the same behavior: gradual approach to exponential growth and fixed ratios of component populations. Section 6.2 developed the basic theory of matrix algebra, with two important results. First, discrete linear population models can be written as $\mathbf{x}_{t+1} = \mathbf{M}\mathbf{x}_t$, where $\mathbf{x}$ is a vector of n component populations and $\mathbf{M}$ is an $n \times n$ matrix. Second, the equation $\mathbf{A}\mathbf{x} = \mathbf{0}$, where $\mathbf{A}$ is a nonzero $n \times n$ matrix, has nonzero solutions if and only if $\det(\mathbf{A}) = 0$. In this section, we combine these two results to develop an efficient mathematical procedure for determining the long-term behavior of discrete linear systems.

6.3.1 Eigenvalues and Eigenvectors

As we saw in Section 6.1, discrete linear population models exhibit growth at a uniform constant rate when the initial conditions are just right. We obtained a set of equations for the growth rate and population ratios by analyzing what happens in the first time step when the initial conditions have the right proportions. This same method can be applied to models written in matrix notation. Suppose we have a discrete linear model $\mathbf{x}_{t+1} = \mathbf{M}\mathbf{x}_t$ and an initial condition $\mathbf{x}_0 = \mathbf{v}$ for which growth occurs at a constant rate λ. Then we can calculate $\mathbf{x}_1$ in two ways. From the mathematical model, we get

$$\mathbf{x}_1 = \mathbf{M}\mathbf{x}_0 = \mathbf{M}\mathbf{v} \,.$$

The exponential growth property, given the assumptions that the initial proportions are just right, gives us

$$\mathbf{x}_1 = \lambda \mathbf{x}_0 = \lambda \mathbf{v} \,.$$

Combining these equations yields a matrix algebra equation,

$$\mathbf{M}\mathbf{v} = \lambda \mathbf{v} \,, \tag{6.3.1}$$

in which both the scalar λ and the vector $\mathbf{v}$ are unknown. This is the *eigenvalue*[11] *problem* of matrix algebra:

Eigenvalue problem for a nonzero $n \times n$ matrix M: *the problem of finding solutions to* $\mathbf{M}\mathbf{v} = \lambda \mathbf{v}$ *with* $\mathbf{v} \neq \mathbf{0}$.

Eigenvalue (of a nonzero $n \times n$ matrix M): *a value of λ for which* $\mathbf{M}\mathbf{v} = \lambda \mathbf{v}$ *has nonzero solutions.*

Eigenvector (of matrix M and corresponding to a particular eigenvalue λ): *a vector* $\mathbf{v}$ *that solves the problem* $\mathbf{M}\mathbf{v} = \lambda \mathbf{v}$.

To solve (6.3.1), we must recast it in a more convenient form. Using the identity matrix $\mathbf{I}$,[12] we have $\mathbf{v} = \mathbf{I}\mathbf{v}$, and this allows us to rewrite $\mathbf{M}\mathbf{v} = \lambda \mathbf{v}$ as

$$\mathbf{M}\mathbf{v} = \lambda \mathbf{v} = \lambda \mathbf{I}\mathbf{v} \,,$$

or

$$\mathbf{M}\mathbf{v} - \lambda \mathbf{I}\mathbf{v} = \mathbf{0} \,.$$

This change allows us to use the distributive property of matrix multiplication (factoring out the $\mathbf{v}$) to get the equivalent equation

$$(\mathbf{M} - \lambda \mathbf{I})\mathbf{v} = \mathbf{0} \,. \tag{6.3.2}$$

---

[11] The pronunciation is "eye-gen-value," with a hard g as in "get."

[12] See Section 6.2.

The key to solving (6.3.2) is to see that it is of the form $\mathbf{Ax} = \mathbf{0}$ By Theorem 6.2.1, the equation has a solution if and only if

$$\det(\mathbf{M} - \lambda\mathbf{I}) = 0. \tag{6.3.3}$$

This equation works out to finding the roots of a polynomial of degree n, called the *characteristic polynomial* of the matrix $\mathbf{M}$.

Example 6.3.1. Let $\mathbf{M}$ be the matrix from Example 6.2.2. We have

$$\mathbf{M} - \lambda\mathbf{I} = \begin{pmatrix} 1 & 20 \\ 0.1 & 0 \end{pmatrix} - \lambda \begin{pmatrix} 1 & 0 \\ 0 & 1 \end{pmatrix} = \begin{pmatrix} 1-\lambda & 20 \\ 0.1 & -\lambda \end{pmatrix}.$$

Thus,

$$0 = \det \begin{pmatrix} 1-\lambda & 20 \\ 0.1 & -\lambda \end{pmatrix} = (1-\lambda)(-\lambda) - (20)(0.1) = \lambda^2 - \lambda - 2 = (\lambda - 2)(\lambda + 1).$$

The matrix $\mathbf{M}$ has eigenvalues 2 and -1.   □

Once the eigenvalues are known, we can determine eigenvectors from (6.3.2). This is best done by rewriting the equation in scalar form and applying the procedure of Example 6.2.9.

Example 6.3.2. With $\mathbf{M}$ from Example 6.3.1 and $\lambda = 2$, we have

$$\mathbf{M} - \lambda\mathbf{I} = \begin{pmatrix} -1 & 20 \\ 0.1 & -2 \end{pmatrix}.$$

The system $(\mathbf{M} - \lambda\mathbf{I})\mathbf{v} = \mathbf{0}$ is equivalent to the scalar equations

$$-J + 20A = 0, \qquad 0.1J - 2A = 0.$$

These equations are redundant, so any solution of one is a solution of the other. Arbitrarily taking $A = 1$, the first equation yields $J = 20$, and this solution also satisfies the second equation. Thus, any vector of the form

$$\mathbf{v} = c_1 \begin{pmatrix} 20 \\ 1 \end{pmatrix}$$

is an eigenvector corresponding to $\lambda = 2$.   □

Check Your Understanding 6.3.1:
Find a formula for all of the eigenvectors of the matrix $\mathbf{M}$ from Example 6.3.2 corresponding to the eigenvalue $\lambda = -1$.

Example 6.3.3. Let

$$\mathbf{M} = \begin{pmatrix} 0 & 104 & 160 \\ 0.01 & 0 & 0 \\ 0 & 0.3 & 0 \end{pmatrix},$$

as in Example 6.2.5. We have

$$0 = \det \begin{pmatrix} -\lambda & 104 & 160 \\ 0.01 & -\lambda & 0 \\ 0 & 0.3 & -\lambda \end{pmatrix} = [-\lambda^3 + (160)(0.01)(0.3)] - [-(104)(0.1)(\lambda)]$$

$$= -\lambda^3 + 10.4\lambda + 0.48.$$

We derived this equation using scalar methods in Example 6.1.5, where we found the root $\lambda = 1.2$. To find the corresponding eigenvectors, we note that the equation $(\mathbf{M} - 1.2\mathbf{I})\mathbf{v} = \mathbf{0}$ is equivalent to the scalar equations

$$-1.2J + 104Y + 160A = 0\,, \qquad 0.01J - 1.2Y = 0\,, \qquad 0.3Y - 1.2A = 0\,.$$

Taking $A = 1$, the third and second equations yield $Y = 4$ and $J = 480$, respectively. Substituting these values into the first equation confirms that the answers are correct.

$\square$

Algorithm 6.3.1 summarizes the procedure for finding eigenvalues and eigenvectors.

Algorithm 6.3.1 *Finding eigenvalues and eigenvectors.*

Eigenvalues and eigenvectors of a matrix $\mathbf{M}$ *are the scalars* λ *and vectors* $\mathbf{v} \neq \mathbf{0}$ *that solve the equation*

$$\mathbf{M}\mathbf{v} = \lambda\mathbf{v}\,.$$

To find eigenvalues and eigenvectors,

1. *Find the characteristic polynomial defined by*

$$p(\lambda) = \det(\mathbf{M} - \lambda\mathbf{I})\,.$$

2. *The eigenvalues are the roots of the characteristic polynomial.*
3. *Substitute an eigenvalue* λ *into the equation*

$$(\mathbf{M} - \lambda\mathbf{I})\mathbf{v} = \mathbf{0}\,,$$

   then rewrite this equation as a set of scalar equations.
4. *Set one of the scalar unknowns to 1 and solve for the remaining unknowns.*[13] *This leaves one unused scalar equation to check the solutions.*
5. *If the last scalar equation does not check, there is an error; most likely the value of* λ *that you used is not actually an eigenvalue.*

6.3.2 Solutions of $x_{t+1} = Mx_t$

The eigenvalues and eigenvectors of a matrix $\mathbf{M}$ can be used to determine the set of all solutions to the matrix model $\mathbf{x}_{t+1} = \mathbf{M}\mathbf{x}_t$, according to the following theorem.

---

[13] This won't work if you choose a scalar unknown whose value needs to be 0, but you can try again with that scalar unknown set to 0 instead. This never happens in stage-structured population models, because the stable structure of a viable population must have positive numbers for each stage. The corresponding mathematical property is guaranteed by the Perron-Frobenius theorem.

> **Theorem 6.3.1 (Solutions of Discrete Linear Systems).** *If* $\lambda_1, \ldots, \lambda_n$ *are distinct eigenvalues of an* $n \times n$ *matrix* $\mathbf{M}$ *with associated eigenvectors* $\mathbf{x}^{(1)}, \ldots, \mathbf{x}^{(n)}$, *then all solutions of the equation*
>
> $$\mathbf{x_{t+1}} = \mathbf{M}\mathbf{x_t}$$
>
> *have the form*
>
> $$\mathbf{x_t} = c_1 \lambda_1^t \mathbf{x}^{(1)} + \cdots + c_n \lambda_n^t \mathbf{x}^{(n)},$$
>
> *where* $c_1, \ldots, c_n$ *are constants.*

Example 6.3.4. Continuing with Example 6.3.2, we have the solution

$$\mathbf{x_t} = c_1 2^t \begin{pmatrix} 20 \\ 1 \end{pmatrix} + c_2(-1)^t \begin{pmatrix} 10 \\ -1 \end{pmatrix}.$$

The values of c_1 and c_2 are determined by the initial data, which is $J(0) = 100$ and $A(0) = 10$. Thus,

$$\begin{pmatrix} 100 \\ 10 \end{pmatrix} = \mathbf{x_0} = c_1 \begin{pmatrix} 20 \\ 1 \end{pmatrix} + c_2 \begin{pmatrix} 10 \\ -1 \end{pmatrix}.$$

This vector equation corresponds to the scalar equations

$$20c_1 + 10c_2 = 100, \qquad c_1 - c_2 = 10.$$

We can solve these equations by adding 10 times the second to the first, with the result $30c_1 = 200$; hence, $c_1 = 20/3$ and then $c_2 = -10/3$. Thus,

$$\mathbf{x_t} = \frac{20}{3} 2^t \begin{pmatrix} 20 \\ 1 \end{pmatrix} - \frac{10}{3}(-1)^t \begin{pmatrix} 10 \\ -1 \end{pmatrix}.$$

$\square$

6.3.3 Long-Term Behavior

In ecological modeling, we are usually concerned with the general characteristics of models rather than the detailed results of specific simulations. The important thing to note about the solution in Example 6.3.4 is *not* the solution formula itself; rather, it is that 2^t dominates $(-1)^t$ as t increases. Regardless of the values of c_1 and c_2 (as long as $c_1 \neq 0$), the solution of the model ultimately approaches

$$\mathbf{x_t} = c_1 2^t \begin{pmatrix} 20 \\ 1 \end{pmatrix}.$$

For this reason, the eigenvalue that matters to the long-term behavior is the one with the largest magnitude. The *value* of c_1, as long as it isn't 0, is not very important because the models are too crude to be used for quantitative prediction. The goal is to predict the qualitative behavior, and for this it is the dominant eigenvalue and its associated eigenvector that matter. The results are summarized in a theorem.

Theorem 6.3.2 (Long-Term Behavior of Discrete Linear Population Models). *Suppose λ_1 has a larger magnitude than the other eigenvalues of a matrix $\mathbf{M}$ and $\mathbf{x}_1$ is an eigenvector corresponding to λ_1. Then almost all solutions of $\mathbf{x}_{t+1} = \mathbf{M}\mathbf{x}_t$ eventually approach*

$$\mathbf{x}_t = c_1 \lambda_1^t \mathbf{x}^{(1)} .$$

Thus, the eigenvalue of largest magnitude determines the long-term growth rate and the corresponding eigenvector determines the long-term solution ratios.

Example 6.3.5. Let

$$\mathbf{M} = \begin{pmatrix} 0 & 104 & 160 \\ 0.01 & 0 & 0 \\ 0 & 0.3 & 0 \end{pmatrix} .$$

The dominant eigenvalue $\lambda = 1.2$ and the corresponding eigenvector were found in Example 6.3.3. From Theorem 6.3.2, we have the qualitative result: the long-term behavior of the model is growth at a rate of 20 % with the stable age distribution of 480:4:1.   □

Check Your Understanding Answers

1. $\mathbf{v} = c_2 \begin{pmatrix} -10 \\ 1 \end{pmatrix}$. This is not the only possible answer. Any 2-vector for which the first component is -10 times the second component works equally well.

Problems

Find the eigenvalue of largest magnitude and a corresponding eigenvector for each of the matrices in Problems 6.3.1–6.3.4.

6.3.1.*

$$\mathbf{M} = \begin{pmatrix} 2 & 3 \\ 2 & 1 \end{pmatrix}$$

6.3.2.

$$\mathbf{M} = \begin{pmatrix} 1 & 1 \\ 2 & 0 \end{pmatrix}$$

6.3.3.

$$\mathbf{M} = \begin{pmatrix} 0 & 9 & 12 \\ 1/3 & 0 & 0 \\ 0 & 1/2 & 0 \end{pmatrix}$$

6.3.4.

$$\mathbf{M} = \begin{pmatrix} 1 & 3 & 2 \\ 2 & 0 & 0 \\ 2 & 2 & 0 \end{pmatrix}$$

6.3.5.* (Continued from Problem 6.2.5.)

(a) Derive the polynomial equation for the eigenvalues of the matrix model of Problem 6.2.5.
(b) Determine the inequality that the parameters must satisfy for a growth rate of at least 1.

6.3.6. (Continued from Problem 6.2.6.)

(a) Derive the polynomial equation for the eigenvalues of the matrix model of Problem 6.2.6.
(b) Determine the inequality that the parameters must satisfy for a growth rate of at least 1.

6.3.7. (Continued from Problem 6.2.7.)

(a) Derive the polynomial equation for the eigenvalues of the matrix model of Problem 6.2.7.
(b) Determine the inequality that the parameters must satisfy for a growth rate of at least 1.

6.3.8. (Continued from Problem 6.2.8.)

(a) Derive the polynomial equation for the eigenvalues of the matrix model of Problem 6.2.8.
(b) Determine the inequality that the parameters must satisfy for a growth rate of at least 1.

6.3.9. The teasel plant (*Dipsacus sylvestris*) has a complicated life cycle consisting of dormant seeds, rosettes (which are vegetative but do not flower), and flowering plants. Rosettes have been subclassified as small, medium, and large by field biologists,[14] and the germination probability of dormant seeds decreases with age. Thus, it makes sense to consider a stage-structured model with two classes of dormant seeds, three classes of rosettes, and one class of flowering plants [1]. Using data obtained by Werner and Caswell [9], the matrix representing teasel populations has been reported as

$$A = \begin{pmatrix} 0 & 0 & 0 & 0 & 0 & 322 \\ 0.966 & 0 & 0 & 0 & 0 & 0 \\ 0.013 & 0.010 & 0.125 & 0 & 0 & 3.45 \\ 0.007 & 0 & 0.125 & 0.238 & 0 & 30.2 \\ 0.008 & 0 & 0 & 0.245 & 0.167 & 0.862 \\ 0 & 0 & 0 & 0.023 & 0.750 & 0 \end{pmatrix}.$$

This model serves as a good example of how careful examination of data and results can yield useful biological information.

(a) Describe the various components of the life history of the teasel plant, as indicated by the pattern of nonzero entries in the matrix. [Hint: Draw a life history graph with a row of nodes for D1, R1, R2, R3, and F and a node for D2 below this row and between D1 and R1. Then draw arrows to represent each possible transition. It is (barely) possible to locate these arrows in such a way that none of them cross. You can then describe the life history in terms of the graph.]
(b) Determine the long-term stable growth rate and the associated ratios of each group population to the adult population. [For a large matrix such as this one, you will want to use mathematical software.]
(c) Note that second-year dormant seeds have a very low probability of germinating and are only capable of producing small rosettes. Suppose their germination probability is changed to 0. How does that change the long-term growth rate?
(d) Based on the result of part (c), it is clear that second-year dormant seeds make no measurable contribution to the teasel population. Reformulate the matrix to omit this group. Note that the new matrix will be 5 × 5.

---

[14] These distinctions are largely arbitrary. Since size is a continuous variable, we could just as easily use two or four classes of rosettes. There are *integral projection* models that deal with continuous size structure and discrete time, but these models are far beyond the scope of this book. They are also impractical, unless there is an enormous amount of data on the effect of rosette size on the future of the plant.

(e) One of the other stages of the teasel plant makes almost no difference to the population dynamics. Try to determine which one this is without doing any calculations. Then check your guess by systematically removing one group at a time. As in part (d), removing one group corresponds to removing one row and one column from the matrix.

(f) Having removed two of the six stages in the model without appreciably affecting the results, we now have a 4×4 matrix. Determine the contribution that the dormant seeds make to the population growth by setting the rate of dormant seed production to 0. You should find that dormant seeds make a noticeable difference but not a major contribution.

(g) As in part (f), determine the contribution to the population growth made by medium rosettes returning the next year as medium rosettes by setting the appropriate matrix entry to 0. [First undo the change in part (f) so that you determine only the contribution from maintenance of medium rosettes.]

(h) Determine the importance of new seeds becoming large rosettes rather than medium rosettes. Note that it is not correct to merely set the appropriate matrix entry to 0; it is better to assume that those seeds become medium rosettes instead of large ones.

(i) Repeat part (a), but include only those stages and transitions that significantly impact the teasel population dynamics. Distinguish between primary and secondary life history elements.

6.3.10. The pea aphid *Acyrthosiphon pisum* exhibits rapid population growth because of its unusual life history.[15] Asexual females hatch from eggs in the spring and find an annual host plant on which to found new colonies. For the remainder of the growing season, the colony consists of wingless asexual females that reproduce by cloning and are born live.[16] Sexual morphs are produced at the end of the growing season; they fly to trees, where they mate and lay eggs for the next year's fundatrices. Experiments conducted by teams of undergraduates at the University of Nebraska-Lincoln measured the vital rates of aphids and aphid population growth [7].

(a) A stage-based model for aphid population dynamics needs six stages: first-, second-, third-, and fourth-instar nymphs, young adults, and mature adults. Aphids progress through each of the first five stages in 1–3 days. Under ideal conditions, experiments have obtained daily survival probabilities of 0.974, 0.952, 0.961, 0.930, 0.955, and 0.903 for the six stages, respectively. The probabilities of surviving and advancing to the next stage in 1 day are 0.421, 0.714, 0.538, 0.379, and 0.455 for the immature stages. There is also a small probability of 0.034 for fourth-instar aphids to become mature adults in 1 day. Young adults have an average of 2.106 offspring per day, and mature adults average 3.630 offspring per day. Use this data to construct a matrix for the aphid population model. Note that the survival probabilities include individuals who stay in the same stage for the whole day as well as individuals who move to the next stage.

(b) Use computer software to determine the long-term growth rate predicted by the model and the fraction of each stage in a population growing at that long-term rate. In particular, what fraction of the population are first-instar nymphs and what fraction are adults?

(c) Use the matrix model to run a simulation of aphid population growth, given a starting population of one reproductive adult and running for 15 days. What is the total population after 15 days?

(d) Use the 15-day total and the long-term growth rate to determine how long it will take for the population to reach 1,000.

(e) Plot N_{t+1}/N_t versus time (starting at $t = 1$), where N_t is the total population at time t.

---

[15] This life history is common among aphid species.

[16] Pea aphids are born pregnant, so the amount of time required for a newborn aphid to become a reproductive adult can be as little as 8 days.

(f) Offer a biological explanation for the shape of the curve in part (c). In particular, explain why the growth rate oscillates and why the amplitude of the oscillation decreases over time, and connect the behavior of the graph at the end of the 2 weeks with the result of part (b).[17]

6.3.11. Matrix population projection models have been used to study the extinction risk and possible conservation strategies for the endangered Serengeti cheetah (*Acinonyx jubatus*) population [2, 8]. The model used for these studies divides cheetah populations into age classes 0–6, 6–12, 12–18, 18–24, 24–30, 30–36, 36–42, and 42+ (months). Here we consider a slightly simplified version in which all of the reproductive age groups (2 years and up) are combined together. In this simplified model, only the oldest of the groups reproduces, with a fertility of 1.277 female offspring per adult female per 6-month period. The survival probabilities for the five age groups in the model are s_1, 0.771, 0.771, 0.920, and 0.888, respectively. (We leave s_1 unspecified to allow for a variety of scenarios.) Note that surviving individuals in the first four groups are in the next group in the following 6-month period, but survivors in the oldest group remain in that group.

(a) Write down the matrix **M** that represents the model.
(b) Use computer software to determine the largest eigenvalue for the case $s_1 = 0.081$, which is the value estimated in [2]. What does this result suggest about the survival chances of the population?
(c) Write the matrix equation that the eigenvector **v** must satisfy if the largest eigenvalue is 1, representing the case where the population is just barely viable. Rewrite the equation as a set of scalar equations. Assuming that the adult population is 1, the system can be solved to determine the other populations and the value of s_1 necessary for viability. Determine this value of s_1.
(d) Use computer software to plot the largest eigenvalue for $0.08 \le s_1 \le 0.2$.
(e) Discuss the outlook for the Serengeti cheetah in the wild, based on the results of your investigation of this model.

6.3.12. One of the best known examples of the use of matrix population projection models in conservation biology is that of Larry Crowder and colleagues' study of loggerhead sea turtles (*Caretta caretta*) [3]. The model is stage-based with five stages: hatchlings, small juveniles, large juveniles, subadults, and adults. The hatchling stage lasts just 1 year, the next three stages last approximately 7 years each, and adult turtles live approximately 32 additional years. This gives adults a long time for population growth to compensate for the low overall survival rate to adulthood. As of 1994, estimates of vital rates of loggerhead turtles in South Carolina yielded the matrix

$$
\mathbf{A} = \begin{pmatrix}
0 & 0 & 0 & 4.665 & 61.896 \\
0.675 & 0.703 & 0 & 0 & 0 \\
0 & 0.047 & 0.657 & 0 & 0 \\
0 & 0 & 0.019 & 0.682 & 0 \\
0 & 0 & 0 & 0.061 & 0.809
\end{pmatrix}.
$$

(a) Use computer software to determine the long-term growth rate for this matrix.
(b) Determine the overall survival rate for turtles at each stage, including both those that move to the next stage and those that don't.
(c) Crowder and colleagues provided evidence that approximately half of the mortality of the large juvenile, subadult, and adult stages (as of 1994) was due to incidental capture and drowning in shrimp trawls [3]. The authors revised the matrix under the assumption that two-thirds of this trawling mortality could be reduced by regulations requiring shrimp trawlers to use turtle-excluder devices. Make these changes to the matrix. [Note that this

---

[17] Experiments corresponding to this scenario consistently show the behavior predicted by the model.

must be done carefully for the large juvenile and subadult stages, since more than one matrix entry is needed to determine the net gain in survivorship.]

(d)* Determine the long-term growth rate for the scenario studied by the authors.

(e) Repeat the process used in parts (c) and (d) to determine the percent mortality reduction (to the nearest 1%) needed for the model to predict long-term viability of the turtle population.

References

1. Caswell H. *Matrix Population Models: Construction, Analysis, and Interpretation*, 2nd ed. Sinauer (2001)
2. Crooks KR, MA Sanjayan, and DF Doak. New insights on cheetah conservation through demographic modeling. *Conservation Biology*, **12**, 889–995 (1998)
3. Crowder LB, DT Crouse, SS Heppell, and TH Martin. Predicting the impact of turtle excluder devices on loggerhead sea turtle populations. *Ecological Applications*, **4**, 437–445 (1994)
4. Deines A, E Peterson, D Boeckner, J Boyle, A Keighley, J Kogut, J Lubben, R Rebarber, R Ryan, B Tenhumberg, S Townley, and AJ Tyre. Robust population management under uncertainty for structured population models. *Ecological Applications*, **17**, 2175–2183 (2007)
5. Edelstein-Keshet L. *Mathematical Models in Biology*. Birkhäuser (1988)
6. Ledder G. BUGBOX-population (2005). http://www.math.unl.edu/~gledder1/BUGBOX/CitedSep2012 http://www.springer.com/978-1-4614-7275-9
7. Ledder G and B Tenhumberg. An interdisciplinary research course in mathematical biology for young undergraduates. In Ledder G, JP Carpenter, TD Comar (eds.) *Undergraduate Mathematics for the Life Sciences: Models, Processes, and Directions*. Mathematics Association of America (2013)
8. Lubben J, B Tenhumberg, A Tyre, and R Rebarber. Management recommendations based on matrix projection models: The importance of considering biological limits. *Biological Conservation*, **141**, 517–523 (2008)
9. Werner PA and H Caswell. Population growth rates and age versus stage-distribution models for teasel (*Dipsacus sylvestris Huds.*). *Ecology*, **58**, 1103–1111 (1977)

Chapter 7
Continuous Dynamical Systems

In Chapter 5, we studied the dynamics of a single variable in continuous time. Now we look at the dynamics of systems of variables in continuous time. The methods available to us are analogous to the methods we used for single variable dynamics. The phase line for one variable scales up to the phase plane for two variables. The analytical method for determining stability with the derivative also scales up to higher dimensions, but with technical complications that rapidly increase as the number of variables increases. When possible, we can greatly simplify the analysis of a model by using an appropriate approximation to reduce the number of dynamic equations.

We begin by introducing two dynamical system models: a pharmacokinetics model in Section 7.1 and an enzyme kinetics model in Section 7.2. Both of these models are very general, and we introduce them at the beginning of the chapter to provide rich examples of manageable linear and nonlinear systems, respectively. Computer simulations give us an idea of the properties of both models.

In Section 7.3, we introduce phase planes and nullcline analysis and use the latter to study the enzyme kinetics model. Nullcline analysis is a powerful tool for understanding systems of two dynamic variables in continuous time.

We develop an analytical (symbolic) method for determining stability of equilibria in dynamical systems in Sections 7.4 and 7.5. In the first of these sections, we restrict our attention to linear systems such as the pharmacokinetics example. In Section 7.5, we extend the method to nonlinear systems such as that for enzyme kinetics.

We bring the chapter to a close in Section 7.6 with a case study of the simplest useful model for HIV immunology. The original model is a nonlinear system of three dynamic variables. However, a preliminary analysis uses nondimensionalization to reduce the model to two dynamic variables with only minimal loss of accuracy. We complete the analysis of the model using the nullcline technique of Section 7.3 and the linearized stability analysis technique of Sections 7.4 and 7.5.

There are several sets of related problems:

Section	7.1	7.2	7.3		7.5	
SIR disease, no births/deaths	7.1.3		7.3.3			
SIR disease, constant population	7.1.4		7.3.4		7.5.1	
SIS disease, constant recruitment	7.1.5		7.3.6		7.5.2	
SIS disease with population change	7.1.6		7.3.7		7.5.3	7.5.4
Plankton		7.2.5	7.3.8		7.5.5	
Self-limiting population		7.2.6	7.3.2			
Chemostat		7.2.7	7.3.9		7.5.6	
Predator-prey type I		7.2.9	7.3.5		7.5.7	
Predator-prey type II			7.3.10	7.3.11	7.5.8	

G. Ledder, *Mathematics for the Life Sciences: Calculus, Modeling, Probability, and Dynamical Systems*, Springer Undergraduate Texts in Mathematics and Technology, DOI 10.1007/978-1-4614-7276-6_7, © Springer Science+Business Media, LLC 2013

7.1 Pharmacokinetics and Compartment Models

After studying this section, you should be able to:

- Sketch compartment diagrams from verbal assumptions.
- Write down systems of differential equations from compartment diagrams.
- Run computer simulations of dynamical systems and interpret the results.
- Nondimensionalize a dynamical system, given a specific set of dimensionless variables.

Pharmacokinetics is the study of the rates of drug interactions in living organisms.[1] While elementary models, such as those in Section 1.1, track a single quantity, it is usually necessary to consider the amounts of the drug stored in distinct components of the organism. Since the various rates at which the drug moves between components or exits the system depend on the amounts of the drug that are present in those different components, pharmacokinetic models appear as systems of differential equations.

7.1.1 Compartment Models

Models for pharmacokinetics usually ignore the chemical reactions of the drug, but focus instead on its movement between different components of the organism. If a chemical reaction occurs in one component, it enters the model by changing the rate at which the drug can move out of that component. As an example, we consider a model for lead poisoning.

When lead is ingested, it is first absorbed by the digestive system into the bloodstream. It is then filtered out of the blood by the kidneys and discarded in the urine. Unfortunately, lead is also absorbed into the body's tissues, to eventually be excreted into the hair or sweat, and it tends to bind with bone. Simple models for lead poisoning can ignore the digestion, which happens very quickly, but they must track the amounts of lead in the blood, bones, and other tissues, and account for the various processes by which it moves among these three areas.

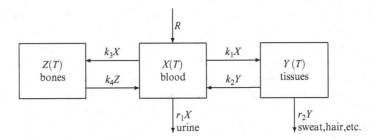

Fig. 7.1.1 Lead transport in a vertebrate body

Example 7.1.1. Figure 7.1.1 is a schematic diagram that depicts the bones, blood, and tissues as three compartments containing quantities $Z(T)$, $X(T)$, and $Y(T)$ of lead, respectively.[2]

---

[1] We use the word *drug* to mean any biochemically reactive substance introduced into an organism from external sources.

[2] There is no common pattern in the choice of symbols used in nondimensionalization. I prefer to make sure that all of the original variables have the same case and give the dimensionless variables the opposite case. Since I am using capital letters for the dimensional concentrations, I use T for dimensional time. The one minor advantage of using capitals for the original model is that most of the work will then be done on the lowercase version.

The arrows indicate rates at which lead enters the body, leaves the body, or moves from one compartment to another. Each of these rates includes a parameter. For example, lead enters the blood compartment from outside the system at the constant rate R and from the Y and Z compartments at the rates $k_2 Y$ and $k_4 Z$, respectively. Lead leaves the blood compartment by three different processes, with the combined rate $(k_1 + k_3 + r_1)X$. Each compartment has a corresponding differential equation, which we construct in the form "overall rate of change is the sum of all rates of increase minus the sum of all rates of decrease." The blood compartment gives us the differential equation

$$\frac{dX}{dT} = R - (k_1 + k_3 + r_1)X + k_2 Y + k_4 Z. \tag{7.1.1}$$

Similarly, the differential equations for the other two compartments are

$$\frac{dY}{dT} = k_1 X - (k_2 + r_2)Y, \tag{7.1.2}$$

$$\frac{dZ}{dT} = k_3 X - k_4 Z. \tag{7.1.3}$$

The differential equations based on the compartment diagram must be supplemented by initial conditions. Two scenarios are of primary interest:

1. Development of lead poisoning, with $R > 0$, $X(0) = Y(0) = Z(0) = 0$.
2. Clearance of lead from the body, with $R = 0$ and $X(0)$, $Y(0)$, and $Z(0)$ determined by the outcome of the first scenario.

Estimates of the parameters are supplied by Rabinowitz et al. [4], who collected the following data from a controlled study of an otherwise-healthy volunteer, with R in micrograms/day and the various rate constants in days^{-1}:

$$R = 49.3, \quad r_1 = 0.0211, \quad r_2 = 0.0162, \tag{7.1.4}$$

$$k_1 = 0.0111, \quad k_2 = 0.0124, \quad k_3 = 0.0039, \quad k_4 = 0.000035. \tag{7.1.5}$$

$\square$

The compartment model of Example 7.1.1 can be used for simulations, which can be conducted through programs built into scientific computing software such as computer algebra systems (Sage, Maple, Mathematica, etc.) or programming environments (Matlab, R, etc.). Such simulations must specify the initial conditions as well as the parameters in the differential equations.

Example 7.1.2.

Q) Suppose a person ingests 49.3 µg of lead per day for 2 years and then stops ingesting any more lead. What does the model of Example 7.1.1 predict for those 2 years and the following 4 years?

A) Figure 7.1.2 displays the results obtained with the ode23t function in Matlab. Other software implementations should yield similar results. Observe what happens each time the lead dose changes. The lead amount in the blood changes very rapidly and approaches a constant value. The amount in the tissues behaves similarly, although the rate of change is not as rapid. The bones accumulate lead at a rate comparable to the tissues during the ingestion phase, but their capacity for storing lead is apparently much larger than that of the blood and tissues. When the lead ingestion ends, the bones yield their store of lead very slowly. $\square$

7.1.2 A Simplified Model

Sometimes the purpose of a model is to create simulations that are as accurate as can be managed. Other times, the goal is to better understand the scientific principles in the scenario being modeled. In this latter case, it can be beneficial to search for a simplified model that still captures the important features of the setting. The simulation result for the three-compartment lead poisoning model suggests a possible simplification. Note that the graphs of

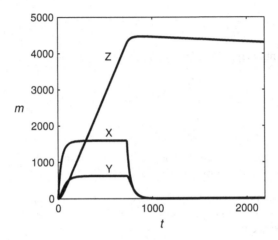

Fig. 7.1.2 The lead poisoning model, with steady doses for 2 years, followed for an additional 4 years

the lead amounts in the blood (X) and the tissues (Y) are very similar. Except for the first few days after a change in the external input rate R, the amount in the tissues seems to be merely a constant multiple of the amount in the blood. If we make the assumption $Y = cX$, where c is a positive parameter, then we can use this simple algebraic equation to replace the differential equation for Y. After algebraic simplification to eliminate Y from the original model, we would have a system that corresponds to the simplified compartment diagram of Figure 7.1.3. The parameters R, k_3, and k_4 are unchanged, but the new parameter r must somehow incorporate both r_1 and the parameters previously associated with the tissue compartment (k_1, k_2, and r_2). The calculation of r is addressed in Problem 7.1.1.

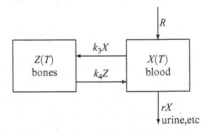

Fig. 7.1.3 A two-compartment model for lead transport in a vertebrate body

Example 7.1.3.

Q) Construct a model corresponding to Figure 7.1.3 and run a simulation using $r = 0.0277$.

A) Following the same procedure as in Example 7.1.1, we have the differential equations

$$\frac{dX}{dT} = R - (k_3 + r)X + k_4 Z , \qquad (7.1.6)$$

$$\frac{dZ}{dT} = k_3 X - k_4 Z . \qquad (7.1.7)$$

Figure 7.1.4 displays the results of the same scenario as Figure 7.1.2. The graphs for X and Z are indistinguishable from those of Figure 7.1.2. ☐

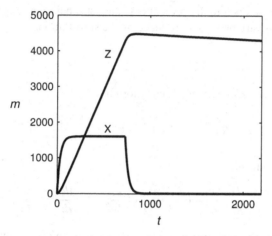

Fig. 7.1.4 The two-compartment lead poisoning model, with steady doses for 2 years, followed for an additional 4 years

Example 7.1.3 illustrates a key principle of mathematical modeling, which we first saw in Chapter 2:

A simpler model can be preferable to a more complex model if the more complex model is only slightly more accurate.

The two-compartment model has three rate parameters (k_3, k_4, and r), compared to six rate parameters in the three-compartment model. If real data were available, we could use the Akaike information criterion[3] to determine which model has greater statistical support. The greater accuracy of the three-compartment model is beneficial only if the parameter values are known to a high degree of accuracy. In this case, as is generally true in biology, they are not well known. The researchers measured the parameters for a single subject, and we have no way to predict the extent to which they might vary from one individual to another. Given the uncertainty in parameters and the tiny difference between the results of the two models, the simpler model is probably the better choice.

---

[3] See Section 2.7.

7.1.3 The Dimensionless Model

If we want to determine the general behavior of the model over a range of possible parameter values, we can obtain further simplification by nondimensionalization.

Example 7.1.4.

Q) Derive the dimensionless two-compartment lead poisoning model using $1/k_3$ as the reference time and R/r as the reference lead mass.[4] This model is specifically intended for the first scenario, in which lead is being ingested; it will need to be modified slightly for the scenario in which lead is initially present but is not still being ingested.

A) The dimensionless variables are introduced by the substitutions

$$T = \frac{1}{k_3}t, \qquad X = \frac{R}{r}x, \qquad Z = \frac{R}{r}z \tag{7.1.8}$$

into (7.1.6) and (7.1.7). The resulting equations are

$$\frac{k_3 R}{r}\frac{dx}{dt} = R - R\left(\frac{k_3 + r}{r}\right)x + \frac{k_4 R}{r}z, \qquad \frac{k_3 R}{r}\frac{dz}{dt} = \frac{k_3 R}{r}x - \frac{k_4 R}{r}z,$$

which we can rewrite as

$$\frac{dx}{dt} = \frac{r}{k_3} - \left(\frac{k_3 + r}{k_3}\right)x + \frac{k_4}{k_3}z, \qquad \frac{dz}{dt} = x - \frac{k_4}{k_3}z.$$

There are apparently two dimensionless parameters, whose values we can estimate from (7.1.4) and (7.1.5):

$$q = \frac{r}{k_3} \approx 7.1, \qquad \varepsilon = \frac{k_4}{k_3} \approx 0.0090. \tag{7.1.9}$$

[5] In terms of these quantities, the model becomes

$$\frac{dx}{dt} = q - (1+q)x + \varepsilon z, \tag{7.1.10}$$

$$\frac{dz}{dt} = x - \varepsilon z. \tag{7.1.11}$$

The initial term q in (7.1.10) is correct for the stage in which lead is accumulating, but should be removed for the stage in which it is no longer being ingested.   □

We will return to the lead poisoning model in Section 7.4, at which point we will see that the presence of the small parameter ε allows us to make predictions about the behavior of the model without even running any simulations.

---

[4] These choices are not obvious. The quantity R/r is the amount of lead that will be in the blood after a long period of ingestion at the rate R. The time $1/k_3$ can be shown to be the expected amount of time that a molecule of lead spends in the blood compartment before being absorbed into the bones.

[5] The Greek letter ε is commonly used in mathematical modeling to represent a dimensionless parameter whose value is typically much less than 1.

Problems

7.1.1. The combined parameter r in the two-component lead poisoning model can be calculated from the data used for the three-component model. The idea is that the approximation that Y reaches equilibrium quickly corresponds to replacing the Y equation in the original model with the algebraic equation

$$k_1 X - (k_2 + r_2)Y = 0 .$$

Derive the two-component model by solving this equation for Y and substituting the result into the original X equation

$$\frac{dX}{dT} = R - (k_1 + k_3 + r_1)X + k_2 Y + k_4 Z$$

to obtain a new X equation. This new equation reduces to the form

$$\frac{dX}{dT} = R - (k_3 + r)X + k_4 Z ,$$

where the quantity inside the parentheses appears as a sum of k_3 and several other terms. Re-define r to be the sum of these other terms and confirm that $r = 0.0277$ is consistent with the given data.

Problems 7.1.2–7.1.6 are concerned with models for SIS (Susceptible-Infective-Susceptible) and SIR (Susceptible-Infective-Recovered) diseases.[6] SIS diseases do not confer immunity and can therefore be modeled with two compartments, labeled S and I, whereas SIR diseases require an additional R compartment. In cases where the total population N is constant, one of the compartments can be omitted from the model. Where the total population can vary, through disease mortality or through long-term population demographics, it is often easier to replace one of the original equations with an equation for the total population.

There are a number of different SIS and SIR models, distinguished by subtle differences. One such difference is in the assumption of how the population size affects the encounter rate of individuals. If the encounter rate per unit time for each person is a constant B, then the rate of encounter with an infective is $B\frac{I}{N}$, since I/N is the probability that a given encounter will be with an infective. If the encounter rate per unit time is proportional to the population size, say CN, then the rate at which one person encounters infectives is $CN\frac{I}{N} = CI$. The choice should be addressed explicitly in models where the population size is not constant, but is irrelevant for a population of constant size. (The former expression is slightly more convenient for the constant population case because it simplifies the nondimensionalization.) SIS and SIR models also vary according to whether they consider a time horizon long enough for population demographics to be necessary and whether the disease can be fatal. One feature that appears in many of these models is a dimensionless parameter R_0 called the *basic reproductive number*. This parameter represents the expected number of secondary infections caused by one infective, given a population comprised of susceptibles.

In all of the SIS and SIR models, we use T for time so that lowercase letters can be used as dimensionless counterparts of the corresponding capital letters. We assume that all parameters are positive; any additional restrictions on parameter values appear in the description of the model.

---

[6] Some problems in Section 2.5 concern SIR models.

7.1.2. Consider an SIS model with no disease-related mortality and population demographics with equal birth and death rates. Such a model is given by the equations

$$\frac{dS}{dT} = bN - pB\frac{I}{N}S + \gamma I - bS,$$

$$\frac{dI}{dT} = pB\frac{I}{N}S - \gamma I - bI,$$

where $p < 1$.

(a) Show that the total population $N = S + I$ must be constant.
(b) Given that $B\frac{I}{N}$ is the rate per person of encounters with infectives, explain what p represents and why the corresponding term has an additional factor of S.
(c) Discuss the assumptions implied by the form γI for the disease recovery rate.
(d) The parameters γ and b are rate constants for the processes of disease recovery and natural death. Which of these should be larger? Should these parameters be approximately equal, or should the larger be considerably larger than the smaller?
(e) Use the algebraic equation $S + I = N$ to eliminate S from the I equation.
(f) Nondimensionalize the new I equation using the dimensionless variables $i = I/N$ and $t = (\gamma + b)T$ and the dimensionless parameter $R_0 = pB/(\gamma + b)$. [The variable i represents the fraction of the total population that is infective at time t. The time scale is chosen to match the mean period of infectiousness, taking the small amount of non-disease-related mortality into account for algebraic convenience.]
(g) Use the algebraic form of the definition of R_0 to justify interpreting it as the expected number of secondary infections caused by one infective given a population in which $S = N$.
(h) Determine the equilibrium solutions for the model, noting any restrictions on R_0 for their existence.
(i) Use the phase line (Section 5.4) and linearized stability analysis (Section 5.5) to determine the stability of the equilibria.

7.1.3. Consider an SIR model without population demographics; that is, the demographic processes of birth and death are ignored. (The model is designed to be used for time periods corresponding to a brief epidemic.) Assuming a contact rate that is independent of population size, the model can be written as

$$\frac{dS}{dT} = -pB\frac{I}{N}S,$$

$$\frac{dI}{dT} = pB\frac{I}{N}S - \gamma I,$$

$$\frac{dR}{dT} = \gamma I,$$

where $p < 1$. The model does not explicitly include disease-based mortality; however, this is no loss of generality. With no possibility of reinfection and no contribution to the birth rate, there is no need to distinguish those who recover from those who die. The model permits disease-based mortality as long as we think of R as representing a "removed" class rather than a "recovered" class.

(a) Justify neglecting the R equation by showing that the total population $N = S + I + R$ must be constant.
(b) Nondimensionalize the model using the dimensionless variables $s = S/N$, $i = I/N$, and $t = \gamma T$ and the dimensionless parameter $R_0 = pB/\gamma$.

(c) Use the algebraic form of the definition of R_0 to justify the claim that it is the expected number of secondary infections caused by one infective given a population in which $S = N$.

(d) Without solving any equations, use the model to determine the time at which the susceptible population is at its maximum.

(e) Use the answer to part (d) to determine the minimum value of $s(0)$ that allows the disease to invade the population, assuming $R_0 > 1$.

(f) Explain what happens if $R_0 < 1$ or if $s(0)$ is less than the critical value found in part (e).

(This problem is continued in Problem 7.3.3.)

7.1.4.* Consider an SIR model with population demographics. For simplicity, we assume that there is no disease-related mortality and that population birth and death rates are equal, so that the population stays constant. Given these assumptions, the S and I equations can be written as

$$\frac{dS}{dT} = bN - pB\frac{I}{N}S - bS \,,$$

$$\frac{dI}{dT} = pB\frac{I}{N}S - \gamma I - bI \,,$$

where $p < 1$.

(a) Use the fact that the population is constant to determine the differential equation for the recovered class.

(b) What assumption does the model make about the status of newborns?

(c) Nondimensionalize the model using the dimensionless variables $s = S/N$, $i = I/N$, and $t = (\gamma + b)T$ and the dimensionless parameters $R_0 = pB/(\gamma + b)$ and $\varepsilon = b/(\gamma + b)$.

(d) Explain why ε is expected to be a small parameter.

(This problem is continued in Problem 7.3.4.)

7.1.5. Consider an SIS model in which the population demographics is determined by natural and disease-related mortality along with recruitment at a constant rate. (This kind of model might be suitable for a population of laboratory mice.) With the additional assumption that the contact rate depends on the population size, we have the model

$$\frac{dS}{dT} = R - pCSI + (1 - \mu)\gamma I - DS \,,$$

$$\frac{dI}{dT} = pCSI - \gamma I - DI \,,$$

where $p, \mu < 1$.

(a) Obtain a differential equation for the total population $N = S + I$.

(b) Explain the biological meaning of the parameter μ.

(c) Nondimensionalize the I and N equations using the dimensionless variables $n = DN/R$, $i = DI/R$, and $t = (\gamma + D)T$ and the dimensionless parameters $R_0 = pCR/D(\gamma + D)$, $m = \mu\gamma/D$, and $\varepsilon = D/(\gamma + D)$.

(d) Explain the significance of the choice R/D of reference quantity for population sizes.

(e) Explain why ε is expected to be a small parameter.

(f) Explain the biological significance of the parameter m. [Hint: Think of the rates represented by the quantities $\mu\gamma$ and D in the definition of m.]

(This problem is continued in Problem 7.3.6.)

7.1.6. Consider an SIS model in which population demographics is determined by a natural carrying capacity along with disease-related mortality. With the additional assumption that the contact rate is independent of the population size (appropriate for a population that expands or contracts its range to maintain constant density), we have the model

$$\frac{dN}{dT} = rN\left(1 - \frac{N}{K}\right) - \mu\gamma I,$$

$$\frac{dI}{dT} = pB\frac{I}{N}S - \gamma I - DI,$$

where $p, \mu < 1$. Note that, when using a natural carrying capacity, it is easier to write the equation for N directly than to compute it from separate equations for S and I. The only disturbance to the standard logistic model is caused by disease-related deaths.

(a) Nondimensionalize the I and N equations using the dimensionless variables $n = N/K$, $i = I/K$, and $t = (\gamma + D)T$ and the dimensionless parameters $R_0 = pB/(\gamma + D)$, $m = \mu\gamma/r$, and $\varepsilon = r/(\gamma + D)$. Note that you must replace S in terms of N and I.
(b) Explain why ε is expected to be a small parameter.
(c) Explain the biological significance of the parameter m. [Hint: Think of the rates represented by the quantities $\mu\gamma$ and r in the definition of m.]

(This problem is continued in Problem 7.3.7.)

7.2 Enzyme Kinetics

After studying this section, you should be able to:

- Explain the basic model of enzyme kinetics.
- Explain the Briggs–Haldane approximation for the initial reaction velocity.

Many important biochemical reactions have the generic form

$$S + E \rightleftharpoons C \rightarrow P + E, \tag{7.2.1}$$

where S is a chemical to be decomposed (often called the *substrate*), E is an enzyme that facilitates the reaction of S without being consumed in the reaction, C is an unstable activated complex formed from the union of S and E, and P represents whatever products result from the reaction. The overall reaction is S $\rightarrow$ P, but the mechanism consists of three separate steps, as indicated in the chemical equation:

1. A forward reaction in which the substrate and enzyme produce the complex.
2. A backward reaction in which the complex decomposes into substrate and enzyme.
3. A completion reaction in which the complex decomposes into products and enzyme.

Reactions of this generic form are called **Michaelis–Menten reactions.**[7]

A mathematical model for the Michaelis–Menten reaction should be based on mechanistic rules that determine the rate of change of each chemical species in terms of the concentrations of all the species. Let S be the concentration of substrate, Y the concentration of enzyme, Z the concentration of complex, and P the concentration of product. We need to determine the rates of the three reactions in terms of these variables.

---

[7] Michaelis is pronounced "mi-KAY-lis."

The backward and completion steps occur spontaneously because of the instability of the complex; thus, it is reasonable to assume that the rates of these reactions are simply proportional to the amount of complex present. Specifically, we assume rates k_2Z for the backward reaction and k_3Z for the completion reaction. The forward reaction requires molecules of substrate and enzyme to encounter each other in the solution. Assuming that the molecules move randomly, the rate is proportional to the concentrations of both[8]; thus, we write the rate of the forward reaction as k_1SY.

Armed with expressions for the reaction rates, we can write down differential equations for the rates of change of each of the quantities in the reaction. Observe that each unit of the forward reaction decreases the concentration of substrate by one unit, while each unit of the backward reaction increases the substrate concentration by one unit. Since the rate of the forward reaction is k_1SY, the rate of increase of substrate via the forward reaction is also k_1SY; similarly, the rate of decrease of substrate via the backward reaction is the same as the reaction rate k_2Z. The overall rate of change of substrate is the sum of the changes caused by the two reactions; therefore, we have

$$\frac{dS}{dT} = -k_1SY + k_2Z . \tag{7.2.2}$$

We can apply similar reasoning to obtain the differential equations for the enzyme, complex, and product:

$$\frac{dY}{dT} = -k_1SY + k_2Z + k_3Z , \tag{7.2.3}$$

$$\frac{dZ}{dT} = k_1SY - k_2Z - k_3Z , \tag{7.2.4}$$

$$\frac{dP}{dT} = k_3Z . \tag{7.2.5}$$

Because the right-hand sides of the equations for Y and Z are identical except for the sign, we can add them together to obtain the simple equation

$$\frac{d(Y+Z)}{dT} = 0 ,$$

which means that the sum of enzyme and complex concentrations is invariant. At any given moment, each enzyme molecule is either free, in which case it counts toward Y, or bound, in which case it counts toward Z. Assuming that the initial amounts of these chemicals are Y_0 and 0, we have

$$Y + Z = Y_0 .$$

We can discard (7.2.3) and substitute $Y = Y_0 - Z$ into (7.2.2) and (7.2.4), obtaining the system

$$\frac{dS}{dT} = -k_1S(Y_0 - Z) + k_2Z , \qquad S(0) = S_0 , \tag{7.2.6}$$

$$\frac{dZ}{dT} = k_1S(Y_0 - Z) - k_2Z - k_3Z , \qquad Z(0) = 0 . \tag{7.2.7}$$

$$\frac{dP}{dT} = V = k_3Z , \qquad P(0) = 0 . \tag{7.2.8}$$

---

[8] This is the *law of mass action*.

Technically, we don't need the extra equation for P either, because the quantity $S + Z + P$ is also invariant; however, we retain this equation because P is the quantity most commonly measured. The rate of change of P is generally called the *reaction velocity*, but it is really the rate of product formation. In keeping with standard notation, we label this quantity V.

7.2.1 Nondimensionalization

To reduce the number of parameters in the model, we can convert it to dimensionless form. We choose reference quantities S_0 for S and P, Y_0 for Z, $k_1 S_0 Y_0$ for V, and $1/(k_1 Y_0)$ for T. Given that the original variables are all uppercase, we use the corresponding lowercase letters for the dimensionless variables. Thus, we apply the substitution formulas

$$S = S_0 s, \quad Z = Y_0 z, \quad P = S_0 p, \quad V = k_1 S_0 Y_0 v, \quad T = \frac{1}{k_1 Y_0} t. \tag{7.2.9}$$

The reference quantities $k_1 S_0 Y_0$ for reaction velocity and $1/k_1 Y_0$ for time are not obvious and should not greatly concern the reader.[9]

With the substitutions of (7.2.9), (7.2.6) becomes

$$k_1 Y_0 S_0 \frac{ds}{dt} = -k_1 S_0 s (Y_0 - Y_0 z) + k_2 Y_0 z.$$

We can remove the common factor of Y_0, yielding

$$k_1 S_0 \frac{ds}{dt} = -k_1 S_0 s (1 - z) + k_2 z.$$

We now have an equation with dimensionless variables, but we still need to arrange the equation to obtain dimensionless parameters. Dividing by $k_1 S_0$ yields

$$\frac{ds}{dt} = -s(1 - z) + \frac{k_2}{k_1 S_0} z.$$

This equation is dimensionless, and for convenience we introduce the dimensionless parameter h defined by $h = k_2/(k_1 S_0)$.

---

[9] The choice for reaction velocity is based on the insight that the forward reaction is the one that drives the overall reaction, and its initial rate is $k_1 S_0 Y_0$. We can approximate the amount of time that corresponds to a unit change of substrate by substituting $\Delta S \approx -S_0$ in the equation

$$\frac{\Delta S}{\Delta T} \approx -k_1 S_0 Y_0$$

and solving for ΔT, with the result $\Delta T \approx 1/k_1 Y_0$. Note that we had other choices. Using the same procedure, we could have found the amounts of time corresponding to a unit change of complex from the forward reaction, a unit change of either substrate or complex from the backward reaction, or a unit change of complex from the completion reaction. Intuitively, the given choice is best, because the substrate is the principal reactant and the forward reaction is the one that uses up substrate.

Treating equations (7.2.7) and (7.2.8) similarly, we eventually obtain the dimensionless model

$$\frac{ds}{dt} = -s(1-z) + hz, \qquad s(0) = 1, \tag{7.2.10}$$

$$\varepsilon \frac{dz}{dt} = s(1-z) - hz - rz, \qquad z(0) = 0, \tag{7.2.11}$$

$$\frac{dp}{dt} = v = rz, \qquad p(0) = 0, \tag{7.2.12}$$

where

$$h = \frac{k_2}{k_1 S_0}, \qquad r = \frac{k_3}{k_1 S_0}, \qquad \varepsilon = \frac{Y_0}{S_0} \ll 1. \tag{7.2.13}$$

The final model has dependent variables s, z, and v, which represent the concentration of substrate, concentration of complex, and rate of product formation, respectively. There are three dimensionless parameters: h, r, and ε. There was some flexibility in defining the dimensionless parameters, but there are some advantages to the form used here. The parameter h represents the strength of the backward reaction relative to the forward reaction. Larger values of h should make the overall reaction slower because molecules of the complex will tend to decompose more quickly than they are formed. Similarly, the parameter r represents the strength of the completion reaction relative to the forward reaction. Larger values of r relative to h mean that the complex is more likely to produce product than to decompose into reactants.

The parameter ε plays a special role in the analysis, because it appears only as a factor on the left side of a differential equation. While h and r represent the relative strengths of the component reactions, ε represents the relative magnitude of the rates of change determined by the set of component reactions. The notation $\varepsilon \ll 1$ means that we expect ε to be very small,[10] which follows from the assumption that the enzyme is usually in short supply relative to the substrate. Insights such as this can simplify the analysis of a model, as we will see later.

7.2.2 Simulation

Our principal question for this model is "How does the reaction velocity $v(t)$ depend on the parameters h, r, and ε?" This question requires analysis because it assumes that these parameters are allowed to vary. Nevertheless, we can obtain some insights by considering simulations with fixed parameter values.

There are no methods that can be used to determine the functions $s(t)$ and $z(t)$ from the differential equations (7.2.10) and (7.2.11). However, numerical methods can be used for satisfactory simulations. A detailed understanding of numerical methods for differential equations is beyond the scope of this presentation. What is important for the reader to understand is that numerical methods involve approximating differential equations with difference equations, thereby turning a continuous model into a discrete model. As we saw in Section 5.5, discrete models sometimes exhibit instabilities that do not occur in the corresponding continuous model. We must recognize that unstable behavior in the simulation for a continuous dynamic system could either be inherent in the system or it could have been introduced by the numerical method used to create the simulation. If the latter, one should seek out numerical methods that are designed to eliminate the danger of introduced instability. Systems such as ours, in which the rates of change of variables are greatly different in magnitude, are said to be *stiff*; these systems are prone to introduced instability. For this reason, we will avoid taking extremely small values of

---

[10] In oral communications, this symbol is generally read as "much less than."

ε in our simulations. If necessary, special methods can be employed for the simulation of stiff problems.

Figure 7.2.1 shows the results of a simulation using the system (7.2.10)–(7.2.12), with parameters $h = 1$, $r = 2$, and $\varepsilon = 0.2$. The curves show a striking feature of Michaelis–Menten

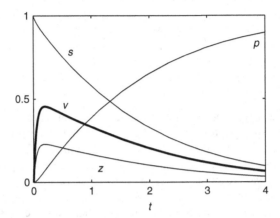

Fig. 7.2.1 The Michaelis–Menten system ((7.2.10)–(7.2.12)), with $h = 1$, $r = 2$, $\varepsilon = 0.2$

reactions: the concentration of the complex (and the reaction velocity) changes rapidly from an initial value of 0 to a maximum value, after which it gradually decreases. Biochemists are particularly interested in the maximum value of the reaction velocity, which we will be able to determine from analysis of the model.

7.2.3 The Briggs–Haldane Approximation

The key to understanding the behavior observed in the graph is the factor of ε in (7.2.11). We can rewrite this equation as

$$\frac{dz}{dt} = \frac{s - (s + h + r)z}{\varepsilon}.$$

At the beginning of the reaction, $s = 1$ and $z = 0$, so $dz/dt = 1/\varepsilon \gg 1$. Thus, dz/dt is initially large because ε is small. The complex increases rapidly, but only for a short interval. There can be only one way for the graph of z to flatten out, and that is for the quantity $s - (s + h + r)z$ to be very small. This occurs at approximately $t = 0.1$ in this example. From this time on, dz/dt is never large, so we must have $s \approx (s + h + r)z$. Solving for z, we have

$$z \approx \frac{s}{s + h + r}. \qquad (7.2.14)$$

If we are willing to ignore the initial transient, we can use the equilibrium equation (7.2.14) in place of the differential equation (7.2.11). Substituting this approximation into (7.2.10) and (7.2.12) yields the simplified equations[11]

$$\frac{ds}{dt} = -\frac{rs}{s + h + r}, \qquad v = \frac{rs}{s + h + r}. \qquad (7.2.15)$$

---

[11] See Problem 7.2.2.

We could at this point have a computer plot the solution of the differential equation for s and use it to obtain a graph of v vs t; however, we already have such a graph from the original model. Instead, the approximate system (7.2.15) is used primarily to estimate the maximum of the reaction velocity curve. This is determined by substituting the maximum value of s[12] into the v equation to get

$$v_0 = \frac{r}{1+h+r},$$

where v_0 is the initial rate of product formation, with "initial" understood to mean "after the brief transient period during which the concentration of the complex reaches equilibrium with the substrate."

Replacing the dimensionless quantities v_0, r, and h in the maximum reaction rate equation with the corresponding dimensional quantities, we obtain[13]

$$V_0 = \frac{V_{max}S_0}{S_0 + K_m}, \tag{7.2.16}$$

where

$$V_{max} = k_3 Y_0, \quad K_m = \frac{k_2 + k_3}{k_1}. \tag{7.2.17}$$

Equation (7.2.16) is the well-known Briggs–Haldane approximation for Michaelis–Menten kinetics.[14] The parameters V_{max} and K_m are usually fit from data for V_0 versus S_0, without attempting to evaluate the k_i or Y_0 independently. However, note that changes in the enzyme concentration lead to a corresponding change in the parameter V_{max}.

Problems

Note: All parameters in these problems are positive.

7.2.1. Nondimensionalize the Michaelis–Menten model using the same definitions of s, z, p, and v as in the text, but using $t = k_1 S_0 T$ for the time. Use the same parameters as in the text. What is the difference in the resulting model? This is another example of equivalent forms of models.[15]

7.2.2. Derive (7.2.15).

7.2.3. Derive (7.2.16).

7.2.4.(a) Use a computer simulation to study the effect of the backward rate constant h on the Michaelis–Menten model. Hold $r = 2$ and $\varepsilon = 0.2$, and plot the simulation results for two values of h larger than 1 and two values of h smaller than 1. Choose values that adequately illustrate the effect of h.
(b) Explain the effect of h in words by reference to the Michaelis–Menten reaction: (1) What does h quantify in the reaction? (2) Given the meaning of h, how should the reaction

---

[12] $s = 1$.

[13] See Problem 7.2.3.

[14] Note that it is the same function, albeit with different notation, as the feeding rate function we derived in Section 2.5.

[15] See Section 2.6.

progress change as h increases? Keep in mind that the graph of z, while interesting, is not important to biochemists. The purpose of the reaction is either to produce product or to break down the substrate.

(c) Use a computer simulation to study the effect of ε on the Michaelis–Menten model. Hold $r = 2$ and $h = 1$ and try $\varepsilon = 0.1, 0.2, 0.4, 0.8$. Plot the simulation results and explain the effect of ε. [Hint: You may need to try different time horizons.]

7.2.5. In the enzyme kinetics problem of this section, we were able to make use of the requirement that each molecule of enzyme had to appear either as unattached enzyme or as part of an activated complex. The same feature appears in a model that connects plankton populations to dissolved nitrogen [1]. The model contains three dynamic quantities: free nitrogen N, phytoplankton (microscopic plants) P, and zooplankton (microscopic animals) Z. The quantities of plankton could be expressed in various units, but we choose to express them in terms of their nitrogen content. In this way, each unit of nitrogen taken up by phytoplankton and eventually consumed by zooplankton counts as one unit of either N, P, or Z.

(a) Suppose the following mechanisms account for nitrogen transfers:

1. Phytoplankton die, returning their nitrogen to the water, at a rate proportional to the population size, with rate constant a.
2. Zooplankton also die and return their nitrogen to the water at a rate proportional to the population size, this time with rate constant b.
3. Free nitrogen is consumed by phytoplankton at a rate proportional to both the nitrogen concentration and the phytoplankton population, with rate constant c.
4. Phytoplankton are consumed by zooplankton at a rate proportional to both populations, with rate constant d.

Write down the appropriate differential equations for the functions $N(T)$, $P(T)$, and $Z(T)$, using T for time.

(b) Show that $N + P + Z$ is constant.
(c) Using A for the constant nitrogen total, eliminate N from the differential equation for P.
(d) Nondimensionalize the P and Z equations using the dimensionless variables $p = P/A$, $z = (c + d)Z/cA$, and $t = cAT$.[16] Choose dimensionless parameters α, β, and δ so that the resulting dimensionless model is

$$p' = p(1 - \alpha - p - z),$$
$$z' = \delta z(p - \beta).$$

(e) Explain the biological significance of the parameters α and β. [Hint: Think of cA as representing the quantity cN.]

(This problem is continued in Problem 7.3.8.)

7.2.6. Many populations are self-limiting in the sense that their waste products decrease the population directly or decrease the capacity of the environment to support them. An example of such a model is [3]

---

[16] The choice of reference quantity for Z is somewhat unusual. Given that the total nitrogen amount is A, it is natural to scale Z as well as P by A. The extra factors of c and $c + d$ make the dimensionless model simpler by having only two parameters that determine the equilibria rather than three. The drawback of this choice is that one unit of zooplankton no longer corresponds to one unit of phytoplankton.

$$\frac{dW}{dT} = AP,$$

$$\frac{dP}{dT} = RP\left(1 - \frac{P}{K}\right) - BPW.$$

(a) Explain the assumptions implicit in the differential equations.
(b) Nondimensionalize the model using dimensionless variables $p = P/K$, $w = BW/R$, $t = RT$, and parameter $a = ABK/R^2$.
(c) Explain the significance of the reference value R/B for W.
(d) Define a new variable $z = p + w$. Obtain a differential equation for z that has the form $z' = pf(z)$.
(e) The equation for z in part (d) contains the unknown population $p(t)$; however, it can still be studied with phase line analysis. Explain why this statement is true.
(f) Use phase line analysis to conclude determine a value that the sum $p + w$ cannot exceed.
(g) Note that w must increase as long as $p > 0$. Explain the consequence of this fact in conjunction with the result of the phase line analysis.
(h) The simple conclusion of this study is only as good as the model from which it was obtained. It serves to explain what happens to yeast in the rising of dough or the brewing of beer, but it obviously does not apply to fish in a pond. What important feature of the pond environment would need to be added to the model to make it realistic for fish?

(This problem is continued in Problem 7.3.2.)

7.2.7. A chemostat is a device consisting of a container filled with water, bacteria, and nutrients, set up so that water and nutrients enter the container at some rate and contents are drained from the container at the same rate, thereby maintaining a constant volume. A simple model for a chemostat is

$$\frac{dR}{dT} = Q(R_0 - R) - \frac{SRC}{A+R},$$

$$\frac{dC}{dT} = \frac{ESRC}{A+R} - QC,$$

where $R(T)$ and $C(T)$ are the amounts of resource and consumer in the container, Q is the flow rate in container volumes per unit time, R_0 is the concentration of the resource in the input stream, measured in resource units per container volume, and E is the amount of consumer biomass that is added through consumption of one unit of resource.

(a) Explain the assumptions implicit in the differential equations. [Hint: Look at the pollution model of Section 2.5 and the resource utilization models of Chapter 5.]
(b) Use the dimensionless variables $r = R/A$, $c = SC/QR_0$, and $t = EST$ to obtain the dimensionless form

$$r' = qr_0\left(1 - \frac{r}{r_0} - \frac{rc}{1+r}\right),$$

$$c' = c\left(\frac{r}{1+r} - q\right).$$

You will need to find the appropriate definitions for the dimensionless parameters q and r_0.
(c) Explain why QR_0/S is a good choice of reference quantity for the consumer biomass. [Hint: Use the resource equation to interpret the quantities QR_0 and SC.]

(d) Explain the biological meaning of the parameters q and r_0. Note that these parameters are largely under control of the experimenter.

(This problem is continued in Problem 7.3.9.)

Problems 7.2.8 and 7.2.9 concern Rosenzweig–MacArthur population models, which have the general form

$$\frac{dV}{dT} = RV \left(1 - \frac{V}{K}\right) - Cf(V),$$

$$\frac{dC}{dT} = ECf(V) - MC,$$

where f has the properties

$$f(0) = 0, f' > 0, f'' \le 0.$$

These models can be used for consumer–resource systems, where V is the biomass of resource and C the biomass of consumer, or for predator–prey systems, where V and C are biomasses of prey species and predator species, respectively. Note that the V equation is the type that we considered in Chapter 5 with fixed C as resource utilization models. In a consumer–resource or predator–prey context, the consumer/predator population is a dynamic variable rather than a fixed parameter.

7.2.8. Explain the assumptions inherent in the general Rosenzweig–MacArthur model.

7.2.9. The simplest Rosenzweig–MacArthur model has a linear predation term, $f(V) = SV$.

(a) Nondimensionalize the model using the dimensionless variables $v = V/K$, $c = SC/R$, and $t = RT$, with parameters $m = M/R$ and $h = ESK/M$.
(b) What is the biological significance of the reference value R/S for the consumer population?

(This problem is continued in Problem 7.3.5.)

7.3 Phase Plane Analysis

After studying this section, you should be able to:

- Plot solution curves in the phase plane using data from simulations.
- Interpret phase portraits of two-dimensional dynamical systems.
- Sketch nullclines for two-dimensional dynamical systems.
- Use nullclines to draw conclusions about the stability of equilibrium points.

Systems of differential equations of the form

$$\frac{dx_1}{dt} = f_1(x_1, x_2, \ldots, x_n)$$

$$\frac{dx_2}{dt} = f_2(x_1, x_2, \ldots, x_n)$$

$$\vdots \quad \vdots$$

$$\frac{dx_n}{dt} = f_n(x_1, x_2, \ldots, x_n)$$

are said to be *autonomous* because the rates of change of the dynamic variables depend only on the state of the system (the values of those dependent variables) and not explicitly on the time. This situation generally occurs when there are no factors external to the mathematical system, such as temperature, that vary in time. It is often helpful to think of the dependent variables as forming an n-dimensional space, called the *state space*.

In this section, we generalize the phase line method for single equations[17] to the phase plane method for systems of two equations.

7.3.1 Equilibrium Solutions

We introduced the concept of an equilibrium solution for a single autonomous differential equation in Section 5.3. This concept can also be generalized to autonomous systems of differential equations.

Equilibrium solution for a system of differential equations: *a constant solution, corresponding to a point in the state space for which all variables have zero rate of change.*

Example 7.3.1.

Q) Find the equilibrium solutions for the dimensionless Michaelis–Menten model[18]

$$s' = -s(1 - z) + hz, \qquad s(0) = 1, \tag{7.3.1}$$
$$\varepsilon z' = s(1 - z) - hz - rz, \qquad z(0) = 0, \tag{7.3.2}$$

where h, r, and ε are all positive and the prime represents the time derivative.
A) Equilibrium solutions for this system must satisfy two equations:

$$-s(1 - z) + hz = 0, \qquad s(1 - z) - hz - rz = 0. \tag{7.3.3}$$

Adding these equations together gives us

$$-rz = 0,$$

from which we conclude that $z = 0$ for any equilibrium solution. Given this conclusion, both of the equilibrium equations reduce to $s = 0$. Hence, the only equilibrium solution is the point $s = 0$, $z = 0$. $\qquad\qquad \square$

7.3.2 Solution Curves in the Phase Plane

In Section 7.2, we plotted the solution of the Michaelis–Menten model with specific parameter values $h = 1$, $r = 2$, and $\varepsilon = 0.2$. An alternative depiction of the solution is obtained by plotting the points in the sz plane. This plane is called the *phase plane*, and a plot showing solutions in the phase plane is called a *phase portrait*. The time series plot and phase portrait for this example appear together in Figure 7.3.1. Each point in the phase portrait corresponds to the

---

[17] See Section 5.4.

[18] See Section 7.2.

values of the state variables at some time; in this way, the plot indicates progression in time, even though the time variable does not appear as a coordinate. The direction of forward time is not apparent in Figure 7.3.1(right), but it is clear from the context, as the time history graph indicates that s is always decreasing. Arrows can be added to a phase portrait to show the direction of time if desired.

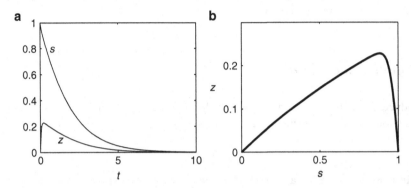

Fig. 7.3.1 The Michaelis–Menten system ((7.3.1)–(7.3.2)), with $h = 1$, $r = 2$, and $\varepsilon = 0.2$, showing (**a**) the time series and (**b**) the phase portrait

Check Your Understanding 7.3.1:
Identify the time in Figure 7.3.1a that corresponds to the peak of the phase portrait. Does the solution move along the curve in Figure 7.3.1b at constant speed?

7.3.3 Nullclines

In Figure 7.3.1, we obtained a phase portrait by plotting a solution curve. The disadvantage of this method is that we have to select a specific set of parameter values for the simulation. Fortunately, it is also possible to get information about solutions by using graphical techniques that work in the general case. The technique is based on curves called *nullclines*.

Nullcline for a two-dimensional system: *a curve in the phase plane composed of points at which one of the variables is not changing.*

A variable is not changing at just those moments when its derivative is 0; hence, the equations of the nullclines are the same as the equilibrium equations. Whereas equilibrium analysis uses all the equilibrium equations together without concern for which is which, the equilibrium equations must be examined separately in nullcline analysis. It is critical to keep track of which of the variables is constant on each particular nullcline.

Example 7.3.2. The first of the equilibrium equations[19] came from setting ds/dt equal to 0; hence, this equation represents an s nullcline. Solving for z puts this equation into the form

$$z = \frac{s}{s+h}, \tag{7.3.4}$$

---

[19] Equation (7.3.3).

which we can graph in the meaningful portion of the sz plane ($s \geq 0$, $z \geq 0$) for a given choice of h. This nullcline partitions the phase plane into regions where s is either increasing or decreasing, as determined by the differential equation for s. For example, (7.3.1) indicates that $ds/dt > 0$ when $z > s/(s+h)$. Hence, $s' > 0$ in the region above the s nullcline and $s' < 0$ in the region below. Figure 7.3.2a shows the s nullcline with $h = 1$ and includes labels to identify the regions where s is increasing and decreasing. Similarly, the second equilibrium equation determines a z nullcline, which we can write as

$$z = \frac{s}{s+h+r} \tag{7.3.5}$$

and graph in the phase plane. A similar analysis of the z differential equation results in the plot of Figure 7.3.2b. □

7.3.4 Nullcline Analysis

While individual nullclines are helpful, the real power of nullclines comes from combining them. The first step is to sketch them on a common set of axes. The nullclines divide the phase plane into regions in which the direction of change of each variable is known, and this information restricts the possible directions for the solution curves to one quadrant of the compass.

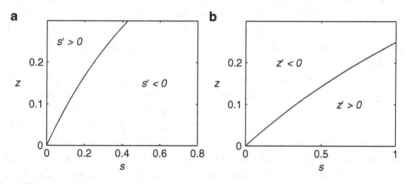

Fig. 7.3.2 (**a**) The s nullcline and (**b**) the z nullcline for the Michaelis–Menten model (using $h = 1$ and $r = 2$), showing the regions where s' and z' are either positive or negative

Example 7.3.3. Figure 7.3.3 shows the full set of nullclines for the Michaelis–Menten system, along with direction arrows. The arrows are created from information recorded in the individual nullcline plots of Figure 7.3.2.

Suppose we begin with the region in the lower right of the plot. We know from Figure 7.3.2 that $s' < 0$ and $z' > 0$ in this region. This means that s is decreasing and z is increasing. Consequently, solution curves in this region must be moving to the left rather than the right and up rather than down. They are thus restricted to directions corresponding to the second quadrant ("northwest"). This restriction is indicated in the figure by the pair of arrows joined at the tail.

Similarly, we can infer from the information in Figure 7.3.2 that solution curves in the middle region must be aimed toward the third quadrant of the compass and curves in the leftmost region must be aimed toward the fourth quadrant. □

We've seen that the direction of a solution curve is restricted to one quadrant of the compass in each region marked out by nullclines. We can draw much stronger conclusions about the

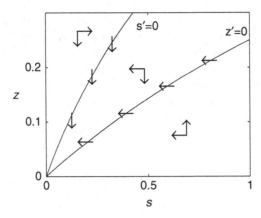

Fig. 7.3.3 The nullcline plot for the Michaelis–Menten model (using $h = 1$ and $r = 2$)

direction of solution curves on the nullclines themselves because one of the rates of change is
known to be 0.

Example 7.3.4. The upper left curve in Figure 7.3.3 is an s nullcline, which means that s is
not changing at points on it. Hence, only z can be changing, which restricts solution curves
to only two possible directions: up and down. Which of these is correct is easily determined
from the adjoining regions. Both of these regions feature arrows that point down rather than up,
so the arrows on the nullcline must point down. Similarly, the arrows on the z nullcline must
be horizontal and pointing to the left. The arrows on the nullclines can be determined directly
from the information in Figure 7.3.2, but it is much easier to do the regions first and then use
the requirement of consistency between a region and its boundaries to determine the directions
on the nullclines.

One can also use the consistency requirement to work across the plot from a starting point
of just one region. Suppose we have done only the rightmost region. The boundary at its left is
a z nullcline; hence the arrows must be horizontal. Consistency demands that they be pointing
to the left, which in turn requires that the horizontal arrow in the middle region be pointing to
the left rather than to the right. The vertical arrow in the middle region must point downward,
because the z nullcline separates two regions with opposite signs for z'. Hence the middle region
arrows point down and to the left. These arrows in turn lead to the conclusion that the arrows on
the s nullcline and the vertical arrows on the region to its left both point down. The horizontal
direction switches in moving across an s nullcline, so the horizontal arrow in the leftmost region
points to the right.   □

Having arrows on the nullclines themselves allows us to identify possible progressions of
solution curves from one region to another.

Example 7.3.5. In the nullcline plot of Figure 7.3.3, the arrows on both nullclines point toward
the center region. Consider a solution curve that begins at the biologically relevant initial point
$(1,0)$. The compass quadrants show that the curve must move up and to the left. Because of
the shape of the nullcline itself, this solution curve must eventually cross the boundary into the
center region. Moreover, all of the arrows on the boundaries of the center region point inward,
so the continuation of the curve must stay in the same region. Because of the shape of the
boundaries, this solution curve must inevitably enter the origin. Note that the same argument
must hold for any other initial point in the rightmost region.   □

Check Your Understanding 7.3.2:
Describe the behavior of solution curves that begin in the leftmost region of Figure 7.3.3.

Nullcline analysis of the Michaelis–Menten system leads to the conclusion that solution curves near the equilibrium point at the origin must move in its direction; hence, the origin is a *locally asymptotically stable* equilibrium point. Moreover, solution curves that start far from the origin must ultimately enter it; this identifies the origin as *globally asymptotically stable*. We will see in Sections 7.4 and 7.5 that analytical methods can usually determine the *local* stability of equilibrium solutions, but not the *global* stability.[20] Thus, we have obtained a very strong conclusion with no calculation other than what was needed to plot the nullclines!

Nullcline analysis does not always yield conclusions about global stability, or even local stability. It all depends on the information in the nullcline plot. In the Michaelis–Menten model, the central region has the special property that all arrows on the boundaries point inward; such a region is called a *no-egress region*. The no-egress boundaries combine with the compass quadrant in the region to significantly constrain the behavior of solution curves. Strong stability conclusions are only possible when the nullcline plot features at least one no-egress region. Even in the absence of such a region, however, nullcline analysis still provides useful information with minimal calculation and without requiring specific values for parameters.

7.3.5 Using Small Parameters

Sometimes the conclusions we can draw from nullclines are strengthened by making use of small parameters in a system of differential equations.

Example 7.3.6. The parameter ε in the dimensionless Michaelis–Menten system represents the ratio of the initial enzyme concentration to the initial substrate concentration, and hence it is always small.[21] Because this parameter multiplies dz/dt, it marks z as a "fast" variable. In Section 7.2, we used this observation to argue that z quickly approaches a point in the phase plane where the right side of the equation is small. In graphical terms, this means that solution curves quickly approach a z nullcline. If the compass quadrant restrictions permit, the solution curves will subsequently follow closely to the nullcline of the fast equation. This information can be used to predict the shape of a solution curve in the Michaelis–Menten system to a considerable degree of detail.

Consider the curve that begins at the point $(1,0)$. This point is not near a z nullcline, so the right side of the z equation is not near 0. Hence, z' must be large. Meanwhile, the s equation is not fast, so s' is never large. The inescapable conclusion is that the solution curve must move almost vertically in the phase plane until it gets close to the z nullcline, at which point the right side of the z equation becomes small and z' stops being large.

Once near the z nullcline, the solution curve must cross to the other side, because the arrows on that nullcline point to the left. The solution curve is subsequently constrained to follow the compass quadrant of the middle region, which means it must move down and to the left. While doing so, it must stay near the z nullcline; otherwise, z must change rapidly. In the middle region, this rapid change in z would be downward, bringing the solution curve back to the nullcline.

The path of the solution curve is now clear. It is in a no-egress region, so it must stay in the middle portion of the plot. As it moves down and to the left, it stays close to the z nullcline. Eventually it must enter the origin. Figure 7.3.4 shows a sample solution curve, with $\varepsilon = 0.2$, along with the nullclines. Our conclusions are clearly confirmed. With a more realistic (smaller) value of ε, the solution will stay even closer to the z nullcline. □

---

[20] There are analytical methods that can sometimes be used to prove global stability, but they are far more sophisticated than anything in this book.

[21] See Section 7.2.

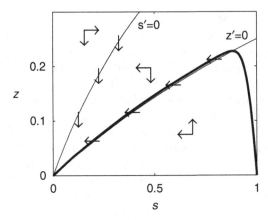

Fig. 7.3.4 The nullcline plot for the Michaelis–Menten model (using $h = 1$ and $r = 2$), along with the solution curve (*heavy*) obtained with $\varepsilon = 0.2$

7.3.6 Nullcline Analysis in General

The principal goal for the analysis of a dynamical system is to understand the phase portrait for the various ranges of parameter values. Nullcline analysis is a graphical tool for addressing this goal. Although the conclusions depend on the details of the system, the procedure is always as presented in the Michaelis–Menten example.

Algorithm 7.3.1 *Nullcline analysis for a dynamical system of variables x and y in continuous time*

1. *Set $x' = 0$ to obtain an equation for the set of points where x is not changing. Plot the curve or curves for this equation and identify each of the regions marked out by the graph as a region where x is either increasing or decreasing.*
2. *Repeat step 1 using the y equation.*
3. *Plot the x and y nullclines together. For each region created by these graphs, use the information from steps 1 and 2 to identify the appropriate compass quadrant for solution curves.*
4. *Use the compass quadrants to determine the direction of solution curves on each nullcline. Generally the direction reverses as the nullcline moves through an equilibrium point.*
5. *Identify any no-egress regions, if possible.*
6. *Use the compass quadrants and no-egress regions to determine the possible paths of solution curves in the phase plane. If either x or y is a fast variable, indicated by a small parameter in front of the derivative term in the dimensionless model, this means that solution curves move in the direction of the fast variable to a corresponding nullcline and remain near that nullcline, as long as doing so does not contradict the compass quadrant in the region.*

The Michaelis–Menten model is a nice first example, but it is important to recognize that some of its features make it easier to analyze than the typical model.

- The nullclines in the Michaelis–Menten model do not cross in the open first quadrant. This means that there are no equilibria with both variables positive. In general, there can be positive equilibria, and these might exist for certain ranges of parameter values and not for others.

- Not all systems have one fast and one slow variable. This means that we can't always find solution curves that follow a nullcline closely.
- For systems that do have a fast and a slow variable, it is not always possible for solutions to follow the fast nullcline. In our system, the nullcline sloped down and to the left, a direction consistent with the arrows in the middle region. When the nullcline slope is not compatible with the direction arrows, the phase portrait becomes complicated.[22]
- Not all systems exhibit no-egress regions. Without such a region, the ultimate fate of solution curves can be unclear from the nullcline plot alone.

Example 7.3.7. Figure 7.3.5 shows the nullcline plot for the system

$$\frac{ds}{dt} = 1 - s - 2vs, \quad s \geq 0,$$

$$\frac{dv}{dt} = 2vs - v, \quad v \geq 0.$$

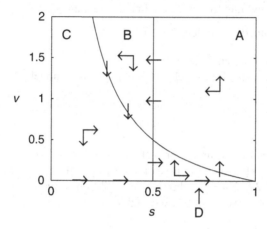

Fig. 7.3.5 The nullcline plot for Example 7.3.7

This nullcline plot lacks no-egress regions, so the analysis is more complicated than in the Michaelis–Menten model. Consider a solution curve that begins just above the point $(1,0)$ in region A. The curve must move up and to the left, and it eventually reaches region B. From there, it must move on to region C and then to region D and finally region A again. In general, rotation like this can spiral either inward or outward, or it can form a closed loop, so we cannot determine the stability of the equilibrium at $(0.5, 0.5)$ by the nullcline method. □

Check Your Understanding Answers

1. The peak is where z reaches its maximum, which occurs at a very small positive time. Thus, the solution moves from the initial point $(1,0)$ to the peak very quickly; it then moves at a progressively slower pace along the curve from the peak toward the point $(0,0)$. We know that the pace is slowing, because the time history graph for s gets flatter as time increases.
2. These solution curves must move down and to the right, cross the boundary into the center region, and continue to the origin.

---

[22] See the problems.

Problems

Note: All parameters in these problems are positive.

7.3.1. This problem demonstrates the limitations of nullcline diagrams.

(a) Sketch the nullcline diagram for the family of systems

$$x' = x - 1.2y,$$
$$y' = k(1.5x - y),$$

where x and y are not required to be nonnegative. [Note that the nullcline plot is the same for all values of k.]

(b) Use computer simulations to plot some solution curves using the value $k = 0.8$. Is the origin stable or unstable?

(c) Repeat part (b) with $k = 1.2$.

(d) Can we conclude anything from the nullcline diagram alone?

(This problem is continued in Problem 7.4.1.)

7.3.2. (Continued from Problem 7.2.6.)
Sketch the nullcline diagram for the self-limiting population model

$$w' = ap,$$
$$p' = p(1 - p - w)$$

with $a = 0.5$. Add in solution curves for $p(0) = p_0$ and $w(0) = 0$, with several values of p_0 satisfying $0 < p_0 \leq 1$. The solution curves must be consistent with the restriction on $p + w$ obtained in Problem 7.2.6. Note that the stable equilibrium value for $p + w$ cannot be reached if p becomes 0, but in no case can it be exceeded.

7.3.3. (Continued from Problem 7.1.3.)

(a) Determine the equilibria for the SIR model with no demographics,

$$s' = -R_0 si,$$
$$i' = R_0 si - i,$$

being careful to indicate any restrictions on existence.

(b) Sketch the nullcline diagram in the si plane for the case $R_0 > 1$.

(c) Add solution curves that begin with very small positive $i(0)$ and various $s(0)$ to the phase plane plot from part (b).

(d) Calculus can be used to find the terminal point of the solution curves. We note first that the paths $i(s)$ in the phase plane must have slope $di/ds = i'/s'$. Use this fact to write a formula for the derivative di/ds.

(e) Integrate the derivative di/ds to obtain a solution formula that contains the parameter R_0 and an integration parameter.

(f) We are interested in solution curves that begin with $s(0) = s_0$ and $i(0) \approx 0$. Use these values to calculate the integration parameter in the solution for $i(s)$.

(g) Identify the value of i at the end of each solution curve. [Hint: Solution curves end at equilibria.] Substitute this value into the equation for $i(s)$ and replace s with s_∞ to indicate that the resulting equation is to determine the final value of s on a solution curve.

(h) Suppose the initial population is fully susceptible ($s(0) = 1$). Solve the equation from part (g) for R_0 and use the result to plot s_∞ as a function of R_0.

(i) The value of R_0 for smallpox is estimated to be at least 4. Given this fact, roughly what fraction of a Native American population newly exposed to smallpox will make it through the epidemic without getting sick?

Problems 7.3.4–7.3.9 call for nullcline plots, but parameters are given only in terms of inequality restrictions. Choose small simple values that meet the requirements. The inequalities that arise in the existence restrictions for equilibria also identify all cases for the nullcline plots in these models. In general, stability analysis[23] is required to fully identify cases. The model in Problems 7.3.10 and 7.3.11 is an example where equilibrium analysis is not sufficient for this purpose.

7.3.4.* (Continued from Problem 7.1.4.)

(a) Determine the equilibria for the SIR model with constant population size,

$$s' = \varepsilon(1 - s) - R_0 si \,,$$
$$i' = R_0 si - i \,,$$

being careful to indicate any restrictions on existence.

(b) Sketch the nullcline diagram in the si plane for the case $0 < R_0 < 1$.

(c) Repeat part (b) for the case $R_0 > 1$.

(d) Determine any conclusions that can be drawn from the nullcline plots.

(This problem is continued in Problem 7.5.1.)

7.3.5. (Continued from Problem 7.2.9.)

(a) Determine the equilibria for the dimensionless Rosenzweig–MacArthur model with Holling type I predator response,

$$v' = v(1 - v - c) \,,$$
$$c' = mc(hv - 1) \,,$$

being careful to indicate any restrictions on existence.

(b) Sketch the nullcline diagram in the vc plane with $h < 1$.

(c) Repeat part (b) for the case $h > 1$.

(d) Determine any conclusions that can be drawn from the nullcline plots.

(This problem is continued in Problem 7.5.7.)

7.3.6. (Continued from Problem 7.1.5.)

(a) Determine the equilibrium points for the SIS model with constant recruitment,

$$n' = \varepsilon(1 - n - mi) \,,$$
$$i' = R_0 i(n - i) - i \,,$$

noting any restrictions on R_0 for their existence.

(b) Verify that the equilibrium value of n is decreased by the long-term presence of the disease.

(c) Sketch the nullcline diagram in the ni plane for the case $0 < R_0 < 1$.

---

[23] See Sections 7.4 and 7.5.

(d) Repeat part (c) for the case $R_0 > 1$.
(e) Determine any conclusions that can be drawn from the nullcline plots.

(This problem is continued in Problem 7.5.2.)

7.3.7. (Continued from Problem 7.1.6.)

(a) Determine the equilibrium points for the SIS model with logistic growth,

$$n' = \varepsilon[n(1-n) - mi],$$
$$i' = R_0 i \left(1 - \frac{i}{n}\right) - i,$$

noting any restrictions on R_0 and m for their existence. You may find it convenient to replace R_0 with the parameter $\alpha = 1 - R_0^{-1}$.
(b) Sketch the nullcline diagram in the ni plane for the case $0 < R_0 < 1$.
(c) Repeat part (b) for the case $R_0 > 1$, $\alpha m > 1$.
(d) Repeat part (b) for the case $R_0 > 1$, $\alpha m < 1$.
(e) Determine any conclusions that can be drawn from the nullcline plots.

(This problem is continued in Problems 7.5.3 and 7.5.4.)

7.3.8. (Continued from Problem 7.2.5.)

(a) Determine the equilibria for the plankton population model,

$$p' = p(1 - \alpha - p - z),$$
$$z' = \delta z(p - \beta),$$

being careful to indicate any restrictions on existence.
(b) Sketch the nullcline diagram in the pz plane for the case $\alpha > 1$.
(c) Repeat part (b) for the case $\alpha < 1 < \alpha + \beta$.
(d) Repeat part (b) for the case $\alpha + \beta > 1$.
(e) Determine any conclusions that can be drawn from the nullcline plots.

(This problem is continued in Problem 7.5.5.)

7.3.9. (Continued from Problem 7.2.7.)

(a) Determine the equilibria for the dimensionless chemostat model,

$$r' = qr_0 \left(1 - \frac{r}{r_0} - \frac{rc}{1+r}\right),$$
$$c' = c \left(\frac{r}{1+r} - q\right).$$

Also show that the equilibrium with $c > 0$ exists if and only if

$$q < \frac{r_0}{1 + r_0}.$$

(b) Sketch the nullcline diagram in the rc plane for the case

$$\frac{r_0}{1 + r_0} < q < 1.$$

(c) Repeat part (b) for the case

$$q < \frac{r_0}{1 + r_0}.$$

(d) Determine any conclusions that can be drawn from the nullcline plots.

(This problem is continued in Problem 7.5.6.)

Problems 7.3.10–7.3.12 concern Rosenzweig–MacArthur models, which have the form[24]

$$\frac{dV}{dT} = RV\left(1 - \frac{V}{K}\right) - Cf(V),$$

$$\frac{dC}{dT} = ECf(V) - MC,$$

where

$$f(0) = 0, f' > 0, f'' \le 0.$$

7.3.10. (a) Determine the equilibria for the dimensionless Rosenzweig–MacArthur model with Holling type II predator response,

$$v' = v\left(1 - \frac{v}{k} - \frac{c}{1 + v}\right),$$

$$c' = \varepsilon c\left(\frac{hv}{1 + v} - 1\right),$$

being careful to indicate any restrictions on h for existence.
(b) Sketch the nullcline diagram in the vc plane with $k = 2$ and $h = 1.4$. Sketch in some solution curves.
(c) Repeat part (b) for the case $k = 2$, $h = 2$.
(d) Determine any conclusions that can be drawn from the nullcline plots.

(This problem is continued in Problem 7.5.8.)

7.3.11. The Rosenzweig–MacArthur model with Holling type II predator response,

$$v' = v\left(1 - \frac{v}{k} - \frac{c}{1 + v}\right),$$

$$c' = \varepsilon c\left(\frac{hv}{1 + v} - 1\right),$$

is capable of predicting a type of dynamics called a *limit cycle*, in which solution curves seem to approach a closed curve rather than a single point.

(a) Run a simulation with $k = 2$, $h = 5$, $\varepsilon = 0.05$, $v(0) = 1$, and $c(0) = 0.1$. Use a time interval of $0 \le t \le 100$. Plot the results as separate curves $v(t)$ and $c(t)$ on common axes. Note that the solutions settle into an oscillatory pattern.
(b) Plot the solution from part (a) in the vc plane.
(c) Add nullclines to the plot of part (b).

---

[24] Also see Problems 7.2.8 and 7.2.9.

(d) The solution curve appears to follow a cycle of four phases:

   1. It moves to the right, almost on a horizontal line, until it just crosses the v nullcline.
   2. It moves up and to the left just above the v nullcline.
   3. It moves sharply to the left almost to the c axis.
   4. It moves down and a little to the right before beginning the cycle anew.

   Use the system of differential equations to explain why the curve moves as it does. In particular, you will need to use the facts that ε is small and that solutions in the region to the right of the c nullcline and above the v nullcline cannot move in a downward direction.

(e) Sketch the nullcline diagram for the case $k = 2$, $h = 2$. Explain why the limit cycle does not occur in this case.

(This problem is continued in Problem 7.5.8.)

7.3.12. One of the best known data sets in ecology consists of counts of hare and lynx pelts collected by the Hudson Bay Company from 1845 to 1935. These data appear to indicate a limit cycle; however, the cycle appears to be moving in the wrong direction. Predator populations should be decreasing when the prey population is small, but the Hudson Bay Company data shows the number of lynx pelts to be increasing when the hare population is small, as though the hares were eating the lynx. Unfortunately, many biologists have inferred from this failure of the Rosenzweig–MacArthur model that mathematical modeling in biology has little value. The real problem, as is always the case when a good model doesn't work, is that the model does not fit the scenario. Bo Deng has developed a better model based on the insight that the pelt data is not a substitute for actual population data unless the number of trappers collecting the data is constant [2]. A proper explanation of the Hudson Bay Company data requires a model that includes the number of trappers as a third dynamic variable. We explore how this can happen with a simplified version of Deng's model. This version is not fit to the actual data, but it does illustrate the resolution of the paradox.

(a) Use biological arguments to explain why the predator population should be decreasing when the prey population is small.

(b) Use mathematical arguments based on the terms in the Rosenzweig–MacArthur model to explain the same fact as in part (a).

(c) Run a simulation for the dimensionless model

$$H' = H\left(1 - H - \frac{3L}{1+3H} - \frac{0.5T}{1+2H+36L}\right), \qquad H(0) = 0.1\,,$$

$$L' = L\left(\frac{3H}{1+3H} - 0.42 - \frac{3T}{1+2H+36L}\right), \qquad L(0) = 0.2\,,$$

$$T' = T\left(\frac{24L}{1+2H+36L} - 0.56\right), \qquad T(0) = 0.3\,,$$

using the time interval $0 < t < 50$. Plot the solutions together on a common graph and confirm that the lynx population decreases when the hare population is small. Note that the trapper population fluctuates also, with a minimum that is only about 60 % of the maximum.

(d) The functions

$$H_T = \frac{0.5HT}{1+2H+36L}, \qquad L_T = \frac{3LT}{1+2H+36L}\,,$$

represent the terms in the hare and lynx equations corresponding to the rate of population loss from trapping. These quantities, rather than the populations H and L, are the ones given in the Hudson Bay Company data. Calculate these for the simulation data and plot the results together on a common graph. Confirm that the lynx pelt data for the model increases during the period when the number of hare pelts is low.

(e) Explain why the trapper population oscillates, including a discussion of the mechanisms for increase and decrease of the number of trappers.

7.4 Stability in Linear Systems

After studying this section, you should be able to:

- Use eigenvalues to determine the stability of two-component linear continuous dynamical systems.
- Use the Routh–Hurwitz conditions to determine the stability of two-component linear continuous dynamical systems.
- Use the Routh–Hurwitz conditions to determine the stability of three-component linear continuous dynamical systems.

In Section 5.5, we obtained the solution formula

$$x = x_0 e^{kx}$$

for the scalar linear differential equation

$$\frac{dx}{dt} = kx, \qquad k \neq 0,$$

which models either exponential growth or exponential decay, depending on the sign of k. From this formula, it is clear that the ultimate fate of the dynamic variable $x(t)$ depends on the algebraic sign of k:

- If $k < 0$, then $\lim_{t \to \infty} x = 0$.
- If $k > 0$, then $\lim_{t \to \infty} x = \infty$.

Alternatively, we can say that the equilibrium point $x = 0$ is asymptotically stable if $k < 0$ and unstable if $k > 0$.

Although this analysis scales up to continuous dynamical systems with more than one component, there is one difficulty: we must find an analog for the constant k for higher dimensional systems.

7.4.1 Linear Systems

As noted in Section 7.3, autonomous systems of differential equations have the general form

$$\frac{dx_1}{dt} = f_1(x_1, x_2, \ldots, x_n)$$

$$\frac{dx_2}{dt} = f_2(x_1, x_2, \ldots, x_n)$$

$$\vdots \quad \vdots$$

$$\frac{dx_n}{dt} = f_n(x_1, x_2, \ldots, x_n).$$

Eventually, we will want to analyze autonomous systems with nonlinear functions f_i; however, in this section we limit our consideration to *linear* autonomous systems, which have a highly structured form.

Linear autonomous system of differential equations: *a system of n differential equations of the form*

$$\frac{dx_1}{dt} = a_{11}x_1 + a_{12}x_2 + \cdots + a_{1n}x_n + b_1$$

$$\frac{dx_2}{dt} = a_{21}x_1 + a_{22}x_2 + \cdots + a_{2n}x_n + b_2$$

$$\vdots \quad \vdots$$

$$\frac{dx_n}{dt} = a_{n1}x_1 + a_{n2}x_2 + \cdots + a_{nn}x_n + b_n \,,$$

where the a_{ij} and b_j are all constants.

Example 7.4.1. The dimensionless two-component lead poisoning model[25]

$$x' = q - (1+q)x + \varepsilon z \,, \tag{7.4.1}$$

$$z' = x - \varepsilon z \,, \tag{7.4.2}$$

where $q, \varepsilon > 0$, is an autonomous two-dimensional system of linear differential equations. If we identify x as x_1 and z as x_2, then

$$a_{11} = -(1+q), \quad a_{12} = \varepsilon, \quad a_{21} = 1, \quad a_{22} = -\varepsilon, \qquad b_1 = q, \quad b_2 = 0 \,.$$

$\square$

Autonomous linear systems are conveniently written in the matrix form

$$\mathbf{x}' = \mathbf{A}\mathbf{x} + \mathbf{b} \tag{7.4.3}$$

by defining the vectors $\mathbf{x}$ and $\mathbf{b}$ and matrix $\mathbf{A}$ as

$$\mathbf{x} = \begin{pmatrix} x_1 \\ x_2 \\ \vdots \\ x_n \end{pmatrix}, \qquad A = \begin{pmatrix} a_{11} & a_{12} & \dots & a_{1n} \\ a_{21} & a_{22} & \dots & a_{2n} \\ \vdots & \vdots & \vdots & \vdots \\ a_{n1} & a_{n2} & \dots & a_{nn} \end{pmatrix}, \qquad \mathbf{b} = \begin{pmatrix} b_1 \\ b_2 \\ \vdots \\ b_n \end{pmatrix}. \tag{7.4.4}$$

Example 7.4.2. The lead poisoning model of Example 7.4.1 has the form (7.4.3), with

$$\mathbf{x} = \begin{pmatrix} x \\ z \end{pmatrix}, \quad A = \begin{pmatrix} -(1+q) & \varepsilon \\ 1 & -\varepsilon \end{pmatrix}, \quad \mathbf{b} = \begin{pmatrix} q \\ 0 \end{pmatrix},$$

$\square$

---

[25] See Section 7.1.

7.4.2 Eigenvalues and Stability

As long as $\det(\mathbf{A}) \neq 0$, there is always a single equilibrium point,[26] which is the unique solution of

$$\mathbf{Ax} + \mathbf{b} = \mathbf{0}, \qquad (7.4.5)$$

and the stability of that equilibrium point can be determined from the eigenvalues of A.

If, for example, the 2×2 matrix A has two real eigenvalues λ_1 and λ_2, then solutions of the linear equation

$$\mathbf{x}' = \mathbf{Ax} + \mathbf{b}$$

have the form

$$\mathbf{x} = \mathbf{x}^\star + c_1 \mathbf{v_1} e^{\lambda_1 t} + c_2 \mathbf{v_2} e^{\lambda_2 t},$$

where $\mathbf{x}^\star$ is the solution of (7.4.5) and $\mathbf{v_k}$ is an eigenvector corresponding to the eigenvalue λ_k.[27] Because asymptotic stability requires

$$\lim_{t \to \infty} \mathbf{x} = \mathbf{x}^\star$$

for all solutions, it only occurs when both eigenvalues are negative. The situation is more complicated than the scalar case, however. The eigenvalue equation is quadratic, so instead of a pair of unique real roots there could either be repeated real roots or a pair of complex roots. These cases make the solution formulas more complicated, but do not affect stability. All real roots, repeated or distinct, yield solutions with the factor $e^{\lambda t}$.[28] Complex roots of the form $\lambda = \alpha \pm i\beta$ yield solutions with the factor $e^{\alpha t}$, where α is the *real part* of λ.[29] This gives us the following general result.

> **Theorem 7.4.1 (Stability of Equilibria for Autonomous Linear Systems).** *The equilibrium solution $\mathbf{x} = \mathbf{x}^\star$ for the equation $\mathbf{x}' = \mathbf{Ax} + \mathbf{b}$, where $\det(\mathbf{A}) \neq 0$ and $\mathbf{Ax}^\star + \mathbf{b} = \mathbf{0}$, is asymptotically stable if and only if all real eigenvalues are negative and all complex eigenvalues have a negative real part.*

Example 7.4.3. For the model of Example 7.4.1, the equilibrium must satisfy

$$q - (1 + q)x + \varepsilon z = 0, \qquad x - \varepsilon z = 0.$$

Adding these equations together produces an equation with only the x variable:

$$q - qx = 0.$$

Since $q > 0$, we have $x = 1$ for all sets of parameter values. The second equilibrium equation then gives the result $z = 1/\varepsilon$. In the context of lead poisoning,[30] the dimensionless variables

---

[26] See Section 5.3.

[27] See Problem 7.4.5.

[28] Repeated roots have to contribute multiple solutions. If the eigenvalue equation has a factor λ^2, then obviously one solution is $e^{\lambda t}$. It can be shown that the other solution is $te^{\lambda t}$. This is important if we need solution formulas, but it is irrelevant to the question of stability.

[29] Complex pairs of roots have to contribute two solutions. The eigenvalue pair $\alpha \pm i\beta$ corresponds to solutions $e^{\alpha t} \cos \beta t$ and $e^{\alpha t} \sin \beta t$. These solutions oscillate, with the rate of oscillation controlled by β; however, only the exponential factor matters toward stability.

[30] See Section 7.1.

x and z represent the amounts of lead in the blood and the bones. With ε small and $z^* = 1/\varepsilon$, the model predicts an equilibrium state with a large amount of lead in the bones. This is why the solutions we saw in Section 7.1 reached equilibrium quickly for X but did not appear to reach equilibrium for Z at all. In order to reach a z value of approximately $1/\varepsilon$, it is necessary to wait for about $1/\varepsilon$ times however long it takes to reach an x value of approximately 1. In Figure 7.1.2, X appears to reach equilibrium in about 100 days. With $\varepsilon \approx 0.01$, it would take about 10,000 days (around 30 years!) for Z to reach equilibrium.

Equilibria play different roles in the behavior of a system, depending on their stability. Only stable equilibria can serve as the long-term behavior of a system. To determine the stability of the lead poisoning equilibrium, we need only find the eigenvalues of the matrix

$$A = \begin{pmatrix} -(1+q) & \varepsilon \\ 1 & -\varepsilon \end{pmatrix} .$$

We have

$$0 = \det(\mathbf{A} - \lambda \mathbf{I}) = \begin{vmatrix} -1 - q - \lambda & \varepsilon \\ 1 & -\varepsilon - \lambda \end{vmatrix} = (-1 - q - \lambda)(-\varepsilon - \lambda) - \varepsilon$$

$$= \lambda^2 + (1 + q + \varepsilon)\lambda + \varepsilon q .$$

Because all eigenvalues have to be negative, we need only consider the root with the positive sign in front of the square root. From the quadratic formula, this root is

$$\lambda_1 = \frac{-(1+q+\varepsilon) + \sqrt{(1+q+\varepsilon)^2 - 4\varepsilon q}}{2} . \tag{7.4.6}$$

Is this root real or complex? If real, is it positive or negative? With the values $q = 7.1$, $\varepsilon = 0.009$, we have

$$\lambda_1 = \frac{-8.1009 + \sqrt{(8.1009)^2 - 0.2556}}{2} \approx -0.0079 < 0 .$$

By Theorem 7.4.1, the equilibrium is asymptotically stable with these parameter values. $\square$

It is more difficult to determine the asymptotic stability without having specific values for the parameters.[31] Fortunately, there is an easier way to determine stability for the general case.

7.4.3 The Routh–Hurwitz Conditions

So far, the general plan for analyzing a two-dimensional system like that of (7.4.1) and (7.4.2) involves two basic computational steps:

1. Rewrite the equation $\det(\mathbf{A} - \lambda \mathbf{I}) = 0$ as a polynomial equation for λ, which is called the *characteristic equation* of the matrix $\mathbf{A}$.
2. Determine the most positive eigenvalue and check to see if it has a negative real part.

---

[31] See Problem 7.4.6.

We can simplify the computation by making use of the relationship between the coefficients of the polynomial in step 1 and the sign of the eigenvalues in step 2. The following theorem supplies the key fact[32]:

Theorem 7.4.2 (Roots of Quadratic Polynomials). *Both roots of the polynomial equation*

$$x^2 + bx + c = 0$$

have negative real parts if and only if

$$b, c > 0.$$

Theorem 7.4.2 allows us to determine stability without having to solve the characteristic equation.

Example 7.4.4. The equilibrium solution for the system of (7.4.1) and (7.4.2) is always asymptotically stable because the coefficients of the polynomial equation

$$\lambda^2 + (1 + q + \varepsilon)\lambda + \varepsilon q = 0$$

are always positive when $q, \varepsilon > 0$. □

Theorem 7.4.2 is a nice improvement on the basic method, but we can do even better by connecting the entries of a matrix and the coefficients of the polynomial that determines the eigenvalues. Using this connection results in the Routh–Hurwitz[33] conditions, which provide stability criteria in terms of properties of the matrix, thereby eliminating the need to actually compute $\det(\mathbf{A} - \lambda\mathbf{I})$.

Theorem 7.4.3 (Routh–Hurwitz Conditions for a System of Two Components). *Let*

$$\mathbf{A} = \begin{pmatrix} a & b \\ c & d \end{pmatrix}.$$

The equilibrium solution $\mathbf{x} = \mathbf{x}^*$ *for the equation*

$$\frac{d\mathbf{x}}{dt} = \mathbf{A}\mathbf{x} + \mathbf{b}$$

is asymptotically stable if

$$\det\mathbf{A} > 0, \qquad tr\mathbf{A} < 0,$$

where $\det\mathbf{A} = ad - bc$ *and the **trace** of the matrix is* $tr\mathbf{A} = a + d$.
The equilibrium solution is unstable if $\det\mathbf{A} < 0$ *or* $tr\mathbf{A} > 0$.

---

[32] See Problem 7.4.7.

[33] Pronounce Routh to rhyme with "mouth."

Example 7.4.5. Given $q, \varepsilon > 0$, the matrix

$$A = \begin{pmatrix} -(1+q) & \varepsilon \\ 1 & -\varepsilon \end{pmatrix}$$

has

$$\det A = \varepsilon q > 0, \qquad \operatorname{tr} A = -(1+q+\varepsilon) < 0;$$

therefore, the equilibrium solution for the system

$$\frac{d\mathbf{x}}{dt} = \mathbf{A}\mathbf{x} + \mathbf{b}$$

is asymptotically stable for any positive values of q and ε. $\square$

7.4.4 The Routh–Hurwitz Conditions for a Three-Dimensional System

There are sets of Routh–Hurwitz conditions for systems of any size, but they get more complicated as the size increases. Here we present the Routh–Hurwitz conditions for 3×3 matrices.

> **Theorem 7.4.4 (Routh–Hurwitz Conditions for a System of Three Components).** *Let*
> **A** *be a* 3×3 *matrix. Let* $\mathbf{A_k}$ *be the* 2×2 *matrix obtained from* **A** *by deleting row* k *and*
> *column* k*. Define* c_1, c_2, *and* c_3 *by*
>
> $$c_1 = -\operatorname{tr}\mathbf{A}, \quad c_2 = \sum_{k=1}^{3} \det \mathbf{A_k}, \quad c_3 = -\det \mathbf{A},$$
>
> *where* $\operatorname{tr}\mathbf{A}$ *is the sum of the diagonal elements of* **A***. Then the equilibrium solution of the*
> *system*
>
> $$\frac{d\mathbf{x}}{dt} = \mathbf{A}\mathbf{x} + \mathbf{b}$$
>
> *is asymptotically stable if all three coefficients are positive and* $c_1 c_2 > c_3$*. The equilibrium*
> *solution is unstable if any of the coefficients is negative or* $c_1 c_2 < c_3$*.*

Example 7.4.6. The matrix A for the three-compartment lead poisoning model (Equations (7.1.1)–(7.1.3)) is

$$A = \begin{pmatrix} -(k_1 + k_3 + r_1) & k_2 & k_4 \\ k_1 & -(k_2 + r_2) & 0 \\ k_3 & 0 & -k_4 \end{pmatrix}.$$

$$(7.4.7)$$

Using the notation of Theorem 7.4.4, we have

$$c_1 = (k_1 + k_3 + r_1) + (k_2 + r_2) + k_4 > 0,$$
$$c_2 = [k_4(k_2 + r_2)] + [k_4(k_1 + k_3 + r_1) - k_3 k_4] + [(k_2 + r_2)(k_1 + k_3 + r_1) - k_1 k_2]$$
$$= k_4(k_2 + r_2) + k_4(k_1 + r_1) + k_2(k_3 + r_1) + r_2(k_1 + k_3 + r_1) > 0,$$

$$c_3 = -[-k_4(k_2+r_2)(k_1+k_3+r_1)+k_1k_2k_4+k_3k_4(k_2+r_2)]$$
$$= k_4(k_2+r_2)(k_1+r_1)-k_1k_2k_4 = k_4(k_2r_1+k_1r_2+r_1r_2) > 0.$$

Finally, $c_1 > k_4$ and $c_2 > (k_2r_1+k_1r_2+r_1r_2)$ combine to make $c_1c_2 > c_3$. The Routh–Hurwitz conditions are satisfied, and the equilibrium solution is asymptotically stable for any parameter values. $\qquad\square$

If your eyes glass over when you see the calculations of Example 7.4.6, there are several points to keep in mind. First, it is much easier to apply the Routh–Hurwitz conditions when at least some of the parameter values are known. Second, the calculations would have been less tedious if we had nondimensionalized the model before doing the stability calculations.[34] The dimensionless three-component model has a Jacobian with three parameters rather than six. Even more importantly, identification of dimensionless parameters as small makes it much easier to check the Routh–Hurwitz conditions.[35] The most important point of all is that identification of small parameters in a system can allow us to reduce the number of variables in the system. Example 7.4.6 is included here to make a valuable point, but it is entirely unnecessary in the study of lead poisoning once we have shown that the two-component model is an adequate approximation to the full model.

Problems

Note: All parameters in these problems are positive.

7.4.1.* (Continued from Problem 7.3.1.)

(a) Determine the stability of the origin for the system

$$x' = x - 1.2y,$$
$$y' = k(1.5x - y).$$

Note that the answer depends on k.
(b) Compare the results of part (a) with Problem 7.3.1.

7.4.2. Use the Routh–Hurwitz conditions to determine the stability of the equilibrium solution of the three-component lead poisoning model (7.1.1)–(7.1.3) using the parameter values given in (7.1.4) and (7.1.5).

7.4.3.* Had we nondimensionalized the three-component lead poisoning model before doing the calculations of Example 7.4.6, we would have obtained the matrix

$$A = \begin{pmatrix} -(1+a_1+b_1) & a_2 & \varepsilon \\ a_1 & -(a_2+b_2) & 0 \\ 1 & 0 & -\varepsilon \end{pmatrix},$$

where $a_1 = k_1/k_3$, $a_2 = k_2/k_3$, $b_1 = r_1/k_3$, $b_2 = r_2/k_3$, $\varepsilon = k_4/k_3$, and ε is expected to be small.

(a) Approximate the parameter c_2 for the Routh–Hurwitz conditions by omitting any terms with the factor ε. Show with minimal calculation that $c_2 > 0$.

---

[34] See Problem 7.4.3.
[35] Compare Problem 7.4.3 with Problem 7.4.2.

(b) Show that all terms in c_3 have a common factor of ε. Explain why it is unnecessary in this case to show that $c_1 c_2 > c_3$ if the other conditions are satisfied.[36]

7.4.4. Figure 7.4.1 shows a very simple two-compartment model for a drug that is absorbed in the digestive tract and then circulated in the blood compartment (including the liver and kidneys).

(a) Write down the appropriate differential equations for the model. Note that the equations must be supplemented by initial conditions $W(0) = W_0$ and $X(0) = X_0$.
(b) Notice that the equations are decoupled, meaning that the W equation does not contain X. This means that we can solve the W equation simply by finding a function that makes it true. No solution technique is needed. From the table of integrals in Section 1.8, we can see that any function Ae^{kt} has derivative equal to k times the function. This fact allows you to identify an appropriate family of solutions for W. Note that the parameter A must be chosen so that $W(0) = W_0$.
(c) The X equation can be simplified with a clever trick. Let $Y(T) = e^{rT} X(T)$. Use the product rule to obtain Y' and then substitute from the differential equation for X to derive the resulting equation

$$Y' = ke^{rT} W(T) .$$

(d) Use the result for W and integration (Section 1.8) to obtain a one-parameter family of functions for Y.
(e) Use the definition of Y to obtain a family of functions for X. Use the initial condition $X(0) = X_0$ to complete the solution for X. In writing your solution, it is advantageous to replace $r - k - b$ by $-(k + b - r)$; under most circumstances, $k > r$.
(f) To study the effect of k, set $b = 0$, $X_0 = 0$, $r = 1$, and $W_0 = 1$ and obtain a simplified formula for X. Now plot $X(T)$ on a single graph using $k = 4$, $k = 3$, and $k = 2$. Discuss the effect of k on the functioning of the drug.
(g) Plot the function X from part (f) with $k = 0.5$. What happens when drug absorption is very slow?
(h) Suppose drug manufacturers can control the value of k without changing the efficacy of the drug. What would be the relative advantages of a larger k and a smaller k?

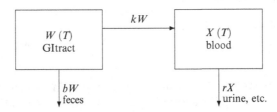

Fig. 7.4.1 A compartment diagram for drug absorption

7.4.5. Let $\mathbf{A}$ be a 2×2 matrix having eigenvectors $\mathbf{v_1}$ and $\mathbf{v_2}$ with corresponding eigenvalues λ_1 and λ_2. Let $\mathbf{b}$ be a two-dimensional vector and let $x^\star$ be the solution of the algebraic equation $\mathbf{Ax} = -\mathbf{b}$. Show that the function

$$\mathbf{x}(t) = \mathbf{x}^\star + c_1 \mathbf{v_1} e^{\lambda_1 t} + c_2 \mathbf{v_2} e^{\lambda_2 t}$$

---

[36] Calculation of c_3 is still messy, but the two hardest parts of the stability demonstration are considerably simplified by the assumption that ε is small.

solves the differential equation

$$\mathbf{x}' = \mathbf{A}\mathbf{x} + \mathbf{b} \, .$$

7.4.6. Show that the eigenvalues of the two-compartment lead poisoning model,

$$\lambda = \frac{-(1+q+\varepsilon) \pm \sqrt{(1+q+\varepsilon)^2 - 4\varepsilon q}}{2} \, ,$$

are either both real and negative or both complex with negative real parts for any $q, \varepsilon > 0$.

7.4.7. For the polynomial equation $x^2 + bx + c = 0$, determine the regions in the bc plane where:

(a) The roots are real,
(b) The roots are real and differ in sign,
(c) The roots are real and positive,
(d) The roots are real and negative,
(e) The roots are complex and have negative real parts,
(f) The roots are complex and have positive real parts.
(g) Sketch a graph in the bc plane showing the regions indicated in parts (b)–(f)

7.5 Stability in Nonlinear Systems

After studying this section, you should be able to:

- Compute the Jacobian to represent a nonlinear system near an equilibrium point.
- Use the Jacobian and the Routh–Hurwitz criteria to determine stability of equilibria for nonlinear dynamical systems.

In Section 7.4, we learned that linear autonomous systems have a single equilibrium solution and that the stability of that solution can be determined by examining the matrix $\mathbf{A}$ that represents the system. For nonlinear systems, there may be any number of equilibria, with different behavior near each. However, we can find a linear approximation for the system near each equilibrium point. The matrices that represent these approximate systems can usually be used to determine the stability of the corresponding equilibrium solution.

7.5.1 Approximating a Nonlinear System at an Equilibrium Point

Suppose we have a two-dimensional nonlinear system of the form

$$x_1' = f_1(x_1, x_2) \, , \tag{7.5.1}$$
$$x_2' = f_2(x_1, x_2) \, . \tag{7.5.2}$$

In general, the linear approximation of a function $f(x_1, x_2)$ near a point $(x_1^\star, x_2^\star)$ is given by[37]

$$f(x_1, x_2) \approx f(x_1^\star, x_2^\star) + \frac{\partial f}{\partial x_1}(x_1^\star, x_2^\star)(x_1 - x_1^\star) + \frac{\partial f}{\partial x_2}(x_1^\star, x_2^\star)(x_2 - x_2^\star) \, .$$

---

[37] Linear approximation for functions of more than one variable requires one term for each independent variable, using partial derivatives rather than ordinary derivatives. See Section 1.3 for partial derivatives and Section 1.4 for linear approximation.

However, we are specifically interested in the case where $(x_1^\star, x_2^\star)$ is an equilibrium solution. Then

$$f_1(x_1^\star, x_2^\star) = 0, \qquad f_2(x_1^\star, x_2^\star) = 0,$$

which simplifies the linear approximation formula for the functions f_1 and f_2 to

$$f_j(x_1, x_2) \approx \frac{\partial f_j}{\partial x_1}(x_1^\star, x_2^\star)(x_1 - x_1^\star) + \frac{\partial f_j}{\partial x_2}(x_1^\star, x_2^\star)(x_2 - x_2^\star).$$

Thus, replacing the functions f_1 and f_2 with their linear approximations yields the system

$$x_1' \approx \frac{\partial f_1}{\partial x_1}(x_1^\star, x_2^\star)(x_1 - x_1^\star) + \frac{\partial f_1}{\partial x_2}(x_1^\star, x_2^\star)(x_2 - x_2^\star), \tag{7.5.3}$$

$$x_2' \approx \frac{\partial f_2}{\partial x_1}(x_1^\star, x_2^\star)(x_1 - x_1^\star) + \frac{\partial f_2}{\partial x_2}(x_1^\star, x_2^\star)(x_2 - x_2^\star), \tag{7.5.4}$$

Now, to simplify the system, we define new variables u_i in place of x_i:

$$u_i = x_i - x_i^\star. \tag{7.5.5}$$

Note that $du_i/dt = dx_i/dt$ because $x_i^\star$ is a constant. Thus, we can rewrite (7.5.3) and (7.5.4) as

$$u_1' = \frac{\partial f_1}{\partial x_1}(x_1^\star, x_2^\star)u_1 + \frac{\partial f_1}{\partial x_2}(x_1^\star, x_2^\star)u_2,$$

$$u_2' = \frac{\partial f_2}{\partial x_1}(x_1^\star, x_2^\star)u_1 + \frac{\partial f_2}{\partial x_2}(x_1^\star, x_2^\star)u_2.$$

In matrix-vector form, we have the system

$$\mathbf{u}' = \mathbf{J}(x_1^\star, x_2^\star)\mathbf{u}, \qquad \mathbf{J} = \begin{pmatrix} \frac{\partial f_1}{\partial x_1} & \frac{\partial f_1}{\partial x_2} \\ \frac{\partial f_2}{\partial x_1} & \frac{\partial f_2}{\partial x_2} \end{pmatrix}. \tag{7.5.6}$$

The matrix $\mathbf{J}$ is called the *Jacobian* of the system.

Jacobian of an n-dimensional system of differential equations: *the $n \times n$ matrix for which the entry in row i and column j is the derivative of the ith function with respect to the jth variable.*

Example 7.5.1. Find the approximate linear system(s) for the dimensionless Michaelis–Menten model

$$\frac{ds}{dt} = -s(1-z) + hz, \tag{7.5.7}$$

$$\varepsilon \frac{dz}{dt} = s(1-z) - hz - rz, \tag{7.5.8}$$

where h and r are positive constants.

As we saw in our earlier analysis, the only equilibrium point is the point $(0,0)$. In the notation of (7.5.1) and (7.5.2), we have

$$f_1(s,z) = -s + sz + hz, \quad f_2(s,z) = \varepsilon^{-1}(s - sz - hz - rz),$$

so

$$\frac{\partial f_1}{\partial s} = -1 + z, \quad \frac{\partial f_1}{\partial z} = s + h,$$

$$\frac{\partial f_2}{\partial s} = \varepsilon^{-1}(1 - z), \quad \frac{\partial f_2}{\partial z} = \varepsilon^{-1}(-s - h - r).$$

These derivatives combine to define the Jacobian matrix as

$$\mathbf{J}(s, z) = \begin{pmatrix} -1 + z & s + h \\ \varepsilon^{-1}(1 - z) & \varepsilon^{-1}(-s - h - r) \end{pmatrix}.$$

At the equilibrium point,

$$\mathbf{J}(0, 0) = \begin{pmatrix} -1 & h \\ \varepsilon^{-1} & -\varepsilon^{-1}(h + r) \end{pmatrix}. \tag{7.5.9}$$

Equation (7.5.9) defines the matrix for the linear approximation of the system near the equilibrium point $(0, 0)$. □

7.5.2 The Jacobian and Stability

Suppose $(x_1^\star, x_2^\star)$ is an equilibrium point for a system of the form given in (7.5.1) and (7.5.2). If we zoom in on the equilibrium point in the phase plane, we approach the system (7.5.6), where the equilibrium $(u_1, u_2) = (0, 0)$ corresponds to the original equilibrium $(x_1^\star, x_2^\star)$. We can certainly determine the stability of the linearized system using Theorem 7.4.3. But does this result also apply to the corresponding equilibrium point for the nonlinear system? Because the behavior of solutions near an equilibrium point in the phase space doesn't change as we zoom in, it seems reasonable to expect that the nonlinear system and its linear approximation should have the same stability properties. This is generally true, although there are exceptions. The important result is summarized in the following theorem:

Theorem 7.5.1 (Stability for Nonlinear Systems of Two Dynamic Variables). *Let $(x_1^\star, x_2^\star)$ be an equilibrium point for the system*

$$\frac{dx_1}{dt} = f_1(x_1, x_2), \quad \frac{dx_2}{dt} = f_2(x_1, x_2).$$

If Theorem 7.4.3 applies to the linearized system having matrix $\mathbf{J}(x_1^\star, x_2^\star)$, then the conclusion from that theorem applies to the corresponding equilibrium $(x_1^\star, x_2^\star)$.

Example 7.5.2. The matrix

$$\mathbf{A} = \mathbf{J}(0, 0) = \begin{pmatrix} -1 & h \\ \varepsilon^{-1} & -\varepsilon^{-1}(h + r) \end{pmatrix}$$

from Example 7.5.1 has

$$\det \mathbf{A} = (-1)(-\varepsilon^{-1}(h + r)) - (h)(\varepsilon^{-1}) = \varepsilon^{-1} r > 0, \quad \operatorname{tr} \mathbf{A} = -1 - \varepsilon^{-1}(h + r) < 0.$$

By Theorem 7.4.3, the origin is asymptotically stable for the linear system $\mathbf{u}' = \mathbf{A}\mathbf{u}$. By Theorem 7.5.1, this result extends to the equilibrium $(s,z) = (0,0)$ for the nonlinear system of (7.5.7) and (7.5.8). □

Note that Theorem 7.4.3 does not apply in certain cases, such as when $\det \mathbf{A} > 0$ and $\operatorname{tr}\mathbf{A} = 0$. This is not very common, but it is important to keep in mind that linearization does not always work.

Example 7.5.3.

Q) Apply linearization to the system

$$\frac{dw}{dt} = ap, \qquad \frac{dp}{dt} = p(1 - p - w), \qquad k > 0,$$

which models the dynamics of a population (p) that is gradually poisoned by its own waste (w).

A) The first equilibrium equation requires $p = 0$, which automatically satisfies the second equation as well. Thus, all points of the form $(w,0)$ are equilibria. The Jacobian is

$$\mathbf{J}(w,p) = \begin{pmatrix} 0 & a \\ -p & 1 - 2p - w \end{pmatrix},$$

so

$$\mathbf{J}(w,0) = \begin{pmatrix} 0 & a \\ 0 & 1 - w \end{pmatrix}, \quad \det \mathbf{J}(w,0) = 0, \quad \operatorname{tr}\mathbf{J}(w,0) = 1 - w.$$

If $w < 1$, then the trace of the Jacobian is positive. In this case, Theorem 7.4.3 indicates that the equilibrium is unstable, and this result carries over to the nonlinear system. However, the theorem does not apply if $w > 1$; we cannot draw a stability conclusion from the linearization in this case.[38] □

We've now seen how the Jacobian provides the matrix needed to determine stability in two-dimensional systems. It works equally well for higher dimensional systems. Theorem 7.5.2 gives the result for three-dimensional systems:

Theorem 7.5.2 (Stability for Nonlinear Systems of Three Components). *Let (x_1^*, x_2^*, x_3^*) be an equilibrium point for the system*

$$x_1' = f_1(x_1, x_2, x_3), \qquad x_2' = f_2(x_1, x_2, x_3), \qquad x_3' = f_3(x_1, x_2, x_3).$$

If Theorem 7.4.4 applies to the linearized system having matrix $\mathbf{J}(x_1^*, x_2^*, x_3^*)$, then the conclusion from that theorem applies to the equilibrium (x_1^*, x_2^*, x_3^*) for the original system.

Example 7.5.4. Consider the one-parameter family of models with one predator and two prey that is given by the system

$$x' = x(1 - x) - xz, \quad y' = ry(1 - y) - yz, \quad z' = z(2x + 2y - 1), \qquad r > 0.$$

---

[38] See Problem 7.3.2 for a conclusive analysis.

The Jacobian for this system is

$$\mathbf{J}(x,y,z) = \begin{pmatrix} 1-2x-z & 0 & -x \\ 0 & r-2ry-z & -y \\ 2z & 2z & 2x+2y-1 \end{pmatrix}.$$

Note that the point $(x,y,z) = (0,1,0)$ is an equilibrium, with Jacobian

$$\mathbf{J}(0,1,0) = \begin{pmatrix} 1 & 0 & 0 \\ 0 & -r & -1 \\ 0 & 0 & 1 \end{pmatrix}.$$

Using the notation of Theorem 7.4.4, we have

$$c_1 = r-2, \qquad c_2 = 1-2r, \qquad c_3 = r > 0.$$

We have $c_1 > 0$ if $r > 2$ and $c_2 > 0$ if $r < 1/2$. These cannot both happen, so the Routh–Hurwitz conditions indicate instability. Thus, the equilibrium $(0,1,0)$ for the original system is unstable.[39]   □

Problems

Note: All parameters in these problems are positive.

7.5.1.* (Continued from Problem 7.3.4.)

(a) Determine the stability of the equilibria for the SIR model with constant population size

$$s' = \varepsilon(1-s) - R_0 si, \qquad \varepsilon > 0,$$
$$i' = R_0 si - i, \qquad R_0 > 0.$$

Note that the results could depend on the parameter values.
(b) Discuss the results with reference to Problems 7.1.4 and 7.3.4.

7.5.2. (Continued from Problem 7.3.6.)

(a) Determine the stability of the equilibria for the SIS model with constant recruitment

$$n' = \varepsilon(1-n-mi), \qquad m, \varepsilon > 0,$$
$$i' = R_0 i(n-i) - i, \qquad R_0 > 0.$$

Note that the results could depend on the parameter values.
(b) Discuss the results with reference to Problems 7.1.5 and 7.3.6.

7.5.3. (Continued from Problem 7.3.7.)

(a) Determine the stability of the equilibrium $(1,0)$ for the SIS model with logistic growth

$$n' = \varepsilon[n(1-n) - mi], \qquad m, \varepsilon > 0,$$
$$i' = R_0 i \left(1 - R_0^{-1} - \frac{i}{n}\right), \qquad R_0 > 0,$$

---

[39] Other equilibria are considered in Problem 7.5.9.

(b) Repeat part (a) in the limit $\varepsilon \to 0$ for the equilibrium $(1 - \alpha m, \alpha(1 - \alpha m))$, where $\alpha = 1 - R_0^{-1}$. Keep in mind that there are restrictions on the parameter values for this equilibrium to exist.

(c) Discuss the results with reference to Problems 7.1.6 and 7.3.7.

7.5.4. (Continued from Problem 7.3.7.)
The equations

$$n' = \varepsilon[n(1-n) - mi], \qquad m, \varepsilon > 0,$$

$$i' = R_0 i\left(1 - R_0^{-1} - \frac{i}{n}\right), \qquad R_0 > 0,$$

for the SIS model with logistic growth are not in a suitable form to determine the stability of the extinction equilibrium $(0,0)$, because the Jacobian is not defined at this point. To get around this problem, we must rewrite the model. For simplicity, assume $R_0 > 1$. (It is intuitively clear that the population will not go extinct if the epidemic cannot get started.)

(a) Let $x = i/n$. Apply the product rule to the equation $i(t) = x(t)n(t)$ and substitute from the n' and i' equations to obtain an equation for x'. With the assumption that ε is very small, this equation reduces to a simple logistic equation for x that does not contain either i or n.

(b) Determine the stable equilibrium solution of the simplified x equation.

(c) Evaluate the Jacobian for the original system, using the stable equilibrium value from part (b) where i/n is needed.

(d) Determine the stability of the extinction equilibrium for the case $R_0 > 1$.

(e) Discuss the results with reference to Problems 7.1.6 and 7.3.7.

7.5.5. (Continued from Problem 7.3.8.)

(a) Determine the stability of the equilibria for the plankton population model

$$p' = p(1 - \alpha - p - z),$$
$$z' = \delta z(p - \beta).$$

Keep in mind that some equilibria exist only for certain ranges of parameter values.

(b) Discuss the results with reference to Problem 7.3.8.

(c) Explain the prediction the model makes for the effect of alpha and beta on the behavior of the system

7.5.6. (Continued from Problem 7.3.9.)

(a) Determine the stability of the $c = 0$ equilibrium for the chemostat model,

$$r' = qr_0\left(1 - \frac{r}{r_0} - \frac{rc}{1+r}\right),$$

$$c' = c\left(\frac{r}{1+r} - q\right).$$

(b) Show that the following pair of conditions guarantees stability for an equilibrium solution in a two-component model:

1. One of the entries in the main diagonal of the Jacobian is negative and the other is 0.
2. The off-diagonal entries of the Jacobian are opposite in sign.

(c) Use the result of part (b) to determine the stability of the $c > 0$ equilibrium solution.

(d) Sketch the region in the $r_0 q$ parameter space for which the consumer population has a stable positive value.

(e) Discuss the results with reference to Problem 7.3.9.

7.5.7.* (Continued from Problem 7.3.5.)

(a) Determine the stability of the equilibria having $c = 0$ for the dimensionless Rosenzweig–MacArthur model with Holling type I predator response,

$$v' = v(1 - v - c),$$
$$c' = mc(hv - 1).$$

(b) Let (v^*, c^*) be the equilibrium in which neither v nor c are 0. Use the equations that these quantities must satisfy to reduce the Jacobian to

$$J(v^*, c^*) = \begin{pmatrix} -v^* & -v^* \\ mhc^* & 0 \end{pmatrix}$$

and use this result to determine the stability of the equilibrium.

(c) Discuss the results with reference to Problems 7.2.9 and 7.3.5.

(d) Explain the prediction the model makes for the affect of h on the behavior of the system.

7.5.8. (Continued from Problems 7.3.10 and 7.3.11.)

(a) Determine the stability of the predator extinction equilibrium $v = k$, $c = 0$ for the dimensionless Rosenzweig–MacArthur model with Holling type II predator response,

$$v' = v\left(1 - \frac{v}{k} - \frac{c}{1+v}\right),$$

$$c' = mc\left(\frac{hv}{1+v} - 1\right),$$

with $k = 2$ and arbitrary h.

(b) Find the Jacobian for the non-extinction equilibrium (v^*, c^*) determined by the equations

$$\left(1 - \frac{v}{k} - \frac{c}{1+v}\right) = 0, \quad \left(\frac{hv}{1+v} - 1\right) = 0.$$

Note that the determinant of this Jacobian must be positive; hence, the stability is determined entirely by the sign of the term in row 1 and column 1.

(c) Set $k = 2$ to simplify the algebra. Then show that the condition that the trace be negative reduces to the requirement $v^* > 1/2$.

(d) Determine the range of h values, given $k = 2$, for which the stability condition $v^* > 1/2$ is satisfied.

(e) Discuss the results with reference to Problems 7.3.10 and 7.3.11.

(f) Explain the prediction the model makes for the affect of h on the behavior of the system.

7.5.9. The one-parameter family of models

$$x' = x(1 - x) - xz, \quad y' = ry(1 - y) - yz, \quad z' = z(2x + 2y - 1),$$

with $r > 0$ was considered briefly in Example 7.5.4, where the Jacobian was found to be

$$\mathbf{J}(x,y,z)= \begin{pmatrix} 1-2x-z & 0 & -x \\ 0 & r-2ry-z & -y \\ 2z & 2z & 2x+2y-1 \end{pmatrix} .$$

(a) Show that the equilibrium in which all populations are 0 is always unstable.
(b) Show that the equilibrium $(1,0,0)$ is always unstable, and similarly, that the equilibrium $(0,1,0)$ is always unstable.
(c) Find the equilibrium in which only z is 0 and show that it is always unstable.
(d) Find the equilibrium in which only y is 0 and determine the range of r for which it is stable.
(e) Repeat part (d) for the case where only x is 0.
(f) Repeat part (d) for the case where none of the variables is 0. [Hint: Leave x, y, and z as factors in all entries in the first, second, and third rows, respectively. The result is a simple matrix with easy calculations to show $c_j > 0$. The final requirement is a little tricky, but much easier if you don't substitute in the formulas for the variables.]
(g) Note that r represents the relative growth rate advantage of the species y as compared to x. Explain why the results of parts (d)–(f) make sense biologically.

7.6 Primary HIV Infection

The simplest model for viral infections of the immune system appears in an article by A.S. Perelson and colleagues that appeared in the year 2000 [6]. We present this model with a minor modification.

T cells are a type of white blood cell. The body produces these cells and uses them to fight infections. Like all cells, T cells do not live forever, so they must be produced continuously to maintain a healthy concentration in the blood.

The HIV virus targets T cells. When a virus particle infects a T cell, it takes over the machinery of the cell to produce more virus particles. The infected cell will continue to produce virus particles until it dies. Infected cells can be expected to die sooner than uninfected cells because of the additional stress put on the cell by the virus.

Consider a dynamical system consisting of three interacting components. There are healthy T cells, infected T cells, and free virus particles, all moving through the bloodstream. We use S (susceptible) for the concentration of healthy T cells, I for the concentration of infected T cells, V for the concentration of free virus particles, and T for time. The three populations interact through several processes, not all of which need appear in our conceptual model. Here are the processes that we do include:

1. Healthy cells are produced by the body at a constant rate R.
2. T cells of both kinds are lost through natural death at a rate proportional to the number of cells: DS for the healthy cells and DI for the infected cells.
3. Infected T cells die as a result of virus infection, with overall rate MI.
4. Healthy cells become infected through encounters with free virus particles. Since this is a chemical reaction, we assume that the rate is given by the law of mass action, which predicts that the rate is proportional to both S and V. Thus, healthy cells are reduced at the rate BVS and infected cells increase at the same rate.
5. Infected cells produce free virus particles at a constant rate, so the overall production rate of virus particles is proportional to the number of infected cells: PI.
6. The body removes free virus particles at a rate proportional to the concentration of such particles: CV.

Combining these assumptions results in the model

$$\frac{dS}{dT} = R - DS - BVS, \tag{7.6.1}$$

$$\frac{dI}{dT} = BVS - DI - MI, \tag{7.6.2}$$

$$\frac{dV}{dT} = PI - CV. \tag{7.6.3}$$

7.6.1 Nondimensionalization

As with the Michaelis–Menten model, we can decrease the number of parameters by nondimensionalization. Specifically, we define dimensionless variables[40] by

$$s = \frac{DS}{R}, \quad i = \frac{(M+D)I}{R}, \quad v = \frac{C(M+D)V}{PR}, \quad t = DT. \tag{7.6.4}$$

Substituting $S = Rs/D$, $I = Ri/(M+D)$, $V = PRv/C(M+D)$, and $\frac{d}{dT} = D\frac{d}{dt}$, and defining the dimensionless parameters

$$b = \frac{BRP}{CD(M+D)}, \quad \theta = \frac{M+D}{C}, \quad \varepsilon = \frac{D}{M+D}, \tag{7.6.5}$$

we have the dimensionless system

$$\frac{ds}{dt} = 1 - s - bvs, \tag{7.6.6}$$

$$\varepsilon \frac{di}{dt} = bvs - i, \tag{7.6.7}$$

$$\theta \varepsilon \frac{dv}{dt} = i - v. \tag{7.6.8}$$

As in the Michaelis–Menten model, we can consider the parameters to be of two types. The parameters ε and θ are ratios of reference times, with ε the lifetime of an infected cell relative to a healthy cell and θ the lifetime of a virus particle relative to that of an infected cell. The parameter b measures the relative strength of the infection process. The factors in the numerator represent the processes that increase the infectivity (B), increase the production of virus (P), and increase the production of target cells (R). The factors in the denominator represent the death rates of virus particles, healthy cells, and infected cells, respectively. We will see in the

---

[40] The reference quantity for healthy cells is chosen by observing that, in the absence of the virus, the equilibrium has $S = R/D$. The reference quantity for infected cells is taken to be $R/(M+D)$ rather than R/D. The former is roughly the number of infected cells lost in one time unit, which will be an advantageous choice. The drawback is that the number of T cells in the dimensionless model will not be $s+i$ on account of the difference in reference quantities; this turns out to be insignificant because there is no important reason why we should want to know the total number of T cells. The reference quantity for the virus is chosen so that the virus equation will yield $v = i$ at equilibrium; in other words, the reference quantity is the number of virus particles that can be maintained by a population of $R/(M+D)$ infected cells. The expected lifetimes of healthy T cells, infected T cells, and virus particles are $1/D$, $1/(M+D)$, and $1/C$, respectively. Any one of these could be taken as the reference time for nondimensionalization. Generally it is best to choose longer times over shorter times, and on the view that the model is mostly about T cells, we choose $1/D$.

subsequent analysis that the model exhibits two qualitatively different behaviors, one for b relatively large and one for b relatively small. A key part of the analysis is to determine the critical value of b where the transition occurs.

7.6.2 Reduction to Two Variables

Experiments have yielded expected lifetimes of approximately 100 days for healthy T cells, 2.5 days for infected T cells, and 1/4 day for virus particles.[41] Thus, the time scale parameters are

$$\varepsilon \approx 0.025, \qquad \theta \approx 0.1 .$$

The assumption that both ε and θ are small allows us to simplify the model for the purpose of analysis, using the same method we used in the Briggs–Haldane approximation. Because both ε and θ are small, the fastest variable in the dynamical system is v. Initially, we should have $i = 0$ and v small. We should expect the v equation to quickly reach equilibrium at $v = i$. Thus, we predict that the virus concentration should very quickly drop to meet the infected cell concentration, while the infected cell concentration begins its increase from 0. This has two consequences: (1) the initial amount of virus is not going to make a big difference in the behavior of the model, since any excess will disappear quickly, and (2) we can ignore the extremely short transient in which $v \neq i$.

This is borne out in Figure 7.6.1, which shows the early history of the infection. Note that the time scale is 100 days, so the plots on the left show approximately the first year of the dynamics and the plots on the right show barely a month. The initial virus load affects the peak of viral concentration, with a smaller load requiring an incubation period. Meanwhile, differences between v and i are barely visible on the shorter time scale. In the lower plot, corresponding to a very large initial load $v(0) = 2$, the virus concentration and infected cell concentration equalize at about $t \approx 0.01$, which corresponds to a real time of about 1 day. With a more realistic initial load, the reduction of v to match i is not even visible on the graph.

Our HIV model shows the great power of nondimensionalization when there are small parameters. The model gives us a considerable amount of detail about the dynamics without any need for nullclines or stability analysis. Most importantly, the combination of analysis and simulation suggests that by approximating the v equation by the relation $v = i$, we can reduce the model to two dimensions with only minimal loss of information. We can then replace v by i in the remaining equations, reducing the model to

$$\frac{ds}{dt} = 1 - s - bis, \qquad b > 0, \tag{7.6.9}$$

$$\varepsilon \frac{di}{dt} = i(bs - 1), \qquad 0 < \varepsilon \ll 1, \tag{7.6.10}$$

$$v = i . \tag{7.6.11}$$

---

[41] Of course, these values differ significantly for different patients; however, we may take them to be good at least to a factor of 10. This may not seem like a useful degree of accuracy, but it is enough to support the mathematical arguments in this section.

7.6.3 Equilibria and Stability

Equilibrium solutions for both the two-dimensional and three-dimensional systems must satisfy the equations:
$$1 - s - bis = 0, \qquad i(bs - 1) = 0, \qquad i = v.$$

The second equation is satisfied either by $i = 0$ (the infection-free equilibrium) or $s = 1/b$ (the disease equilibrium). In the first case, we get $s = 1$ and $v = 0$; in the second case, we get $i = v = 1 - 1/b$, which only has biological meaning if $b > 1$.

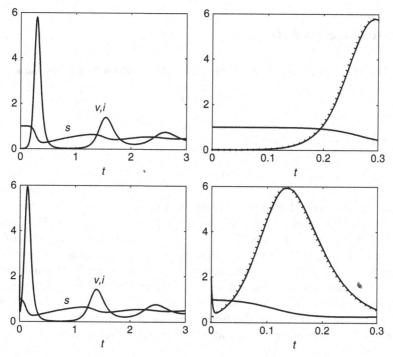

Fig. 7.6.1 HIV dynamics, from (7.6.6) to (7.6.8), with $b = 2$, $\varepsilon = 0.025$, $\theta = 0.1$; the *upper* plots have $v(0) = 0.01$, the *lower* plots have $v(0) = 2$, and in both cases i is *dotted*, although not necessarily visible. The plots on the *right* cover the first 10 % of the time interval shown on the *left*

Using the two-dimensional model, the Jacobian is

$$\mathbf{J}(s,i) = \begin{pmatrix} -1 - bi & -bs \\ bi & bs - 1 \end{pmatrix}.$$

Thus,

$$\mathbf{J}(1,0) = \begin{pmatrix} -1 & -b \\ 0 & b-1 \end{pmatrix}, \quad \mathbf{J}(b^{-1}, 1 - b^{-1}) = \begin{pmatrix} -b & -1 \\ b-1 & 0 \end{pmatrix}.$$

We have

$$\det \mathbf{J}(1,0) = 1 - b, \qquad \operatorname{tr} \mathbf{J}(1,0) = b - 2,$$
$$\det \mathbf{J}(b^{-1}, 1 - b^{-1}) = b - 1, \qquad \operatorname{tr} \mathbf{J}(b^{-1}, 1 - b^{-1}) = -b.$$

Note that the two determinants are negatives of each other, so there is no scenario in which both equilibria are stable. The infection-free equilibrium $(1,0)$ is asymptotically stable whenever $b < 1$, as both the trace and determinant requirements are satisfied. The disease equilibrium is asymptotically stable whenever $b > 1$. To summarize:

- If $0 < b < 1$, the asymptotically stable equilibrium $(1,0)$ is the only equilibrium.
- If $b > 1$, the infection-free equilibrium $(1,0)$ is unstable and the disease equilibrium $(b^{-1}, 1 - b^{-1})$ is asymptotically stable.

The three-dimensional model yields the same stability results.[42]

7.6.4 Phase Plane Analysis

Equation (7.6.9) indicates that the s nullcline is $bis = 1 - s$, which we can rewrite as

$$i = \frac{1-s}{bs} .$$ (7.6.12)

From (7.6.10) we find two i nullclines:

$$i = 0, \qquad s = \frac{1}{b} .$$ (7.6.13)

We saw that the set of equilibrium points is different for the cases $0 < b < 1$ and $b > 1$. The value $b = 1$ marks the boundary between two qualitatively different model behaviors. Such a value is called a *bifurcation* value. When there is a bifurcation, we need distinct nullcline sketches for each of the cases. Figure 7.6.2 illustrates the two cases, using b values of 0.8 and 2. In the $b = 0.8$ case, we show two hypothetical solution curves (dotted). Although these curves illustrate the behavior expected of solution curves, the initial conditions are unrealistic. The actual solution starts near the stable equilibrium point at $(1,0)$ and moves quickly to that point. In the $b = 2$ case, the solution curve that begins near the $(1,0)$ equilibrium exhibits oscillatory behavior, as shown in the corresponding time series graphs of Figure 7.6.1.

Problems

7.6.1. Determine the stability of the equilibria for the dimensionless three-component HIV model.

The immune system utilizes a variety of different components to combat infections, foreign bodies, and other dangers to the organism. A complete model of the immune system would be very complicated; however, there are a number of simple models that deal with specific aspects of the immune system. We consider one of these models here[43]:

---

[42] See Problem 7.6.1.

[43] These are adapted from [5].

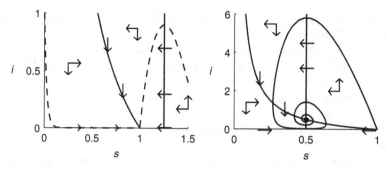

Fig. 7.6.2 Nullcline sketches for the HIV model, with $b = 0.8$ (*left*) and $b = 2$ (*right*), along with solution curves obtained with $\varepsilon = 0.025$, $\theta = 0.1$

$$\frac{dP}{dT} = RP\left(1 - \frac{P}{K}\right) - QMP - SNP\,,$$

$$\frac{dM}{dT} = L - DM - AMP\,,$$

$$\frac{dN}{dT} = \frac{CP}{H+P} - BN\,,$$

where P is the pathogen and M and N are two different kinds of macrophages (white blood cells), one non-specific and the other a cell that is "trained" to recognize and attack a specific invader, and all other quantities are positive parameters. The model can be written in dimensionless form as

$$p' = p(1 - p - qm - sn)\,,$$

$$m' = a[\delta(1 - m) - mp]\,,$$

$$n' = b\left(\frac{p}{h+p} - n\right)\,.$$

7.6.2. (a) Explain the three terms in the (dimensional) pathogen equation. In particular, what assumptions does the model make about pathogen population growth and about interaction with macrophages?

(b) Explain the three terms in the non-specific macrophage (M) equation. In particular, what are the processes by which the body maintains a population of these cells in the absence of infection, and what is the outcome on these cells of interaction with the pathogen?

(c) The first term in the specific macrophage (N) equation is simplified from a model that involves the training of cells from a reservoir of untrained cells. While it does not show all the features of that model, it does indicate how the rate at which these trained cells are created depends on the pathogen population. Explain this dependence.

(d) Explain the biological process that removes trained cells from the body. In particular, what significant advantage is conferred by these cells as compared to the non-specific cells?

(e) The parameter δ is always small. What is the biological meaning of this fact?

7.6.3. In some circumstances, the non-specific macrophages are sufficient to clear an infection, in which case we can ignore the N equation. For a specific example, assume $q = 1.5$ and $\delta = 0.2$.[44]

---

[44] The value given for q is realistic, while the value for δ is about 10–20 times larger than what usually occurs. This larger value has been chosen because it makes the nullcline diagram easier to interpret, without changing the qualitative features of the system appreciably.

(a) Plot the nullcline diagram for the *pm* subsystem with the given parameter values. Determine the number and stability of the equilibria.
(b) Run a simulation with the given parameter values and also $a = 0.5$, $m(0) = 1$, and $p(0) = 1$, with a total time of 100 units. [Experiment with the number of time steps. What happens if you don't make the time intervals short enough?]
(c) Repeat part (b) with $p(0) = 0.9$.
(d) Repeat part (b) with $p(0) = 0.2$.
(e) Add the data from parts (b)–(d) to the nullcline plot.
(f) Discuss the results. What should we expect a realistic value of $p(0)$ to be, given that $p = 1$ would be the maximum capacity for the pathogen in a person without a functioning immune system?
(g) Explain why the assumption that N can be omitted from the model is reasonable with small $p(0)$ and the other parameters as given above.

7.6.4. (a) Show that the *pm* subsystem always has exactly one chronic infection equilibrium state when $q < 1$ and determine the range of possible p values. [Keep in mind that δ is small, so it is certainly less than 1.]
(b) Determine the Jacobian for the chronic infection state. To keep the result as simple as possible, do not try to substitute in the formulas for p^* and m^*. Instead, use the equilibrium equations to eliminate m^* from the top row and p^* from the bottom row.
(c) Use the Routh–Hurwitz conditions to show that the chronic infection equilibrium is stable if and only if qm^{*2} is less than some critical value.
(d) Use the equilibrium equations to eliminate first q and then m^* from the stability criterion of part (c). Then show that the resulting stability criterion, which includes only p^* and δ, is always met when $q < 1$.[45]

7.6.5. In this problem, we explore the predictions the model makes about the effect of the pathogen-specific macrophages.

(a) Determine the stability of the disease-free equilibrium for the *pm* subsystem in terms of the parameters q, δ, and a. When the stability condition is not met, the pathogen is guaranteed to persist over time.
(b) Determine the stability of the disease-free equilibrium for the full system in terms of the parameters. [Hint: In this case, it is easier to compute the eigenvalues than to apply the Routh–Hurwitz conditions.]
(c) Suppose $q = 0.5$ and $\delta = 0.2$ in the *pm* subsystem. Determine the equilibrium value of p for this case. (The stability of this equilibrium is confirmed in Problem 7.6.4.)
(d) Take $s = 1$ and $h = 0.06$ with the same values for q and δ as in part (a). Determine the equilibrium value of p for the full system.
(e) Use the Routh–Hurwitz conditions to confirm the stability of the equilibrium of part (d).
(f) Run (separate) simulations for the *pm* subsystem and the full system using $p(0) = 0.2$, $m(0) = 1$, $n(0) = 0$, $b = 0.1$, and other parameter values as given in parts (c) and (d). Plot the results on a common graph. [Choose an appropriate total time to make a meaningful plot.]
(g) Explain the model's prediction for the benefit of having pathogen-specific macrophages.

---

[45] This problem is yet another example of the general principle that computer algebra systems are no substitute for human algebra capability. It would be possible to get a CAS to do the manipulations to obtain the results of this problem, but only by anticipating the manipulations needed. There is no CAS command for "Use human ingenuity to simplify this formula so as to obtain a result with a simple biological interpretation."

(h) What qualitative prediction of this model is not realistic?[46] [Hint: Think about what it means to have an equilibrium with $p > 0$ and what advantages the pathogen-specific macrophages ought to add to the functioning of the immune system.]

7.6.6. (a) Show that the full system always has exactly one positive equilibrium value for p under the assumption that δ is arbitrarily small (in other words, take $\delta = 0$).

(b) Use the Routh–Hurwitz conditions to show that the chronic disease equilibrium of the full system with parameters

$$q = 0.5, \quad \delta = 0.2, \quad s = 1, \quad h = 0.06, \quad a = 0.4, \quad b = 0.1$$

is stable (see Problem 7.6.5d).[47]

References

1. Britton NF. *Essential Mathematical Biology*. Springer, New York (2004)
2. Deng B. An inverse problem: Trappers drove hare to eat lynx. in preparation for publication
3. Ledder G. *Differential Equations: A Modeling Approach*, McGraw-Hill, New York (2005)
4. Rabinowitz M, GW Wetherill, and JD Kopple. Lead metabolism in the normal human: Stable isotope studies. *Science*, **182**, 725–727 (1973)
5. Reynolds A, J Rubin, G Clermont, J Day, Y Vodovotz, and GB Ermentrout. A reduced mathematical model of the acute inflammatory response: I. Derivation of model and analysis of anti-inflammation. *Journal of Theoretical Biology*, **242**, 220–236 (2006)
6. Stafford MA, L Corey, Y Cao, ES Daar, AS Perelson. Modelling viral and immune system dynamics. *Journal of Theoretical Biology*, **203**, 285–301 (2000)

---

[46] Making one unrealistic prediction does not mean the model is a failure. It makes a lot of accurate predictions and could be made more realistic by adding more features.

[47] With a considerable amount of algebra, it can be shown that the chronic disease equilibrium is stable for any realistic set of parameter values.

Erratum to:

Continuous Dynamical Systems

Glenn Ledder

Department of Mathematics, University of Nebraska-Lincoln, Lincoln, NE, USA

G. Ledder, *Mathematics for the Life Sciences: Calculus, Modeling, Probability, and Dynamical Systems*,
Springer Undergraduate Texts in Mathematics and Technology, DOI 10.1007/978-1-4614-7276-6_7,
© Springer Science+Business Media, LLC 2013

DOI 10.1007/978-1-4614-7276-6_8

The paperback and online versions of the book contain some errors, and the corrections to these versions are given below:

Page 379, please replace Reference 6

"Stafford MA, L Corey, Y Cao, ES Daar, AS Perelson. Modelling viral and immune system dynamics. *Journal of Theoretical Biology*, **203**, 285–301 (2000)"

by

"Perelson AS and PW Nelson, Mathematics analysis of HIV-1 dynamics *in vivo. SIAM Review*, **41**, 3–44 (1999)"

The online version of the original chapter can be found at
http://dx.doi.org/10.1007/978-1-4614-7276-6_7

G. Ledder, *Mathematics for the Life Sciences: Calculus, Modeling, Probability, and Dynamical Systems*,   E-1
Springer Undergraduate Texts in Mathematics and Technology, DOI 10.1007/978-1-4614-7276-6_8,
© Springer Science+Business Media, LLC 2013

Appendix A
Additional Topics in Discrete Dynamical Systems

This appendix contains three topics that are thematically related to Chapter 6. Discrete nonlinear systems (Section A.1) are analogous to the continuous nonlinear systems of Chapter 7, but are more complicated. These models are overused in practice. Discrete models are indicated when life history events are synchronous, such as for many plants and fish that reproduce once per year; continuous models should be used in other cases. Markov chains (Section A.2) are dynamical systems in which the variables are probabilities rather than populations. The development focuses on the problem of measuring the extent to which two different species are genetically related. Boolean algebra models (Section A.3) provide a simplified setting for dynamical systems that is useful for analyzing complicated systems, such as those in gene regulatory networks.

A.1 Discrete Nonlinear Systems

After studying this section, you should be able to:

- Run simulations for discrete nonlinear systems.
- Analyze the stability of fixed points for discrete nonlinear systems.

On conceptual grounds, it would have made sense to present discrete nonlinear systems alongside their linear counterparts in Chapter 6. There are two reasons why this topic appears in a separate appendix instead. First, the analysis of discrete nonlinear systems is easier to understand after studying continuous nonlinear systems.[1] Second, one should only use discrete nonlinear systems when absolutely mandated by the synchrony of life events. There are several reasons for this. As we saw in Chapter 5, discrete models have some mathematical complications that are absent from continuous models. Here again, we will see complications arising in a model with only weak nonlinearities in Example A.1.4. A further problem with discrete models is the weakness of graphical techniques, which are much more complicated and far less general than those for continuous models. In our study of single dynamic models in Chapter 5, we ran simulations and used linearization techniques to determine stability for both discrete and continuous models. These techniques generalize to higher order systems in both cases.[2] The very powerful phase line method for continuous equations[3] generalizes to phase plane methods

---

[1] See Section 7.5.

[2] See Chapter 7 for continuous systems.

[3] Section 5.4.

G. Ledder, *Mathematics for the Life Sciences: Calculus, Modeling, Probability, and Dynamical Systems*, Springer Undergraduate Texts in Mathematics and Technology, DOI 10.1007/978-1-4614-7276-6,
© Springer Science+Business Media, LLC 2013

for continuous systems.[4] In contrast, the graphical technique of cobweb analysis for discrete equations,[5] which is less powerful than the phase line method, does not generalize to multi-component discrete systems, with the exception of a limited class of discrete systems that can be recast as single-component models. We begin our study with this limited class for which Chapter 5 methods are applicable.

Suppose a population is divided between two age classes, young (Y) and adults (A), with the additional requirements that all surviving young become adults in the next year and all adults die after reproducing. With these restrictive assumptions, the populations can be modeled by equations of the form

$$Y_{n+1} = f(A_n),$$
$$A_{n+1} = g(Y_n).$$

Because generations do not overlap, we can combine the two equations into a single one that spans a discrete interval of 2 years; for example,

$$Y_{n+2} = f(A_{n+1}) = f(g(Y_n)).$$

We can then define the composite function $h(Y) = f(g(Y))$ and rewrite the system as

$$Y_{n+2} = h(Y_n).$$

This is a single discrete dynamic equation, albeit with a time interval of 2 years rather than 1, and we can apply all of the methods for dealing with such equations from Chapter 5.

Example A.1.1. Suppose the adults of a population occupy an ecological niche that allows for density-independent reproduction, while the young occupy a niche in which resources are limited. If we assume that the survival/recruitment of the young follows the saturation curve used for Holling type II dynamics, we have the model

$$Y_{n+1} = fA_n, \qquad f > 0,$$
$$A_{n+1} = \frac{Y_n}{b + Y_n}, \qquad b > 0.$$

These combine to make a 2-year discrete model

$$Y_{n+2} = \frac{fY_n}{b + Y_n} \equiv h(Y_n).$$

Fixed points satisfy $Y_{n+2} = Y_n = Y$, resulting in the equation

$$Y = h(Y) = \frac{fY}{b + Y}.$$

Thus, $Y = 0$ is a fixed point, and a second fixed point $Y = f - b$ exists if $f > b$. The fixed point $Y^\star$ is stable if and only if $|h'(Y^\star)| < 1$. We have

$$h'(Y) = \frac{fb}{(b + Y)^2} \, ;$$

---

[4] See Section 7.3.

[5] See Section 5.2.

thus,

$$h'(0) = \frac{f}{b}, \qquad h'(f-b) = \frac{b}{f}.$$

The fixed point $Y = 0$ is stable if $f < b$, while the fixed point $Y = f - b$ is stable whenever it exists. Cobweb analysis confirms these conclusions and also demonstrates that the long-term behavior of the population always converges to the stable fixed point.[6]   □

A.1.1 Linearization for Discrete Nonlinear Systems

In Chapter 6, we saw that the quantity λ in the discrete model $N_{t+1} = \lambda N_t$ generalizes to the eigenvalues for a discrete linear system. We also saw in Chapter 7 that the stability of an equilibrium solution of a continuous nonlinear system is determined by the eigenvalues of the Jacobian matrix at the corresponding equilibrium point.[7] The same connections hold for discrete nonlinear systems. Near a fixed point, a discrete nonlinear system can be approximated by a linear system represented by the Jacobian matrix.[8] The eigenvalues of the Jacobian determine the behavior near the corresponding fixed point, with $|\lambda| < 1$ for all λ required for stability. It is often more convenient to use the equivalent criterion $|\lambda|^2 < 1$ for all λ; this generalizes to the case of complex eigenvalues, with the magnitude of a complex number defined by $|a+ib|^2 = a^2 + b^2$.

Example A.1.2. The Jacobian for the system of Example A.1.1 is

$$J = \begin{pmatrix} 0 & f \\ \frac{b}{(b+Y)^2} & 0 \end{pmatrix}.$$

Thus, the eigenvalues are given by

$$0 = \det \begin{pmatrix} -\lambda & f \\ \frac{b}{(b+Y)^2} & -\lambda \end{pmatrix} = \lambda^2 - \frac{fb}{(b+Y)^2}.$$

The stability requirement $\lambda^2 < 1$ yields the inequality

$$\frac{fb}{(b+Y)^2} < 1.$$

This is the same requirement that we obtained in Example A.1.1 using the method for single discrete equations.   □

Usually it requires significant algebraic calculation to compute eigenvalues for a model with arbitrary parameters. Some of this calculation can be avoided by making use of the Jury conditions for stability, which are a set of inequalities written in terms of quantities calculated directly from the Jacobian matrix.[9]

---

[6] See Problem A.1.1.

[7] See Section 7.5.

[8] See Problem A.1.2.

[9] These correspond to the Routh–Hurwitz conditions for continuous systems.

Theorem A.1.1 (Jury Conditions for a System of Two Components). *Let* **J** *be the Jacobian matrix that represents a nonlinear system of two components near a fixed point* $\mathbf{x}^\star$. *The fixed point is asymptotically stable if*

$$|tr\,\mathbf{J}| < 1 + \det\mathbf{J} < 2,$$

where $tr\,\mathbf{J}$ *is the sum of the main diagonal entries of the Jacobian. The fixed point is unstable if any one of the inequalities*

$$tr\,\mathbf{J} > 1 + \det\mathbf{J}, \qquad tr\,\mathbf{J} < -1 - \det\mathbf{J}, \qquad \det\mathbf{J} > 1$$

is true.

Example A.1.3. For the model of Examples A.1.1 and A.1.2, we have

$$\mathrm{tr}\,\mathbf{J} = 0, \qquad \det\mathbf{J} = -\frac{fb}{(b+Y^\star)^2}.$$

The first of these results reduces the Jury conditions to $-1 < \det\mathbf{J} < 1$, and the second of this latter pair is automatically satisfied because the determinant is negative. The remaining condition, $\det\mathbf{J} > -1$, reduces to $(b+Y^\star)^2 > fb$, which is equivalent to the condition $fb/(b+Y^\star)^2 < 1$ that we derived in Example A.1.2. $\qquad\square$

As with the Routh–Hurwitz conditions for continuous systems, there are Jury conditions for any size, but they get more complicated as the size increases. Here we present the Jury conditions for 3×3 matrices.

Theorem A.1.2 (Jury Conditions for a System of Three Components). *Let* **A** *be the Jacobian matrix for a three-component discrete system at a fixed point. Let* $\mathbf{A_k}$ *be the* 2×2 *matrix obtained from* **A** *by deleting row k and column k. Define* c_1, c_2, *and* c_3 *by*

$$c_1 = -tr\,\mathbf{A}, \; c_2 = \sum_{k=1}^{3} \det\mathbf{A_k}, \; c_3 = -\det\mathbf{A},$$

where $tr\,\mathbf{A}$ *is the sum of the diagonal elements of* **A**. *Then the fixed point of the original system is asymptotically stable if*

1. $1 + c_1 + c_2 + c_3 > 0$,
2. $1 - c_1 + c_2 - c_3 > 0$, *and*
3. $|c_2 - c_1 c_3| < 1 - c_3^2$.

The fixed point is unstable if any of the inequalities is reversed.

A.1.2 A Structured Population Model with One Nonlinearity

In Chapter 7, we examined three-component continuous models in which each equation is nonlinear. Here we consider a three-component discrete model with only one nonlinear equation. This is enough to produce some very complicated behavior.

Example A.1.4. Consider a species whose life history encompasses three yearly stages, with the two older stages occupying the same ecological niche, but only the oldest stage reproducing. With linear recruitment for the two older classes and Beverton–Holt[10] recruitment for the youngest class, we have the model

$$L_{t+1} = \frac{fA_n}{b + Y_n + A_n},$$

$$Y_{n+1} = s_1 L_n,$$

$$A_{n+1} = s_2 Y_n,$$

where $f, b > 0$ and $0 < s_1, s_2 < 1$. Figure A.1.1 shows plots of Y and A for three simulations using common values $f = 10,000$, $b = 15$, $s_1 = 0.03$, along with different values for s_2. These plots show some of the behaviors that the system can exhibit. □

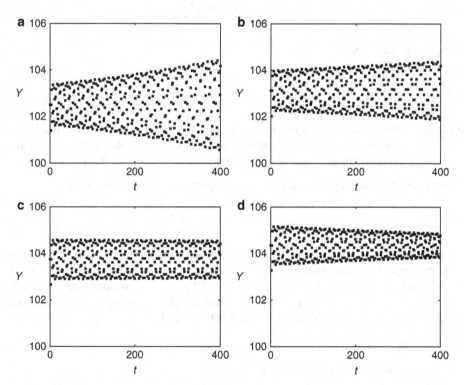

Fig. A.1.1 Y values for the model of Example A.1.4, using $f = 10,000$, $b = 15$, $s_1 = 0.03$, with (**a**) $s_2 = 0.595$, (**b**) $s_2 = 0.600$, (**c**) $s_2 = 0.605$, (**d**) $s_2 = 0.610$

How were the parameter sets for the plots of Figure A.1.1 chosen so as to illustrate different behaviors for nearly identical parameter values? Not by trial and error. The only realistic way to identify cases such as these is to do a thorough analysis of the model.[11]

---

[10] See Section 5.1.

[11] Stumbling across a good set of parameter values by chance is like throwing a dart at a fog-obscured board and managing to hit the bullseye. After the analysis, finding good parameter values is like walking up to the board and stabbing the bullseye with the dart.

Example A.1.5. Consider the model of Example A.1.4. The Jacobian has the form

$$J = \begin{pmatrix} 0 & J_{12} & J_{13} \\ s_1 & 0 & 0 \\ 0 & s_2 & 0 \end{pmatrix},$$

where

$$J_{12} = \frac{-fA}{(b+Y+A)^2}, \qquad J_{13} = \frac{f(b+Y)}{(b+Y+A)^2}.$$

With c_1, c_2, c_3 defined as in Theorem A.1.2, we have

$$c_1 = 0, \quad c_2 = -s_1 J_{12} = \frac{s_1 fA}{(b+Y+A)^2} \geq 0, \quad c_3 = -s_1 s_2 J_{13} = \frac{-s_1 s_2 f(b+Y)}{(b+Y+A)^2} < 0. \quad \text{(A.1.1)}$$

For convenience, we define

$$k_3 = -c_3 = \frac{s_1 s_2 f(b+Y)}{(b+Y+A)^2} > 0, \qquad \text{(A.1.2)}$$

whereupon the Jury conditions become

1. $1 + c_2 - k_3 > 0$,
2. $1 + c_2 + k_3 > 0$, and
3. $c_2 < 1 - k_3^2$.

The second of these conditions is automatically satisfied, given $c_2, k_3 > 0$. The third requires $k_3 < 1$, which guarantees that the first condition is satisfied as well. Hence, the stability hinges on the third condition. The subsequent analysis has to be done for each fixed point.

1. There is an obvious fixed point in which all components are 0. For this point,

$$c_2 = 0, \qquad k_3 = \frac{s_1 s_2 f}{b}.$$

Thus, the extinction point is stable if $s_1 s_2 f < b$. The quantity $s_1 s_2 f / b$ has a clear biological interpretation. If we start with L_0 individuals in year 0, this cohort yields $Y_1 = s_1 L_0$, $A_2 = s_1 s_2 L_0$, and $L_3 = s_1 s_2 f L_0 / (b + A_2)$. In the limit of a small population, we have $L_3 / L_0 = s_1 s_2 f / b$. Thus, $s_1 s_2 f / b$ represents the maximum possible population growth rate, and of course the population dies out if this quantity is less than 1.

2. Now assume $b < s_1 s_2 f$.

   a. Any non-extinction fixed points must satisfy

   $$Y^\star = s_1 L^\star = \frac{s_1 fA^\star}{b + Y^\star + A^\star} = \frac{s_1 s_2 fY^\star}{b + Y^\star + A^\star}.$$

   With $Y^\star > 0$, this equation reduces to

   $$b + Y^\star + A^\star = s_1 s_2 f. \qquad \text{(A.1.3)}$$

   Given $A^\star = s_2 Y^\star$, this equation reduces to

   $$b + (1 + s_2)Y^\star = s_1 s_2 f,$$

from which we see that there is one fixed point with $Y^\star > 0$, given by

$$Y^\star = \frac{s_1 s_2 f - b}{1 + s_2}, \qquad A^\star = \frac{s_2(s_1 s_2 f - b)}{1 + s_2}. \tag{A.1.4}$$

Note that this fixed point exists under the condition for which the extinction fixed point is unstable.

b. Analysis of this fixed point requires some messy algebra. We can make it a little less messy by defining a new parameter

$$\beta = \frac{b}{s_1 f} < s_2. \tag{A.1.5}$$

In terms of this parameter, we have

$$Y^\star = \frac{s_1 f (s_2 - \beta)}{1 + s_2}. \tag{A.1.6}$$

From here, some careful algebra yields[12]

$$c_2 = \frac{Y^\star}{s_1 s_2 f}, \qquad 1 - k_3 = \frac{Y^\star}{s_1 f}, \qquad 1 + k_3 = 2 - (1 - k_3) = 2 - \frac{Y^\star}{s_1 f}. \tag{A.1.7}$$

By rewriting the stability requirement (the third Jury condition) as

$$c_2 < 1 - k_3^2 = (1 - k_3)(1 + k_3),$$

and substituting from (A.1.7), we get one factor of $Y^\star$ to cancel, leaving the inequality

$$\frac{1}{s_2} < 2 - \frac{Y}{s_1 f}.$$

Substituting the formula for $Y^\star$ and doing more algebra[13] eventually results in the inequality

$$s_2^2 + (1 + \beta)s_2 - 1 > 0. \tag{A.1.8}$$

See Problem A.1.5 for a complete analysis of this condition; here we simply note that the condition is obviously not satisfied if s_2 is too small, given the additional restriction $\beta < s_2$. This is an interesting result. The model is similar to the one-component Beverton–Holt model discussed in the Sections 5.1, 5.2, and 5.5 problem sets; however, there is a wide range of parameter values for which the three-component model has no stable fixed points.[14] As is frequently the case with discrete models, the behavior can be very complicated even when the model is fairly simple.   □

---

[12] See Problem A.1.3a.

[13] See Problem A.1.3b.

[14] See Problem A.1.4 for more simulations of model behavior.

Problems

A.1.1. Plot cobweb diagrams for the cases $f < b$ and $f > b$ for the model

$$Y_{n+2} = \frac{fY_n}{b + Y_n}$$

and use them to confirm the conclusions of Example A.1.1.

A.1.2. Suppose $(X^\star, Y^\star)$ is a fixed point for a system

$$X_{n+1} = f(X_n, Y_n), \qquad Y_{n+1} = g(X_n, Y_n).$$

Near the fixed point, we can replace X and Y by

$$X = X^\star + \varepsilon x, Y = Y^\star + \varepsilon y.$$

Use the method of Section 7.5 to linearize the system near the fixed point. Conclude that the system is approximately

$$\mathbf{z}_{n+1} = \mathbf{J}(X^\star, Y^\star)\mathbf{z}_n,$$

where $\mathbf{z}$ is the vector whose components are x and y and $\mathbf{J}(X^\star, Y^\star)$ is the Jacobian matrix evaluated at the fixed point.

A.1.3. (a) Derive Equation (A.1.7).
(b) Derive Equation (A.1.8).

A.1.4.*

(a) Run numerical simulations for the model

$$L_{t+1} = \frac{fA_n}{b + Y_n + A_n},$$
$$Y_{n+1} = s_1 L_n,$$
$$A_{n+1} = s_2 Y_n,$$

with parameter values $f = 10,000$, $b = 15$, $s_1 = 0.03$, and $s_2 = 0.604$ and initial conditions $L_0 = 3,482$, $Y_0 = 106$, and $A_0 = 61$. Plot graphs of Y for 1,000 time steps and for 10,000 time steps. [Do not connect the points!] Is there a stable fixed point?
(b) Repeat part (a) with $s_2 = 0.605$.
(c) Repeat part (b) with $b = 210$ and just 100 time steps.
(d) Do the results of these simulations agree with the analysis of Example A.1.5?

A.1.5. This problem completes the general analysis begun in Example A.1.5. Suppose $0 < \beta < s_2$.

(a) Use the case $\beta = 0$ to find a value s_2^+ big enough so that the inequality

$$s_2^2 + (1 + \beta)s_2 - 1 > 0$$

is satisfied for any β values.
(b) Use the case $\beta = s_2$ to find a value s_2^- small enough so that the inequality from part (a) is not satisfied for any β values.

(c) Plot the triangle formed by the lines $\beta = 0$, $\beta = s_2$, and $s_2 = 1$ in the $s_2\beta$ plane. Solve the equation

$$s_2^2 + (1 + \beta)s_2 - 1 > 0$$

for β on the interval $s_2^- < s_2 < s_2^+$ and add this curve to the plot.

(d) The region inside the triangle in the graph of part (c) represents the permissible sets of parameter values, and the curve divides this region into two portions. Explain the system behavior that results from points (s_2, β) in each portion of the plot.

Parasitoids[15] are animals whose life history combines a free-living stage and a parasitic stage. Many wasps and flies, for example, lay eggs in a caterpillar or other insect host. When the eggs hatch, the wasp larvae consume the host from the inside. These animals are of significant interest in ecology because they are common in nature and because they can be useful as bio-control agents. In many cases, parasitoids have a synchronized life cycle, with one generation per year; in these cases, a discrete nonlinear model is most appropriate. Models for host-parasitoid systems are explored in Problems A.1.6–A.1.8.

A.1.6. In this problem, we develop a general model for host-parasitoid systems. Assume that the host population dynamics is given by $H_{n+1} = R_0 H_n$ in the absence of parasitoids, where $R_0 > 1$. Let f be the fraction of hosts not parasitized, so that we have $H_{n+1} = R_0 f H_n$.

(a) Assume that each parasitized host results in c parasitoids in the next generation. Write down the equation for the parasitoid population dynamics.

(b) Assume the generic form $f(aP)$ for the fraction of hosts that are parasitized. Nondimensionalize the resulting model by multiplying the H equation by ac and the P equation by a and then using the substitutions $p = aP$, $h = acH$.

(c) Compute the Jacobian matrix for a fixed point $(h, p) = (h^\star, p^\star)$ with both components positive.

(d) The H equation yields a simple equation for the parasitoid population at a fixed point in terms of the function f. Use this equation to eliminate f from the Jacobian. Note that it still contains f'.

(e) Compute $\mathrm{tr}\,\mathbf{J}$ and $\det \mathbf{J}$. Use the assumption $f' < 0$ to show that $\mathrm{tr}\,\mathbf{J}$ is positive and that the first of the two conditions in Theorem A.1.1 is always satisfied. Hence, stability is determined by the requirement $\det \mathbf{J} < 1$.

(f) Use the other fixed point equation to eliminate $h^\star$ from the formula for the determinant. Conclude that the fixed point determined by $f(p^\star) = 1/R_0$ is stable if and only if $\det \mathbf{J} < 1$, where

$$\det \mathbf{J} = -p^\star f'(p^\star) \frac{R_0{}^2}{R_0 - 1} > 0.$$

A.1.7.* The dimensionless Nicholson–Bailey host-parasitoid model is

$$h_{n+1} = R_0 h f(p_n), \quad R_0 > 1,$$
$$p_{n+1} = h_n[1 - f(p_n)],$$

with the host survival function $f(p) = e^{-p}$ (see [1]).

(a) Calculate the fixed point $p^\star$ and the corresponding value of $\det \mathbf{J}$.

(b) Show that this fixed point is unstable by putting the stability requirement into the form $g(R_0) > 1$ and using calculus to show $g(R_0) \le 1$.

---

[15] This is pronounced PAR-uh-si-toid.

(c) Run a simulation using $R_0 = 1.1$, $h_0 = 1$, $p_0 = 0.4$, and a total of 120 years.

(d) Are the results consistent with part (a)? Are they biologically realistic?

A.1.8. The Nicholson–Bailey model corresponds to the assumption of a Poisson distribution for the number of parasitoid attacks in a given amount of time. A different model is obtained from the assumption of a different distribution, called the negative binomial distribution. This model is

$$h_{n+1} = R_0 h f(p_n), \qquad R_0 > 1,$$
$$p_{n+1} = h_n[1 - f(p_n)],$$

with the host survival function

$$f(p) = \left(1 + \frac{p}{k}\right)^{-k}.$$

The parameter $k > 0$ measures the similarity to the Poisson distribution, with the latter achieved in the limit $k \to \infty$.

(a) Calculate the fixed point $p^\star$ for $k = 2$ and determine the stability of the fixed point.

(b) Repeat part (a) for $k = 0.5$.

(c) Run a simulation for the model using $k = 2$, $R_0 = 1.1$, $h_0 = 1$, $p_0 = 0.4$, and a total of 120 years.

(d) Repeat part (c) with $k = 0.5$.

(e) Are the results of the simulations consistent with the theoretical results?

The flour beetle *Tribolium confusum* is often used as a model insect species for both theory and experiment. The best known model for this population is a discrete stage-structured model with a time step of 2 weeks [2, 3]. Flour beetles are larvae for about 2 weeks; then they go through three life stages (nonfeeding larvae, pupae, and callow adults) in approximately 2 weeks before becoming adults that can live for several years.[16] The model assumes that adults lay an average of b eggs, but these numbers lead to far fewer larvae because larvae and adults eat eggs. A fraction s of larvae survive to become "pupae," the pupae are either eaten by adults or survive to become adults, and a fraction m of adults die in each 2-week period. The full model is

$$L_{t+1} = bA_t e^{-\alpha L_t - \beta A_t},$$
$$P_{t+1} = sL_t,$$
$$A_{t+1} = P_t e^{-\gamma A_t} + (1 - m)A_t.$$

Typical parameter values are given in Table A.1.1.[17] The value of γ can be manipulated experimentally by removing additional pupae by hand at each census. The flour beetle model is explored in Problems A.1.9–A.1.11.

---

[16] It is common in practice to use discrete models for cases such as this, where the stage durations are approximately comparable. My view is that discrete models should only be used when life history events are synchronous, which is not the case for flour beetles. Discrete models impose synchronicity, which adds complexity that is not part of the actual biological setting.

[17] Note that large numbers of measurements allow for relatively small confidence intervals; for example, the 95 % confidence interval for γ for [3] is $(0.004446, 0.004792)$. However, these results are not reproducible, as the value of γ for [2] indicates. One should not put too much faith in reported parameter values, and it does not make sense to use values that appear to indicate a high degree of precision. Two significant digits is as much as is ever warranted for ecological data.

Table A.1.1 Reported parameter values for the *Tribolium* model

Reference	b	s	m	α	β	γ
Cushing et al. [2]	11.6772	0.4871	0.1108	0.0093	0.0110	0.0178
Dennis et al. [3]	10.45	0.8000	0.007629	0.01731	0.01310	0.004619

A.1.9. One of the interesting features of the discrete flour beetle model is that it can exhibit chaotic solutions when there are no stable fixed points. In this problem, we show that cannibalism of eggs by larvae is required for this to occur. To that end, we take $\alpha = 0$ and show that there is always a unique stable fixed point. The result is not quite general, but it can be obtained with the reasonable restrictions $1 < bs < 10$ and $m < 1/8$.

(a) Define additional parameters

$$k = \ln \frac{bs}{m}, \qquad q = mk,$$

which will turn out to be useful in the analysis. Show that q is an increasing function of both bs and m, so that the maximum value of q occurs at $bs = 10$ and $m = 1/8$. Conclude that $q < 0.6$. Note also that $k > 0$.

(b) Write down the system of three equations to determine the fixed points.

(c) Determine the Jacobian matrix for the general case. Simplify the last entry by using one of the equations from part (a) to eliminate P in favor of A. The Jacobian should then include all of the parameters as well as the unknown value $A^\star$ for the fixed point.

(d) Show that the extinction fixed point is unstable with the assumed parameter value restrictions. Conclude that the population persists.

(e) Derive the formula

$$A^\star = \frac{k}{\beta + \gamma}$$

for the unique positive fixed point. We will show that this fixed point is stable.

(f) Use the Jacobian matrix and the result for $A^\star$ to obtain

$$c_1 = \frac{\gamma}{\beta + \gamma} q + m - 1, \quad c_2 = 0, \quad c_3 = \frac{\beta}{\beta + \gamma} q - m,$$

where the c_i are those used in the Jury conditions.

(g) Compute $c_1 + c_3$ and show that the first two Jury conditions are satisfied, using the result of part (a).

(h) Find the maximum values of $|c_1|$ and $|c_3|$ and show that

$$|c_1 c_3| + c_3^2 < 1.$$

Conclude that the third Jury condition is also satisfied.

A.1.10. Consider the discrete flour beetle model for the special case $\gamma = 0$, in which there is no cannibalism of pupae by adults. Use average parameter values of $b = 11$, $s = 0.64$, $m = 0.06$, $\alpha = 0.013$, $\beta = 0.012$.

(a) Determine the Jacobian matrix for the extinction fixed point and use the Jury conditions to show that it is unstable.

(b) Determine the Jacobian matrix for a fixed point with positive values. This matrix can be simplified with some algebraic substitutions. Rewrite the entries in the top row in the form kL, where k is whatever is left over after dividing the entry by L. The values of k do not include any exponential functions if you substitute for L from the first fixed point equation.

(c) Solve the fixed point equations to find the unique fixed point with positive population values.
(d) Determine the stability of the positive fixed point by computing the eigenvalues of the Jacobian or by using the Jury conditions.
(e) Run a simulation showing the behavior of the system starting with a population that consists of 10 adults and running for 2 years (each time step represents 2 weeks). What is the long-term stable behavior of the system? Is this consistent with the stability analysis?

A.1.11. Consider the discrete flour beetle model with average parameter values $b = 11$, $s = 0.64$, $m = 0.06$, $\alpha = 0.013$, $\beta = 0.012$, $\gamma = 0.011$.

(a) Determine the Jacobian matrix for a fixed point with positive values, simplifying it as in part (b) of Problems A.1.9 and A.1.10.
(b) Eliminate the pupae from the three fixed point equations to obtain a pair of fixed point equations for L and A.
(c) Solve one of the remaining fixed point equations for L in terms of A and substitute it into the other to obtain a single nonlinear equation for A. Solve this equation numerically and then determine the corresponding fixed point values for L and P.
(d) Use the Jury conditions or compute eigenvalues to determine the stability of the positive fixed point.
(e) Run a simulation showing the behavior of the system starting with a population that consists of 10 adults. What is the long-term stable behavior of the system? Is this consistent with the stability analysis?

A.2 Markov Chains

After studying this section, you should be able to:

- Construct Markov chain models for stochastic processes in which the state at some future time depends solely on the current state.
- Find steady-state probability distributions for Markov chains.

In Chapter 6, we used equations of the form

$$\mathbf{x}_{t+1} = \mathbf{M}\mathbf{x}_t$$

to model changes in a set of dynamic variables that represent population sizes. The same mathematical structure applies to models that track dynamic changes in probabilities. These mathematical models have become useful tools in molecular biology, leading to important discoveries about the development of species in evolutionary history. The full subject is very complicated, but we can get a general sense of the possibilities by examining the simplest mathematical model for genetic change.

In this section, we consider the problem of estimating the *phylogenetic distance* between species, a concept that refers to the overall amount of genetic difference between genomes. Phylogenetic distance is important because it has caused significant changes in our understanding of the evolutionary relationships of species. For example, scientists had long thought that chimpanzees were more closely related to gorillas than to humans. We now know that the phylogenetic distance between the chimpanzee and human genomes is less than that between the chimpanzee and gorilla genomes, and therefore chimpanzees are more closely related to humans.[18]

---

[18] We are thinking specifically of the total amount of genetic difference in the genomes. One could still argue that the smaller number of differences between chimpanzee and human are more important than the larger number between chimpanzee and gorilla.

A.2.1 Some Scientific Background

We've seen earlier that DNA carries information in the pattern of four types of nucleotides, labeled A, G, C, and T.[19] DNA containing these nucleotide sequences is arranged in long molecules called chromosomes; taken together, these chromosomes constitute the genome of the individual. We can think of the DNA in a genome as falling broadly into three categories:

1. Essential DNA is in the form of genes that are crucial to species survival, such as the genes that determine the network of blood vessels or the function of organs. These genes are largely resistant to change because such changes from the norm tend to be harmful. The corresponding DNA may be different between species, but will likely be almost the same for individuals in the same species.
2. Some DNA is in the form of genes that play at best a small role in species survival, such as the genes that determine hair color. The corresponding DNA shows significant variation within a population. This DNA is useful for identification of individuals in a species.
3. There is also non-essential DNA, which does not affect the characteristics of the organism but is merely a residue of the evolutionary past. At one time it was thought that most DNA is non-essential, but scientists now estimate that this category encompasses about 20% of a genome.

The genome of a species can be thought of as being defined by the combination of its essential and non-essential DNA. Although these portions are largely inherited intact from one's parents, there are two important processes that cause them to change over time: natural selection and mutation. Essential DNA is subject to natural selection. If there are individual variations in a portion of this DNA, then some individuals will be more successful at survival and reproduction than others; over time, the population will be dominated by those individuals who have the more successful variation. In contrast, non-essential DNA is not subject to natural selection.

Natural selection must, of course, have individual variation to work with. This variation results from genetic mutations that occur in the production of sperm and egg cells of organisms that reproduce sexually. Mutations are rare events,[20] and those that alter the individual's fitness are either removed by natural selection or replace earlier versions. Successful mutations and mutations in non-essential DNA can accumulate over evolutionary time. This is what allows us to associate the differences in similar regions of DNA between species with their phylogenetic distance. While there is not a simple linear relationship between number of mutations and evolutionary time, it seems reasonable that more evolutionary time should result in more mutations. Thus, a larger difference between species A and B than between species B and C indicates a more recent common ancestor for B and C. The challenge is to use mathematical models to quantify the relationship between genome differences and phylogenetic distance.

The full story is actually much more complicated.

1. Substitutions appear to account for only 35–50% of mutations [4]. There are a number of other types of mutations, but the most common appear to be insertions and deletions, in which a small bit of DNA is inserted between two formerly adjacent nucleotides, or a small bit is lost from a section of DNA. These mutations are much harder to identify over long

---

[19] See Section 3.2.

[20] See Problem 3.7.15d.

periods of evolutionary time and harder to quantify.[21] Methods that consider only substitutions can only be used on portions of a chromosome in which any insertions or deletions are known.

2. Non-essential DNA is subject to mutation without natural selection, which raises the question of how mutations in non-essential DNA could be identical for individuals of the same species.

3. Natural selection occurs at the level of genome function, not genome structure. Changes in individual nucleotides do not change the function in cases where both the new and old codons make the same protein. A glance at the genetic code[22] shows that there is a lot of redundancy. For example, the third nucleotide is irrelevant in 8 of the 16 possibilities for the first two codons.

4. Natural selection is based on preferential survival of some mutations over others. The accumulation rate of a mutation depends on how much survival difference that mutation makes. For example, HIV is particularly insidious because one mutation for resistance to a drug can quickly become dominant in a population. We need a new vaccine for influenza every year, but the chicken pox vaccine is the same now as it was when first created. Hence, the molecular clock that connects mutations with time does not tick at a constant rate across species or even within species. The molecular clock is close to constant for species that are closely related and for genome portions that have the same or no function.

A.2.2 A Model for DNA Change

Suppose we leave the details of what portion of a genome to study to the molecular biologists. Assume that there are J nucleotides in a strand that has had no insertions or deletions and let N be the unknown number of generations that have passed between the ancestral strand and the contemporary strand. For any position in the sequence, the nucleotide must be either A, G, C, or T. By comparing the ancestral and contemporary strands, we can measure the fraction of DNA sites that are different between the two. This value, commonly called β, is a measure of the difference between genomes.

Now let α be the probability of a mutation in one site over one generation. Over N generations, we expect the total number of mutations to be αN for each site, yielding a total of $\alpha N J$ for the strand. The product $d = \alpha N$ is the number of mutations per site. This is the phylogenetic distance, which we tentatively assume to be proportional to evolutionary time. Our goal is to infer d from β.

At first thought, this sounds easy. The total number of differences between the strands is βJ and the total number of mutations is $\alpha N J = dJ$. These should be equal, so $d = \beta$. However, this reasoning is flawed. If a site starts as A, mutates to G, and then mutates back to A, with no further changes, then both of these mutations are counted toward dJ. However, the two strands are identical because the second mutation reversed the first one, so neither of them contributes to βJ. Thus, $d > \beta$, because some mutations actually *decrease* the number of differences between the strands. We need a nuanced mathematical model to connect the unknown phylogenetic distance with the known fraction of sequence differences.

---

[21] One extreme case similar to insertion and deletion can be seen in a comparison of the human and chimpanzee genomes. Humans have 23 pairs of chromosomes, while chimpanzees have 24, which seems to refute the claim that the two species are closely related. However, a careful study of human chromosome 2 shows that it appears to consist of two formerly distinct chromosomes that have been joined together, with the two portions corresponding to two of the chimpanzee chromosomes.

[22] See Table 3.2.1.

Let $p_A(n)$, $p_G(n)$, $p_C(n)$, and $p_T(n)$ be the probabilities of having each given nucleotide at a particular site in the nth generation. Mutations from one generation to the next change these probabilities, and we must quantify these changes. The simplest assumption is that all possible changes are equally likely. Since α is the probability of change, and each nucleotide has three possible changes, the probability of any particular change is $\alpha/3$. Of course the probability of no change is $1 - \alpha$. With these assumptions, the probability that a site will contain the nucleotide A at time $n + 1$ is the sum of the probabilities of starting with A and not changing plus the probabilities of starting with one of the others and then changing to A:

$$p_A(n+1) = (1 - \alpha)p_A(n) + \frac{\alpha}{3}p_G(n) + \frac{\alpha}{3}p_C(n) + \frac{\alpha}{3}p_T(n). \tag{A.2.1}$$

Similar equations can be written for the other probabilities in generation $n + 1$, and the four equations can be combined into a single matrix equation of the form

$$\mathbf{x}_{n+1} = \mathbf{M}\mathbf{x}_n.^{23} \tag{A.2.2}$$

Each row in the matrix corresponds to the coefficients in one of the four equations. Since we have arbitrarily chosen the order A, G, C, T, the coefficients of (A.2.1) are the first row of the transition matrix $\mathbf{M}$. The full probability vector $\mathbf{x}$ and the matrix $\mathbf{M}$ are given by

$$\mathbf{x} = \begin{pmatrix} p_A \\ p_G \\ p_C \\ p_T \end{pmatrix}, \quad \mathbf{M} = \begin{pmatrix} 1 - \alpha & \alpha/3 & \alpha/3 & \alpha/3 \\ \alpha/3 & 1 - \alpha & \alpha/3 & \alpha/3 \\ \alpha/3 & \alpha/3 & 1 - \alpha & \alpha/3 \\ \alpha/3 & \alpha/3 & \alpha/3 & 1 - \alpha \end{pmatrix}. \tag{A.2.3}$$

Any model consisting of a matrix equation that represents dynamic changes in vectors of probabilities is called a *Markov chain* model. The specific model we are examining, defined by the assumption that the transition probabilities are all the same, is called the *Jukes–Cantor* model.[24]

A.2.3 Equilibrium Analysis of Markov Chain Models

There are some fundamental mathematical differences between the matrices obtained in structured population models and those obtained in Markov chain models. These differences lead to different features for the corresponding dynamical systems.

1. Structured population models have matrices in which the ij entry represents the contribution of population component j at time n to population component i at time $n + 1$. Thus, none of the entries can be negative. Nonnegative matrices have three special properties: (1) the eigenvalue of largest magnitude is always positive, (2) there is a one-parameter family

---

[23] Given that we have always done matrix-vector multiplication with the matrix on the left and the vector on the right, this is the natural way to proceed. Unfortunately, most of the literature on Markov chains makes the opposite choice. In our matrix $\mathbf{M}$, the entry in row i and column j represents the probability of a transition from state j to state i. In the more common representation of Markov chains, the matrix is written so that the entry in row i and column j represents the probability of a transition from state i to state j. This sounds more natural, but it means that the probability vectors must be written as rows rather than columns and the matrix multiplication must have the vector on the left. This necessitates changes in the definition of eigenvectors, which is an unfortunate complication.

[24] Other models make more sophisticated assumptions about the relative probabilities of specific substitutions. The Jukes–Cantor model illustrates the important features of Markov chain models and phylogenetic distance while keeping complications to a minimum.

of eigenvectors corresponding to this dominant eigenvalue, and (3) the eigenvector corresponding to the dominant eigenvalue is positive. These properties guarantee that solutions will approach an asymptotic growth rate that corresponds to a stable distribution of component populations.

2. Markov models have matrices in which the ij entry represents the probability of being in state i at time $n + 1$ after having been in state j at time n. Thus, each entry is between 0 and 1. Moreover, there is always exactly one state at the end of each time step, so the total of the probabilities for any time step must be 1. This means that the sum of entries in each column is 1. Most Markov matrices share the same properties as nonnegative matrices, with the dominant eigenvalue $\lambda = 1$.[25] This means that Markov models generally have an equilibrium solution that represents the stable distribution of probabilities.

Example A.2.1. Let $\mathbf{M}$ be the matrix of (A.2.3). Finding eigenvalues of a matrix this size is outside the scope of our presentation; however, we can start with the assumption that $\lambda = 1$ is an eigenvalue. If $\mathbf{x}$ is an eigenvector corresponding to $\lambda = 1$, then it satisfies the equation

$$(\mathbf{M} - \mathbf{I})\mathbf{x} = \mathbf{0}, \quad \mathbf{M} - \mathbf{I} = \begin{pmatrix} -\alpha & \alpha/3 & \alpha/3 & \alpha/3 \\ \alpha/3 & -\alpha & \alpha/3 & \alpha/3 \\ \alpha/3 & \alpha/3 & -\alpha & \alpha/3 \\ \alpha/3 & \alpha/3 & \alpha/3 & -\alpha \end{pmatrix}.$$

The components of $\mathbf{x}$ must satisfy a system of four equations, each corresponding to a row of $\mathbf{M} - \mathbf{I}$. Such a system would normally be difficult to solve, but here we can observe that the entries in each row sum to 0. If all four components of the vector are the same, then the products of coefficients and components will also sum to 0. The stable distribution of probabilities has to be an eigenvector, and as a set of probabilities it also has to sum to 1, which means that each probability is 1/4. This should not be surprising, as the symmetry in the rule that determines the probability of each possible mutation represents a process in which none of the nucleotides is favored over the others.   □

A.2.4 Analysis of the DNA Change Model

The initial goal of our analysis is to connect the measured value of β with the phylogenetic distance $d = \alpha N$.[26] The equilibrium distribution discovered in Example A.2.1 is of no help in accomplishing this goal; by definition, this is the distribution we expect to see as $N \to \infty$. Instead, we proceed by a method that follows the strategy of calculating some quantity in two different ways, one involving β and the other involving α and N. The method requires us to use another eigenvector in addition to the one for $\lambda = 1$. This calculation is beyond the scope of our treatment, so we simply present the result.

---

[25] There are some additional requirements that guarantee these properties; further discussion of this topic is outside the scope of this presentation.

[26] We used the same strategy in Section 6.1 to calculate eigenvalues.

The vectors

$$\mathbf{v_1} = \begin{pmatrix} 1/4 \\ 1/4 \\ 1/4 \\ 1/4 \end{pmatrix}, \qquad \mathbf{v_2} = \begin{pmatrix} 3/4 \\ -1/4 \\ -1/4 \\ -1/4 \end{pmatrix}$$

are eigenvectors of the matrix $\mathbf{M}$ of (A.2.3) corresponding to the eigenvalues $\lambda_1 = 1$ and $\lambda_2 = 1 - \frac{4}{3}\alpha$.

Check Your Understanding A.2.1:
Verify that the vector $\mathbf{v_2}$ is an eigenvector of $\mathbf{M}$ corresponding to the eigenvalue $\lambda_2(\alpha) = 1 - \frac{4}{3}\alpha$.

Define the vector $\mathbf{u}$ by

$$\mathbf{u} = 3\mathbf{M}^N(\mathbf{v_1} + \mathbf{v_2}).[27] \tag{A.2.4}$$

We now proceed to calculate $\mathbf{u}$ by two different methods, taking advantage of two facts:

1. $\mathbf{v_1}$ and $\mathbf{v_2}$ are eigenvectors, which means that multiplication by $\mathbf{M}$ yields a simple result.
2. The sum $\mathbf{v_1} + \mathbf{v_2}$ is also very simple.

Calculating u in Terms of N and α

The calculation of $\mathbf{u}$ is somewhat tedious, so we leave much of it as a problem. The essential idea is that repeated use of the eigenvector equation $\mathbf{Mv} = \lambda\mathbf{v}$ leads to a more general result,

$$\mathbf{M}^N\mathbf{v} = \lambda^N\mathbf{v}, \tag{A.2.5}$$

with which we eventually obtain the answer

$$\mathbf{u} = \frac{3}{4}\begin{pmatrix} 1 + 3\lambda_2^N \\ 1 - \lambda_2^N \\ 1 - \lambda_2^N \\ 1 - \lambda_2^N \end{pmatrix}, \qquad \lambda_2 = 1 - \frac{4}{3}\alpha. \tag{A.2.6}$$

Estimating u in Terms of β

The matrix $\mathbf{M}^N$ represents the overall transition probabilities for N successive generations. We can't calculate this matrix directly, but we can estimate it. Given that β is the measured fraction of sites that have changed nucleotides over N generations, we can approximate $\mathbf{M}^N$ by

---

[27] There is no obvious reason why this should be helpful. It is always more satisfying when methods have a clear conceptual motivation, but occasionally mathematicians must resort to methods that appear simply as clever tricks.

$$\mathbf{M}^N = \begin{pmatrix} 1-\beta & \beta/3 & \beta/3 & \beta/3 \\ \beta/3 & 1-\beta & \beta/3 & \beta/3 \\ \beta/3 & \beta/3 & 1-\beta & \beta/3 \\ \beta/3 & \beta/3 & \beta/3 & 1-\beta \end{pmatrix}. \tag{A.2.7}$$

This is not entirely correct, as it assumes both that the fraction of changed sites is the same, no matter what the original nucleotide, and that the changed sites are equally distributed among the three possible nucleotides. These assumptions are no worse than the basic Jukes–Cantor assumption about the structure of $\mathbf{M}$, however. Combining (A.2.4) and (A.2.7) yields the result

$$\mathbf{u} = \begin{pmatrix} 3(1-\beta) \\ \beta \\ \beta \\ \beta \end{pmatrix}. \tag{A.2.8}$$

The Jukes–Cantor Distance

Equations (A.2.6) and (A.2.8) provide two different results for the same quantity. Comparing them yields the equation

$$\beta = \frac{3}{4} - \frac{3}{4}\lambda_2^N = \frac{3}{4} - \frac{3}{4}\left(1 - \frac{4}{3}\alpha\right)^N, \tag{A.2.9}$$

which predicts the fraction of sites with changes in terms of the mutation rate and the number of generations. We can solve this equation for N, with the elegant result

$$N = \frac{\ln\left(1 - \frac{4}{3}\beta\right)}{\ln\left(1 - \frac{4}{3}\alpha\right)}, \tag{A.2.10}$$

from which we have

$$d = \alpha\frac{\ln\left(1 - \frac{4}{3}\beta\right)}{\ln\left(1 - \frac{4}{3}\alpha\right)} = \frac{\alpha}{\ln\left(1 - \frac{4}{3}\alpha\right)}\ln\left(1 - \frac{4}{3}\beta\right). \tag{A.2.11}$$

This result still appears to depend on α, which is difficult to measure. In practice, this dependence is meaningless. Given the realistic assumption that α is very small, we can approximate[28] the Jukes–Cantor distance as

$$d = -\frac{3}{4}\ln\left(1 - \frac{4}{3}\beta\right). \tag{A.2.12}$$

This simple result is a reasonable approximation of the amount of genetic change corresponding to a particular net substitution probability β. The properties of this function match reasonable expectations.[29] It increases as β increases, with $d \approx \beta$ if β is small and $d \to \infty$ as $\beta \to 3/4$.[30]

---

[28] See Problem A.2.1b.

[29] See Problem A.2.2.

[30] Note that $\beta = 3/4$ means that the system has reached equilibrium; theoretically this requires infinite time.

Problems

A.2.1. (a) Derive (A.2.6).

(b) Verify that

$$\beta = \frac{3}{4} - \frac{3}{4}\left(1 - \frac{4}{3}\alpha\right)^N$$

satisfies the equation

$$\begin{pmatrix} 3(1-\beta) \\ \beta \\ \beta \\ \beta \end{pmatrix} = \mathbf{u} = \frac{3}{4}\begin{pmatrix} 1+3\lambda_2^N \\ 1-\lambda_2^N \\ 1-\lambda_2^N \\ 1-\lambda_2^N \end{pmatrix},$$

where

$$\lambda_2 = 1 - \frac{4}{3}\alpha.$$

A.2.2. (a) Plot the Jukes–Cantor phylogenetic distance function

$$d = -\frac{3}{4}\ln\left(1 - \frac{4}{3}\beta\right).$$

Be careful to restrict β to values that make sense biologically.

(b) Use linear approximation to show that $d \approx \beta$ for small genome changes. Why does this make sense?

(c) Compute $\lim_{\beta \to 3/4} d$. Explain the meaning of the result.

(d) Use linear approximation to derive (A.2.12) from (A.2.11).

A.2.3. Let $\mathbf{M}$ be a 2×2 Markov chain matrix with entries a and b as shown below.

$$\mathbf{M} = \begin{pmatrix} & a \\ b & \end{pmatrix}.$$

(a)* Fill in the blanks to complete the matrix.

(b) Show that $\begin{pmatrix} 1 \\ 1 \end{pmatrix}$ is an eigenvector for $\lambda = 1$ if and only if the entries in each row of $\mathbf{M}$ sum to 1. What must be true about a and b in this case?

A.2.4. The Kimura model of genetic change assumes that the rates for the AG, GA, CT, and TC substitutions are faster than those for the other substitutions. (There is a biochemical basis for why this should be the case.)

(a) Construct the matrix $\mathbf{M}$ for the Kimura model, using α for the faster rate and β for the slower rate.

(b) Show that $\begin{pmatrix} 1 \\ 1 \\ 1 \\ 1 \end{pmatrix}$ is an eigenvector for the Kimura model for $\lambda = 1$ and conclude that all nucleotides are equally likely.

A.2.5. The Felsenstein model of genetic change assumes that rates of change depend on the nucleotide being changed to, but not the nucleotide being changed from.

(a) Assume that other nucleotides change to A at rate a, G at rate g, and so on. Construct the matrix $\mathbf{M}$.

(b) Show that $\begin{pmatrix} a \\ g \\ c \\ t \end{pmatrix}$ is an eigenvector for the Felsenstein model for $\lambda = 1$.

A.3 Boolean Algebra Models

After studying this section, you should be able to:

- Compute Boolean functions.
- Find fixed points of Boolean networks.

Only a small fraction of an organism's genes are active in any particular cell; genes for the heart muscle, for example, are only expressed in heart cells. Even so, the number of chemical compounds that could be present in a cell, including messenger RNA molecules, proteins, and enzymes, is large. Each of these could be considered as a variable in a dynamical system that represents the cell; at any given time, nearly all are at a concentration of 0. This suggests a modeling strategy of focusing strictly on the distinction between zero and nonzero values, ignoring the specific amounts of those quantities that are not zero. A special type of mathematics, called Boolean algebra,[31] is ideal for this kind of modeling. In this section, we develop the basic principles of Boolean algebra and briefly indicate how Boolean models can be used to study the regulation of gene expression. For a more advanced introduction to gene regulation networks and their Boolean models, the reader should consult the outstanding paper on this topic by Martins et al. [5], from which this section draws heavily.

A.3.1 Boolean Algebra

A *Boolean variable* is a variable that can take on only two values, 0 and 1. These can be combined into dynamical systems, called Boolean networks. As we saw in Chapters 5–7, dynamical systems are defined by a set of formulas that calculate the new state of the system in terms of the old state. For a Boolean network, this means that we need functions that define Boolean dependent variables in terms of Boolean independent variables; in other words, we must first develop the machinery of Boolean algebra.

Algebra with number systems is based on the arithmetic operations of addition and multiplication. Algebra with Boolean variables is instead based on logic. Think of the Boolean variables 1 and 0 as representing the logical constants TRUE and FALSE. Suppose A is a species of animal and B is a group of species that can be divided into mutually exclusive subgroups B1 and B2. Consider the statements

1. A is a species of type B1.
2. A is a species of type B2.
3. A is a species of type B.

Suppose we use the Boolean variables X, Y, and Z to indicate the truth of these statements. Given a particular A, we need assess the values of only two of them; the third can be calculated

---

[31] Named for its creator, the nineteenth century English mathematician George Boole.

by rules of logic. For example, statement 3 is TRUE when either of statements 1 or 2 is TRUE. In terms of the variables, $Z = 1$ requires either $X = 1$ or $Y = 1$. We can define a Boolean operation X OR Y that is TRUE if either $X = 1$ or $Y = 1$ and FALSE otherwise. Using the notation $\vee$ for this operation,[32] we can write

$$Z = X \vee Y.$$

Note that the definition of Z means $Z = 1$ in the event that $X = 1$ *and* $Y = 1$, although the setting rules out this possibility.

Suppose we want to indicate how to calculate Y from knowledge of X and Z. This is a bit tricky. Certainly $Z = 1$ is necessary for $Y = 1$. In addition, $X = 1$ precludes $Y = 1$. So two separate conditions are necessary for $Y = 1$: we must have $Z = 1$ and $X = 0$. There is an AND operation that is TRUE only when both of its operands are TRUE, but we want Y to be TRUE when one operand is TRUE and the other FALSE. The solution is the negation operation; the symbol $\neg X$ (read "not X") is TRUE when X is FALSE, and vice versa. Thus, we want to require Z and $\neg X$ to be TRUE. Using the symbol $\wedge$ for the AND operation,[33] the correct notation is

$$Y = Z \wedge \neg X.$$

The OR, AND, and negation operations are sufficient to define all necessary logical constructions in Boolean algebra, but there is an additional operation that we define for convenience. Suppose we want to indicate that Z is TRUE whenever one of X and Y is TRUE, but not when both are TRUE. The OR operation does not do this by itself. We can denote this with the EXCLUSIVE OR operation, written as

$$Z = X \oplus Y.$$

Table A.3.1 summarizes the binary operations of Boolean algebra.

Table A.3.1 Binary Boolean operations

AND	$0 \wedge 0 = 0$	$0 \wedge 1 = 1 \wedge 0 = 0$	$1 \wedge 1 = 1$
OR	$0 \vee 0 = 0$	$0 \vee 1 = 1 \vee 0 = 1$	$1 \vee 1 = 1$
EXCLUSIVE OR	$0 \oplus 0 = 0$	$0 \oplus 1 = 1 \oplus 0 = 1$	$1 \oplus 1 = 0$

As in the motivating example, the names of the operations describe how they work. Thus, $a \wedge b$ is TRUE only when both a and b are TRUE, $a \vee b$ is TRUE whenever either a or b is TRUE, and $a \oplus b$ is TRUE when one of a and b is TRUE, but not both. The EXCLUSIVE OR operation could be omitted from the list, as it can be constructed from the other operations in several different ways. However, these constructions are sufficiently complicated to justify thinking of EXCLUSIVE OR as an independent operation.

Example A.3.1. To demonstrate the identity

$$x \oplus y = [x \wedge (\neg y)] \vee [(\neg x) \wedge y],$$

we calculate the complicated expression on the right side for each possible pair of x and y values and check that the answer is the same as $x \oplus y$:

---

[32] The symbol $\cup$, which represents the union of two sets, is probably more familiar to most readers. Think of the OR operation as similar to a union of sets.

[33] Think of the symbol as being similar to the symbol $\cap$ used for the intersection of sets.

$$[0 \wedge (\neg 0)] \vee [(\neg 0) \wedge 0] = [0 \wedge 1] \vee [1 \wedge 0] = 0 \vee 0 = 0 = 0 \oplus 0 \, ;$$
$$[0 \wedge (\neg 1)] \vee [(\neg 0) \wedge 1] = [0 \wedge 0] \vee [1 \wedge 1] = 0 \vee 1 = 1 = 0 \oplus 1 \, ;$$
$$[1 \wedge (\neg 0)] \vee [(\neg 1) \wedge 0] = [1 \wedge 1] \vee [0 \wedge 0] = 1 \vee 0 = 1 = 1 \oplus 0 \, ;$$
$$[1 \wedge (\neg 1)] \vee [(\neg 1) \wedge 1] = [1 \wedge 0] \vee [0 \wedge 1] = 0 \vee 0 = 0 = 1 \oplus 1 \, .$$

$\square$

Check Your Understanding A.3.1:
Verify the identity $x \oplus y = \neg[(x \wedge y) \vee (\neg x \wedge \neg y)]$.

A.3.2 Boolean Functions and Boolean Networks

A *Boolean function* is a function $f(\mathbf{x})$ that uses the values of the Boolean input vector $\mathbf{x}$ to compute a Boolean variable output. For convenience, we often write the argument $\mathbf{x}$ as a list of the scalar components.

Example A.3.2. Let $\mathbf{x}$ be a Boolean variable with three components, called X, Y, and Z. Define the Boolean function

$$f(\mathbf{x}) = f(X, Y, Z) = X \wedge Y.$$

Then, for example, $f(1,1,0) = 1$, while $f(0,1,1) = 0$. In this particular function, the state of Z does not matter, but it may be convenient to include Z as an independent variable. $\square$

Check Your Understanding A.3.2:
Verify the results $f(1,0,0) = 0$ and $f(1,1,1) = 1$ for the Boolean function of Example A.3.2.

Now suppose $\mathbf{x}$ is a vector of n Boolean variables and $\mathbf{f}$ is a vector of n Boolean functions $f_1, f_2, \ldots, f_n$. Then we can use the function $\mathbf{f}$ to define a **Boolean network** using the dynamic equation

$$\mathbf{x}_{t+1} = \mathbf{f}(\mathbf{x}_t). \tag{A.3.1}$$

Example A.3.3. Suppose the Boolean function $\mathbf{f}$ of three variables has components defined by

$$f_1(X, Y, Z) = \neg Z, \qquad f_2(X, Y, Z) = X \wedge Y, \qquad f_3(X, Y, Z) = X \oplus Y.$$

Suppose further that $\mathbf{x}_0 = (1, 1, 1)$. Then

$$f_1(1,1,1) = 0, \quad f_2(1,1,1) = 1, \quad f_3(1,1,1) = 0,$$

so $\mathbf{x}_1 = (0, 1, 0)$. $\square$

Check Your Understanding A.3.3:
Verify that $\mathbf{x}_2 = (1, 0, 1)$ and $\mathbf{x}_3 = (0, 0, 1)$ for the Boolean sequence defined in Example A.3.3.

Boolean networks are autonomous, meaning that changes depend only on the state and not on the time. For example, the result $\mathbf{f}(1,1,1) = (0,1,0)$, which we calculated for the Boolean network in Example A.3.3 means that the state $(1,1,1)$ is always followed by the state $(0,1,0)$. We could display this fact as a graph consisting of two nodes for the states $(1,1,1)$ and $(0,1,0)$ along with an arrow that goes from the former to the latter. The network of Example A.3.3 has only eight states, so it is not a difficult matter to construct the complete graph of the Boolean network.

In the discrete models of Chapter 6, the index t had the specific meaning of time. This is not strictly necessary with Boolean networks. The use of dynamical systems notation is often merely an artifice used to identify the long-term behavior of a network from any given starting point. It does not matter what length of time is needed for the system to move from one state to another; all that matters is that states can be classified according to the way they appear in the graph of the network. Many states are *transient*, which means that systems that start in these states never return. Some states are *fixed points*, meaning that the system stops changing once it reaches that state. In some networks, there are *recurrent* states that occur as part of a periodic movement through two or more states.

Example A.3.4. Let $\mathbf{f}$ be the Boolean function defined in Example A.3.3. The dynamics of the Boolean network defined by $\mathbf{f}$ is illustrated in Figure A.3.1. The state $(1,1,0)$ is a fixed point that cannot be attained unless the system is initially at that point. There is a cycle of four states, and there are three transient states, each of which eventually reaches the cycle rather than the fixed point. Note that graphs are defined by the pattern of nodes and arrows. The same graph could be displayed with any number of different layouts.   □

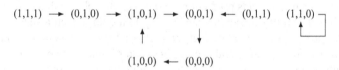

Fig. A.3.1 A graph of the Boolean network of Examples A.3.3 and A.3.4

One important difference between Boolean networks and dynamical systems based on differential equations or difference equations is that the notion of stability does not apply in the same way. In a dynamical system, it makes sense to consider what happens when the initial state is arbitrarily close to a fixed point; in particular, it is possible for the system to evolve away from the fixed point. In a Boolean network, variables can only be 0 or 1, so the system cannot start arbitrarily close to a fixed point. In a sense, fixed points and cycles in a Boolean network are always stable. This limits the kind of questions that can be addressed with Boolean models.

A.3.3 Using a Boolean Network for a Consistency Check

The primary use of Boolean models is to serve as a consistency check on a proposed mechanism. Before we consider using Boolean models for this purpose, it is helpful to consider an alternative notation that streamlines the discovery of fixed points and cycles. The key idea is that each variable in a Boolean network is either present or absent, and the state of the system is determined by the set of variables that are present. We could therefore name states by the list of variables whose value is 1. For example, if the Boolean variables are X, Y, and Z, then the state XZ corresponds to the point $(1,0,1)$, in which X and Z are both present while Y is absent. Similarly, the state $(0,0,1)$ can be denoted as Z. The null state $(0,0,0)$ can be denoted by the symbol $\emptyset$, which is used in set theory to indicate a set that has no elements. Figure A.3.2 reproduces the graph of the Boolean network of Examples A.3.3 and A.3.4 with this alternative labeling scheme. The advantage of this method is that each of the arrows is easy to check with the definition of the function. One can see at a glance, for example, that X is present at the tip of an arrow precisely when Z is not present at the tail, in keeping with the definition $f_1(X,Y,Z) = \neg Z$.

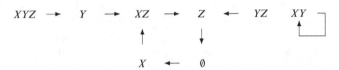

Fig. A.3.2 A graph of the Boolean network of Examples A.3.3 and A.3.4 and defined by the function $\mathbf{f}(X,Y,Z) = (\neg Z, X \wedge Y, X \oplus Y)$

Suppose the quantities X, Y, and Z represent chemicals that could be present in a cell and interact according to the following rules:

1. X forms whenever Z is absent, but never when Z is present.
2. Y cannot be produced in the system, but it can be maintained if X is present.
3. Either X or Y is necessary to produce Z, but the combination of X and Y suppresses formation of Z.

This set of rules corresponds exactly to the Boolean network of Examples A.3.3 and A.3.4. Hence, the network analysis yields predictions about what the physical system will do. Specifically, it predicts that two things can happen. The system could reach a state in which both X and Y are always present and Z is always absent. Alternatively, the system could reach a state in which Y is always absent and X and Z cycle between present and absent in such a way that Z follows X, with some overlap, while X reappears only when Z is absent. These are very specific predictions about what states can be seen in the physical system. If the system exhibits fixed states or cycles that are not in agreement with the network analysis, then the proposed mechanism must be false.

Boolean networks cannot be used to discover mechanisms because for any particular set of fixed points and/or cycles one can find a very large number of possible underlying mechanisms. However, they do find wide employment in biochemistry in cases where some features of the chemical mechanism are known and others are not. Many incorrect mechanisms can be rejected because their Boolean networks do not have the right behavior. For systems with large numbers of components, it is far easier to identify fixed points in Boolean networks than in dynamical systems such as the ones examined in Chapters 6 and 7.

Problems

A.3.1.* Consider the Boolean network $\mathbf{x}_{t+1} = \mathbf{f}(\mathbf{x}_t)$ defined by the functions

$$f_1(X,Y,Z) = Y \vee Z, \qquad f_2(X,Y,Z) = \neg(X \vee Z), \qquad f_3(X,Y,Z) = \neg(X \wedge Z).$$

(a) Determine the action of the function $\mathbf{f}$ on each of the eight possible states.
(b) Arrange the information from part (a) as a graph similar to that in Figure A.3.2.

Problems A.3.2 and A.3.3 consider a situation in which the three components of a Boolean network represent interacting chemical species in a cell.[34]

A.3.2. Suppose we know some elementary facts about the biochemistry:

1. The presence of Y inhibits formation of X, which is naturally present.
2. The presence of X is necessary for the formation of Y.
3. The presence of X promotes the formation of Z.
4. The presence of Y promotes the formation of Z.

(a) Determine the correct Boolean functions for the formations of X and Y at time $t + 1$ from the Boolean state at time t. These are unambiguous.
(b) The list of facts is not sufficient to determine the correct function for the formation of Z. In fact, three possibilities are consistent with the data:

   1. It could be that both X and Y must be present together.
   2. It could be that either X or Y is sufficient, regardless of the presence of the other.
   3. It could be that formation of Z requires either X or Y, but not both.

   Determine the correct Boolean function for Z in the first case. Use the complete set of Boolean functions to determine the graph of states. Which states are persistent?
(c) Repeat part (b) for the second case.
(d) Repeat part (b) for the third case.

A.3.3. Repeat problem A.3.2 given the following set of facts.

1. The presence of Y promotes formation of X.
2. The presence of Z inhibits formation of X.
3. The presence of X inhibits the formation of Y.
4. The presence of Y promotes the formation of Z.

References

1. Britton NF. *Essential Mathematical Biology*. Springer, New York (2004)
2. Cushing JM, B Dennis, and RF Constantino. An interdisciplinary approach to understanding nonlinear ecological dynamics. *Ecological Modelling*, **92**, 111–119 (1996)
3. Dennis B, RA Desharnais, JM Cushing, SM Henson, and RF Constantino. Estimating chaos and complex dynamics in an insect population. *Ecological Monographs*, **71**, 277–303 (2001)
4. Denver DR, K Morris, M Lynch, and WK Thomas. High mutation rate and predominance of insertions in the Caenorhabditis elegans nuclear genome. *Nature*, **430**, 679–682 (2004)
5. Martins A, P Vera-Licona, and R Laubenbacher. Computational systems biology: discrete models of gene regulation networks. In Ledder G, JP Carpenter, TD Comar (eds.) *Undergraduate Mathematics for the Life Sciences: Models, Processes, and Directions*. Mathematics Association of America (2013)

---

[34] These problems use Boolean networks that the author obtained from Dan Hrozencik of Chicago State University and Timothy Comar of Benedictine Universtiy.

Appendix B
The Definite Integral via Riemann Sums

To estimate the total amount of a quantity V that is produced in an interval $a \le t \le b$, given a production rate $Q(t)$, we can divide the interval $[a,b]$ into n equal subintervals of duration $h = (b-a)/n$. Let $Q_0 = Q(a)$, $Q_1 = Q(a+h)$, $Q_2 = Q(a+2h)$, and so on, with $Q_n = Q(b)$ at the end. For the first subinterval, $a \le t \le a+h$, we can estimate the rate for the entire subinterval using $Q_0 = Q(a)$. Similarly, we can use $Q_1 = Q(t+a)$ for the interval $a+h \le t \le a+2h$. The result is the left-hand sum

$$V_L = Q_0 h + Q_1 h + \cdots + Q_{n-1} h = \sum_{j=0}^{n-1} Q_j h.$$

Similarly, we could use Q_1 for the interval $a \le t \le a+h$, and so on, and generate a right-hand sum

$$V_R = Q_1 h + \cdots + Q_n h = \sum_{j=1}^{n} Q_j h.$$

In general, we cannot say that either V_L or V_R is a maximum estimate or a minimum estimate, since the function Q might not be always increasing or always decreasing. However, we can always say that the difference between the two sums is

$$V_R - V_L = [Q(b) - Q(a)]h = \frac{(b-a)[Q(b) - Q(a)]}{n},$$

so the difference between the two estimates approaches 0 in the limit as $n \to \infty$. This means that the left and right sums converge to the same result, which must be the exact answer. We may summarize the result mathematically as follows:

$$V = \lim_{n \to \infty} \sum_{j=1}^{n} Q(t_j)h, \quad \text{where} \quad a + (j-1)h \le t_j \le a + jh. \tag{B.1}$$

Note that we need not specify a left-hand sum or a right-hand sum. The result is the same if we alternate left endpoints and right endpoints, or if we use the midpoint of each subinterval instead. All that matters is that we use some value of t in each subinterval to represent the rate Q over the entire subinterval.

G. Ledder, *Mathematics for the Life Sciences: Calculus, Modeling, Probability, and Dynamical Systems*, 407
Springer Undergraduate Texts in Mathematics and Technology, DOI 10.1007/978-1-4614-7276-6,
© Springer Science+Business Media, LLC 2013

The result of (B.1) motivates the formal definition of the *definite integral*, the second of the two principle concepts of calculus.[35]

Definite integral of a function $Q(t)$ **over**   **an interval** $a \leq t \leq b$: *the quantity defined by the formula*

$$\int_a^b Q(t)\,dt = \lim_{n \to \infty} \sum_{j=1}^{n} Q_j h, \tag{B.2}$$

where $Q_j = Q(t_j)$ *for any* t_j *in the interval* $a + (j-1)h \leq t_j \leq a + jh$ *and* $h = (b-a)/n$.

---

[35] The other is, of course, the derivative.

Appendix C
A Runge–Kutta Method for Numerical Solution of Differential Equations

It is easy to program simulations for discrete dynamical systems because the equations of the model are all that is needed. For continuous dynamical systems, we need a special numerical method that uses a discrete approximation of the system of differential equations. The simplest of these is Euler's method, which we used in Example 5.3.1. More sophisticated methods work far better. The simplest of these methods is the Runge–Kutta[36] method of order 4.

Like Euler's method, Runge–Kutta methods work by computing approximate values x_j of the dependent variable[37] at a collection of times t_j. Thus, they are based on discrete approximations to a continuous equation. The most obvious way to discretize a differential equation $x' = f(t,x)$ is to evaluate the differential equation at the point (t,x) using a forward difference approximation for the derivative:

$$\frac{x_{j+1} - x_j}{h_j} = f(t_j,x_j), \; h_j = t_{j+1} - t_j,$$

which leads to the Euler approximation

$$x_{j+1} = x_j + h_j f(t_j,x_j).$$

The drawback of Euler's method is the use of the value $f(t_j,x_j)$ to represent dx/dt over the entire interval $[t_j,t_{j+1}]$. As the actual solution curve moves through this interval, its slope changes and the approximation becomes less accurate. A better approximation for the average rate of change over the interval could in theory be obtained by averaging $f(t_j,x_j)$ and $f(t_{j+1},x_{j+1})$, the latter being the slope at the end of the interval. This does not work in practice because the value x_{j+1} is not known. The idea of Runge–Kutta methods is to obtain several approximations for slopes in the interval and average them together to obtain the slope to be used to compute x_{j+1}. The details of how best to do this are beyond the scope of this presentation, so we merely present the algorithm.

---

[36] This is pronounced Run-ga Kut-ta.

[37] For convenience, we are assuming a single differential equation. Numerical methods for single differential equations also work for systems.

G. Ledder, *Mathematics for the Life Sciences: Calculus, Modeling, Probability, and Dynamical Systems*, 409
Springer Undergraduate Texts in Mathematics and Technology, DOI 10.1007/978-1-4614-7276-6,
© Springer Science+Business Media, LLC 2013

Algorithm C.1 *Runge–Kutta order 4 for $x' = f(t,x)$.*

Given (t_j, x_j) and $h = t_{j+1} - t_j$, x_{j+1} is given by

$$k_1 = hf(t_j, x_j),$$
$$k_2 = hf(t_j + 0.5h, x_j + 0.5k_1),$$
$$k_3 = hf(t_j + 0.5h, x_j + 0.5k_2),$$
$$k_4 = hf(t_j + h, x_j + k_3),$$
$$x_{j+1} = x_j + \frac{1}{6}(k_1 + 2k_2 + 2k_3 + k_4).$$

Hints and Answers to Selected Problems

Chapter 1

Section 1.1

1.1.1 $t = \dfrac{\ln 3}{0.3}$

1.1.2 $t_z = \dfrac{\ln 650 - \ln z}{0.3}$

1.1.7 (a) $\left(\dfrac{b}{2}, -\dfrac{b^2}{4} \right)$

1.1.12

(a) $y = Ae^{-1.6}$
(b) $y = 650 + Ae^{-1.6}$
(c) $A = \dfrac{650}{1 - e^{-1.6}} \approx 814.4, \quad y_{\min} = Ae^{-1.6} \approx 164.4$

1.1.16

(a) $y(t) = e^{-mt}, \quad s = e^{-m}$
(b) $n = -\dfrac{\ln s}{s - s^2}$. Since $0 < s < 1$, $n > 0$.

Section 1.2

Interval	1960–1970	1970–1980	1980–1990	1990–2000	2000–2010
Average rate of change	0.877	1.300	1.548	1.524	2.038

1.2.1 There is a general increase.

G. Ledder, *Mathematics for the Life Sciences: Calculus, Modeling, Probability, and Dynamical Systems*, 411
Springer Undergraduate Texts in Mathematics and Technology, DOI 10.1007/978-1-4614-7276-6,
© Springer Science+Business Media, LLC 2013

1.2.5

(a) -156 on $[0.5, 1]$, -134 on $[1, 1.5]$, -145 on $[0.5, 1.5]$

(b) -144.5

(c) The central difference approximation is clearly best.

1.2.8 We get 3.2, 10, 32, and 100. The slope appears to be infinite, and the tangent line is vertical.

1.2.9 $0.22, 0.125$, and 0.08

Section 1.3

1.3.1 $f'(x) = 2x$

1.3.3 $f'(x) = \cos x$

1.3.7 $f'(x) = -5e^{-5x} + \pi \cos \pi x$

1.3.11 $g'(x) = 2x \ln x + x$

1.3.13 $f'(x) = \sec^2 x = \dfrac{1}{\cos^2 x}$

1.3.16 $g'(t) = \dfrac{e^v(v-1)}{(e^v + v)^2} v'(t)$

1.3.17 $g'(t) = \dfrac{2t - t^4}{(t^3 + 1)^2}$

1.3.22 $g'(x) = \left(2e^{2x} + \dfrac{1}{x}\right) \cos\left(e^{2x} + \ln x\right)$

1.3.25 $f'(x) = \pi \cos x e^{\pi \sin x}$

1.3.27 $g'(t) = +\dfrac{2\pi t}{(t^2 + 1)^2} \sin\left(\dfrac{\pi}{t^2 + 1}\right)$

1.3.33 The answer is 1. To get there, you must simplify the denominator, replace x with $y = -x$, and factor the resulting limit using one factor that matches the appropriate formula in (1.3.3).

Section 1.4

1.4.6 0 is a local minimum.

1.4.8 0 is a local maximum, 2 and -2 are local minima.

1.4.10 -1 is a local minimum and 2 is not a local extremum.

1.4.14

(a) 0 is a local maximum.

(b) $f' = 4x(x^2 - 3x - b)$; the critical points $x = \dfrac{3 \pm \sqrt{9 + 4b}}{2}$ require $b > -9/4$.

(c) $f''(x^\star) = 4x^\star(2x^\star - 3)$

(d) The value $x^\star$ is a local maximum if it is between 0 and 3/2, and a minimum if it is less than 0 or greater than 3/2.

(e) The larger point is a local minimum, while the smaller is a local maximum.

1.4.16 $\pm 0.01x_0$

Section 1.5

1.5.1 $r = 1/e$

1.5.2 $r = q$

1.5.3 (d) $v = \dfrac{3}{2}u$

1.5.4

(a) $w = 1$

(b) $F' > 0$ for all p, so the global maximum is at $p = 1$.

(c) $p = 1/(2 - w)$

(e) $(1 - p)^2 = \left(\dfrac{1 - w}{2 - w}\right)^2$

1.5.5

(a) $t(x) = X - x + \dfrac{\sqrt{x^2 + Y^2}}{s}$

(b) $x^\star = Y\dfrac{s}{\sqrt{1 - s^2}}$

(c) $x^\star < X$

(f) $\cos\phi^\star = s$

1.5.9

(a) $x = e^{-kt}$

(b) $f + x$ is constant because resources lost to the patch are gained by the forager. Thus, $f(t) = 1 - D - e^{-kt}$.

(d) $D = 1 - x^\star(1 - \ln x^\star)$

Section 1.6

1.6.1 $30\pi D \,\mathrm{mm}^2/\mathrm{day}$

1.6.6

(a) $\dfrac{dX}{dT} = X - \dfrac{QX}{1 + X}$

(b) It will be easier to analyze a model that has one parameter than a model with three parameters.

Section 1.7

1.7.1

(a) $y_L = 5$, $y_R = 14$, $\Delta y = 9$
(b) $y_L = 6.875$, $y_R = 11.375$, $\Delta y = 4.5$
(c) $\Delta y = 27/n$
(d) 270
(e) $y_L = 8.95006$, $y_R = 9.05006$
(f) 9.00006
(g) 8.99997
(h) $S_3 = 9.0$, $S_6 = 9.0$

1.7.6 $\displaystyle\int_0^{100} Lf(x)\,dx$

1.7.9

(a) 2,000
(b) 1.000850
(c) 1.000838
(d) ≈ 1.000834
(e) 0.000016 and 0.000004
(f) The midpoint approximation is generally much better than the left and right approximations; also, the midpoint approximation is more accurate when the function is close to linear.

Section 1.8

1.8.1 $-4/5$

1.8.4 $(e^2 + 1)/2$

1.8.7 $1/\sqrt{2} + 1/2$

Section 1.9

1.9.1 $\dfrac{2}{3}\left(2^{3/2} - 1\right)$

1.9.2 $1/8$

1.9.7 0

Chapter 2

Section 2.2

2.2.3

(a) $q^2, \quad 2pq, \quad p^2$

(b) $0.36, 0.41$

2.2.7 (d) $ne^{-ra} \int_0^1 (1-x) \left(e^{5rx} + e^{-5rx} \right) dx = 1$

Section 2.3

2.3.3 The slopes are $m_1 = 1 - 0.4c$ and $m_2 = 1 - 0.2c$. Errors at the edges of the graph matter more than errors near the middle.

2.3.5

(a) $C = -15.6 + 4.22t$

(b) There is a lot of scatter in the data, which limits our confidence in the choice of a linear model.

Section 2.4

2.4.1 $N = 6.01e^{0.400t}$

2.4.2 $N = 6.18e^{0.390t}$

2.4.5 $z = 1007e^{-0.0827t}$

Section 2.5

2.5.6

$$\frac{dS}{dT} = -pBSI + QR, \quad \frac{dI}{dT} = pBSI - KI, \quad \frac{dR}{dT} = KI - QR.$$

Section 2.6

2.6.3 $Y = e^{-T}$, where $Y = y/y_0$ and $T = kt$

2.6.8 $\dfrac{dc}{dt} = 1 - c - rc, \quad r = \dfrac{RV}{F}.$

The quantities F/V and R represent the rates for the two processes that eliminate pollutant from the lake. Hence, r is the ratio of the magnitude of the decay process to the magnitude of the flushing process.

Section 2.7

2.7.2

(a) $y = -0.11667x + 0.04x^2 - 0.00083x^3$
(c) The cubic polynomial is too "curvy" for a good fit.

2.7.6

(a) $B = b + m\bar{x} + a\bar{x}^2 - \bar{y}$, $M = m + 2a\bar{x}$
(b) $m = M - 2a\bar{x}$, $b = B + \bar{y} - m\bar{x} - a\bar{x}^2$

Chapter 3

Section 3.1

3.1.2 See Figure A.1.

Section 3.2

3.2.1 1/4

3.2.2

(a) 5/36
(b) The experiment is to roll two dice and add the results. The sample space is the set of integers from 2 to 12.

3.2.3 0.15

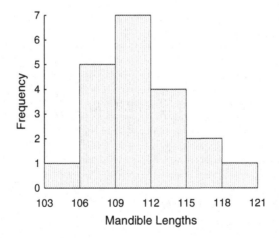

Fig. A.1 Problem 3.1.2

Section 3.3

3.3.1 $\mu = 1.016$, $\sigma = 0.1340$

3.3.3 1/8

3.3.5

(a) $P[X = 2] = 0.1$, $P[X = 1] = 0.6$, $P[X = 0] = 0.3$
(b) $\mu = 0.8$, $\sigma^2 = 0.36$, $\sigma = 0.6$

Section 3.4

3.4.3 Needing at least three tries is equivalent to having two consecutive failures; thus, the probability is $b_{2,\,0.3}(0) = 0.49$.

Section 3.5

3.5.1

(a) $b_{2n,\,0.5}(n) = \dfrac{(2n)!}{(2^n\, n!)^2}$

(b) 0.176
(c) 0.080

3.5.2

(a) See Table A.1
(b) See Figure A.2
(c) See Figure A.3
(d) See Figure A.4

Table A.1 Problem 3.5.2

Measurement	32	34	36	38	40	42	44	46	48
Frequency	1.5	61	434	1488.5	2080.5	1270	349.5	44.5	2.5

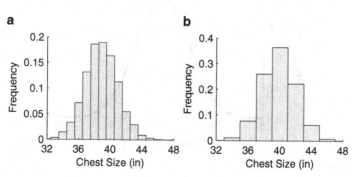

Fig. A.2 Problem 3.5.2b

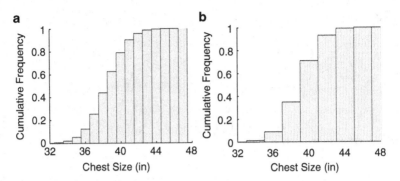

Fig. A.3 Problem 3.5.2c

Section 3.6

3.6.2

(a) 0.0668
(b) 0.3795

3.6.6 See Figure A.5.

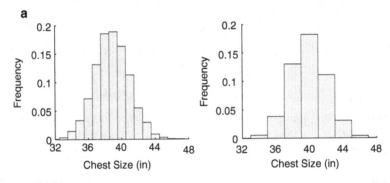

Fig. A.4 Problem 3.5.2d

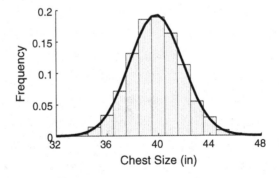

Fig. A.5 Problem 3.6.6

Section 3.7

3.7.2

(a) $f_1(1) = e^{-1} \approx 0.3679$
(b) $f_2(2) = 2e^{-2} \approx 0.2707$
(c) $f_n(n) = \dfrac{n^n}{n!} e^{-n}$, so $f_3(3) = 0.2240$, $f_4(4) = 0.1954$, $f_5(5) = 0.1755$, $f_6(6) = 0.1606$. The probably of hitting the mean goes down as the mean increases.

3.7.3 $E_4(1) = 1 - e^{-4} = 0.9817$

3.7.7 (a) $\mu = 0.425$, $\sigma^2 = 0.453$
(b), (c) See Figure A.6.
(d) $\sigma^2 \approx \mu$ and the shape looks right.

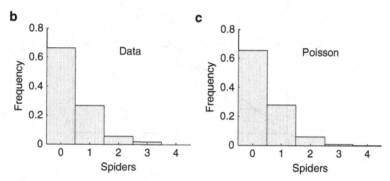

Fig. A.6 Problem 3.7.7

Chapter 4

Section 4.1

4.1.4 $B_{80,0.05}(2) = 0.2306$

4.1.6

(a) 0.0761
(b) $50 \times \dfrac{300}{5651} \times 0.0761 = 0.2020$
(c) 0.2619
(d) 0.2209

Section 4.2

4.2.1

(a) See Figure A.7.
(b) 0.036

(c) 0.028

(d) The fit to a normal distribution is unusually good, but not quite so good as to arouse suspicion.

4.2.9

(a) 0.036

(b) 0.073 and 0.148 respectively. Larger samples score worse. This makes sense. Generally we expect the residual sum of scores to increase with sample size, as it does here. It doesn't increase if the distribution is truly normal because the accuracy improves as the data is sorted into increasing order.

(c) About 0.125 is typical. Note that it is larger than the value for consecutive integers.

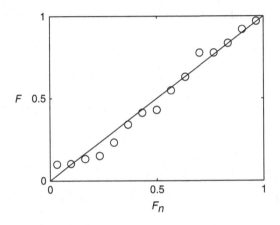

Fig. A.7 Problem 4.2.1

(d) About 0.2 is typical. This is not quite twice that of part (c). Apparently the difference between consecutive and random stays roughly the same (about 0.05) as the sample size increases.

(e) The average residual should be about the same, but there are twice as many points as in part (b).

Section 4.3

4.3.3

(a) Means are all 2; standard deviations are 0.671, 0.335, and 0.168.

(c) 0.183, 0.090, 0.066

Section 4.4

4.4.2

(a) Yes. While we cannot know the distribution from which the data was drawn, the value of the Cramer–von Mises statistic is well within the range of normally distributed data.

(b) The probability of getting the psychotic data from a population of nonpsychotics is 6×10^{-8}, so we can safely conclude that psychotics have higher dopamine levels than nonpsychotics. Given the large difference in the means, we can confidently claim biological significance as well as statistical significance.

4.4.8

(a) 0.0014
(b) 0.0046
(c) The normal distribution is not a good model for such a small number of Bernoulli trials. The binomial distribution probability is correct.
(d) 0.0384 and 0.0500. The handedness of presidents in the period between Roosevelt and Obama, exclusive, is only 5 % likely in the population at large, which (barely) meets the usual requirement for an interpretation of statistical significance. There are two possible explanations. It could be that there is some significant characteristic difference between left-handed and right-handed people that accounts for the difference, or it could merely be an example of the high probability of unrestricted coincidences. Unless more data, such as a high number of left-handed presidents prior to Roosevelt, can strengthen the case for a real difference, the latter explanation seems to be the better tentative conclusion.

Section 4.5

4.5.1 $0.0140 < \mu < 0.0188$ and $0.0133 < \mu < 0.0195$

4.5.6

(a) 0.3377
(b) $L(p) = \left(\dfrac{p}{0.3377}\right)^{106602} \left(\dfrac{1-p}{0.6623}\right)^{209070}$
(c) See Figure A.8
(e) The correct value is $L = 0.146$ rather than $L = 0.15$
(f) $L = 0.036$

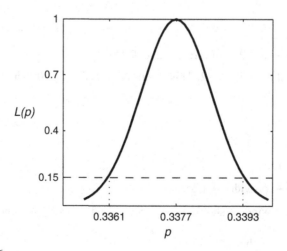

Fig. A.8 Problem 4.5.6

Section 4.6

4.6.1

(a) 0.9670
(b) 0.8523
(c) 0.8242
(d) does not make sense

4.6.3

(a) 13.89 %
(b) 0.33 %

Section 4.7

4.7.2

(a) $P[B|A] = 0.2, \quad P[B|A^c] = 0.3$
(b) $P[A] = 3/7, \quad P[B] = 9/35$
(c) $P[A^c|B] = 2/3$

4.7.3

(a) $P[E] = 0.012$
(b) $P[E|S] = 0.05$
(c) $P[E|S^c] = 0.0025$

Chapter 5

Section 5.1

5.1.1

(a) $N^\star = \dfrac{R \pm \sqrt{R^2 - 4}}{2}$. Both roots are positive if $R > 2$.

(b) There are no real roots if $R < 2$. This means $N_t^2 - RN_t + 1 > 0$, which in turn means that $N_{t+1} - N_t < 0$.

5.1.2

(a) $S - 1 + \dfrac{A}{1 + N_t}$

(b) Adults do not survive.

(c) $N^\star = \dfrac{A - (1 - S)}{1 - S}$, provided $A + S > 1$.

(d) The plot viewing window should be $0 \le S \le 1$ and $0 \le A \le \hat{A}$, where $\hat{A}$ is somewhere around 3–5.

(e) $N \to 0$, consistent with predictions.

(f) $N \to 2$, consistent with predictions.

(g) $N \to 0.6$, consistent with predictions.

Section 5.2

5.2.1 (c) See Figure A.9.

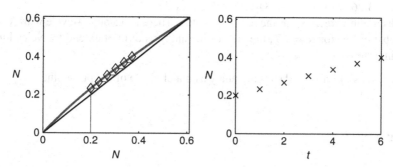

Fig. A.9 Problem 5.2.1

Section 5.3

5.3.4

(a) 0 and 8.9
(b) 0, 0.7, 2.0, and 7.3
(c) 0 and 0.36

5.3.9

(b) The term $\dfrac{cv^2}{1+v^2}$
(c) Use the equilibrium equation to find an alternative that is equivalent to the answer of part (a).
(d) $k/2$
(f) $\dfrac{1}{k}+\dfrac{k}{4}$

Section 5.4

5.4.1 0 and K are stable and T is unstable.

Section 5.5

5.5.1

(a) 0 is stable if $S+A<1$; $\dfrac{A-(1-S)}{1-S}$ is stable if $S+A>1$.
(b) The results of different methods are fully consistent.

5.5.5

(a) $f' = 1 - (2X - B)^2$

(b) $X = 1.64$ and $X = 2.56$, with $f' = 0.154$; asymptotically stable 2-cycle.

(c) For $R = 2.5$: $X = 1.5$ and $X = 3$, with $f' = -1.25$; unstable 2-cycle. For $R = 2.7$: $X = 1.44$ and $X = 3.26$, with $f' = -2.312$; unstable 2-cycle.

(e) Results are in agreement to the extent possible. There is a stable 4-cycle with $R = 2.5$ and no stable cycles for $R = 2.7$; the eigenvalue analysis did not extend to cycles longer than two time steps.

5.5.6 The results of linearized stability analysis match those of the phase line.

Chapter 6

Section 6.1

6.1.1 (b), (c) See Figure A.10.

(d) Steady growth with a fixed population ratio.

(e) Growth rate 1.1, ratio 22:1

(f) The model predicts the population will grow without bound. This is alright in the short term. In the long term, some mechanism must be present to limit population.

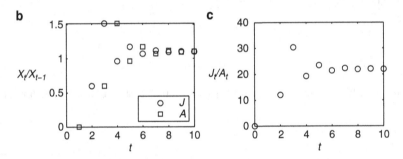

Fig. A.10 Problem 6.1.1

6.1.5 $s + rf = 1$

6.1.8 (b) $\gamma = 1$

Section 6.2

6.2.1 4

6.2.5 $\mathbf{x}_{t+1} = \mathbf{M}\mathbf{x}_t$, where $\mathbf{x} = \begin{pmatrix} J \\ A \end{pmatrix}$ and $\mathbf{M} = \begin{pmatrix} 0 & f \\ r & b \end{pmatrix}$. $\det \mathbf{M} = -rf$

6.2.9

(a) $\lambda = 3, -1$
(b) The ratios of the components of the vectors must be 1:3 and $-1:1$, respectively.

Section 6.3

6.3.1 $\lambda_1 = 4$. The eigenvector has ratio 3:2.

6.3.5

(a) $\lambda^2 - b\lambda - rf = 0$
(b) $rf + b = 1$

6.3.12 (d) 1.024

Chapter 7

Section 7.1

7.1.4

(a) $\dfrac{dR}{dT} = \gamma I$
(b) They are susceptible.
(c) $\dfrac{ds}{dt} = \varepsilon(1 - s) - R_0 si, \quad \dfrac{di}{dt} = R_0 si - i.$
(d) The parameter ε is the ratio of the time scale for the disease to the time scale for births. These are on the order of days and years, respectively.

Section 7.3

7.3.4 (a) $(1, 0)$ is always an equilibrium point; $\left(\dfrac{1}{R_0}, \varepsilon \left(1 - \dfrac{1}{R_0} \right) \right)$ is an equilibrium point when $R_0 > 1$.
(b), (c) See Figure A.11.
(d) $(1, 0)$ is stable when $R_0 < 1$ and unstable when $R_0 > 1$. The stability of the other equilibrium point cannot be determined from the nullcline plot.

Section 7.4

7.4.1

(a) The origin is stable for $k > 1$ and unstable for $k < 1$.
(b) The nullcline plot is unable to distinguish these cases.

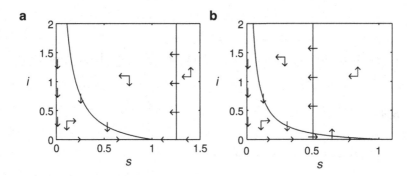

Fig. A.11 Problem 7.3.4

7.4.3

(a) $c_2 \approx (1 + a_1 + b_1)(a_2 + b_2) - a_1 a_2 = (1 + b_1)a_2 + (1 + a_1 + b_1)b_2 > 0$
(b) Under the assumption that ε is arbitrarily small, we have $c_3 \to 0$ while $c_1 c_2 > 0$. Hence, $c_1 c_2 > c_3$ for ε small enough.

Section 7.5

7.5.1

(a) The equilibrium $(1,0)$ is stable when $R_0 < 1$. The equilibrium $(R_0^{-1}, \varepsilon(1 - R_0^{-1}))$ is stable when it exists $R - 0 > 1$.
(b) The result for $(1,0)$ is the same as in Problem 7.3.4. The stability of the other point could not be determined from nullcline analysis. All results are consistent with the biological setting. It makes sense that the disease-free state should be stable if R_0 is below some threshold and unstable if it is higher.

7.5.7

(a) $(0,0)$ is unstable; $(1,0)$ is stable if $h < 1$.
(b) Stable when it exists $(h > 1)$.
(c) Results are consistent. Nullcline analysis was unable to determine the stability of the equilibrium with $c > 0$.
(d) Increasing h decreases the vegetation and increases the consumer population. Decreasing h too much drives the consumer to extinction.

Appendix A

Section A.1

A.1.4

(a) There are no stable fixed points.
(b) There is a stable fixed point with positive values.
(c) The origin is stable.
(d) All results are consistent with the analysis.

A.1.7

(a) $p^* = \ln R_0$, $\quad \det \mathbf{J} = \dfrac{R_0 \ln R_0}{R_0 - 1}$

(b) $g(R_0) = R_0(1 - \ln R_0)$

Section A.2

A.2.3

(a) There is only one way to complete the matrix that is consistent with the requirements for Markov chain matrices.

Section A.3

A.3.1

(a)

$$f(0,0,0) = (0,1,1), \quad f(0,0,1) = (1,0,1), \quad f(0,1,0) = (1,1,1), \quad f(0,1,1) = (1,0,1),$$
$$f(1,0,0) = (0,0,1), \quad f(1,0,1) = (1,0,0), \quad f(1,1,0) = (1,0,1), \quad f(1,1,1) = (1,0,0).$$

(b) See Figure A.12

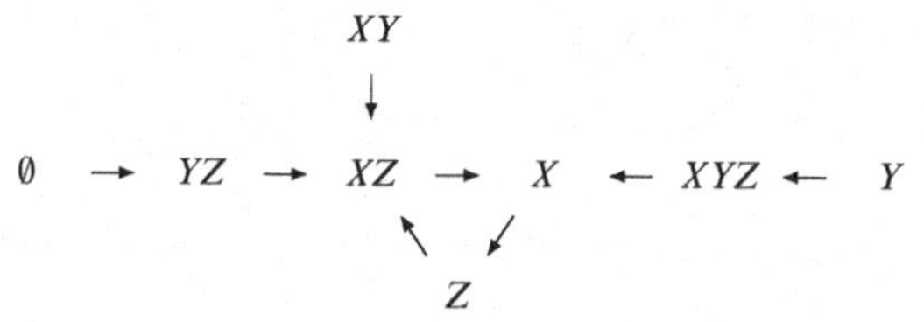

Fig. A.12 Problem A.3.1

Index

G. Ledder, *Mathematics for the Life Sciences: Calculus, Modeling, Probability, and Dynamical Systems*, 429
Springer Undergraduate Texts in Mathematics and Technology, DOI 10.1007/978-1-4614-7276-6,
© Springer Science+Business Media, LLC 2013

Printed in the United States
By Bookmasters